Fundamentals of Signals and Systems

This innovative textbook provides a solid foundation in both signal processing and systems modeling using a building block approach. The authors show how to construct signals from fundamental building blocks (or basis functions), and demonstrate a range of powerful design and simulation techniques in MATLAB®, recognizing that signal data are usually received in discrete samples, regardless of whether the underlying system is discrete or continuous in nature.

The book begins with key concepts such as the orthogonality principle and the discrete Fourier transform. Using the building block approach as a unifying principle, the modeling, analysis and design of electrical and mechanical systems are then covered, using various real-world examples. The design of finite impulse response filters is also described in detail.

Containing many worked examples, homework exercises, and a range of MATLAB laboratory exercises, this is an ideal textbook for undergraduate students of engineering, computer science, physics, and other disciplines.

Accompanying the text is a CD containing the MATLAB m-files for MATLAB generated figures, examples, useful utilities and some interesting demonstrations. The files can be easily modified to solve other related problems.

Further resources for this title including solutions, lecture slides, additional problems and labs, are available at www.cambridge.org/9780521849661.

Philip D. Cha is a Professor in the Department of Engineering at Harvey Mudd College in Claremont, California. He obtained his Ph.D. in mechanical engineering from the University of Michigan. Prior to joining Harvey Mudd, he worked as a senior research engineer at the Ford Motor Company Research Laboratory. He has been a visiting Professor at the University of Michigan and Tsinghua University in China. He has also been a participating guest at the Lawrence Livermore National Laboratory. In 2000, he received the Ralph R. Teetor Educational Award from the Society of Automotive Engineers as one of the top ten mechanical engineering educators in the USA.

John I. Molinder is a Professor in the Department of Engineering at Harvey Mudd College in Claremont, California. He received his Ph.D. in electrical engineering from the California Institute of Technology and worked as a senior engineer at the Jet Propulsion Laboratory before joining Harvey Mudd College. For more than 20 years he held a part-time position at the Jet Propulsion Laboratory. He has also worked as a principal engineer at Qualcomm in San Diego, California and, most recently, as a contractor for Boeing Satellite Systems in El Segundo, California. Prior to receiving his Ph.D. he served as a project officer in the USAF in San Bernardino, California.

Fundamentals of Signals and Systems

A building block approach

Philip D. Cha

Harvey Mudd College, Claremont, California

and

John I. Molinder

Harvey Mudd College, Claremont, California

CAMBRIDGE UNIVERSITY PRESS

Cambridge, New York, Melbourne, Madrid, Cape Town, Singapore, São Paulo

Cambridge University Press
The Edinburgh Building, Cambridge CB2 2RU, UK

Published in the United States of America by Cambridge University Press, New York

www.cambridge.org
Information on this title: www.cambridge.org/9780521849661

First published 2006

Printed in the United Kingdom at the University Press, Cambridge

A catalog record for this publication is available from the British Library

ISBN-13   978-0-521-84966-1 hardback
ISBN-10   0-521-84966-7 hardback

Contents

Figures

Tables

Preface

This text results from a new approach to teaching the sophomore engineering course entitled *Introduction to Engineering Systems* at Harvey Mudd College. Since the course is required of all students regardless of major, the goal is to provide a clear understanding of concepts, tools, and techniques that will be beneficial in engineering, physics, chemistry, mathematics, biology, and computer science. Cha, Rosenberg, and Dym's *Fundamentals of Modeling and Analyzing Engineering Systems*,[1] written specifically for this course and providing an excellent introduction to the modeling and analysis of systems from a wide variety of disciplines, was used for several years. This text complements that text's focus on modeling with a strong emphasis on representation of continuous-time and discrete-time signals in both the time and frequency domains, modeling of mechanical and electrical systems, and the design of finite impulse response (FIR) discrete filters. *Fundamentals of Modeling and Analyzing Engineering Systems* and this text can be used together in a two-semester sequence to provide a thorough introduction to both signal processing and modeling, or selected topics from both can be used as the basis for a one-semester course. The sampling theorem, continuous-to-discrete and discrete-to-continuous converters, the discrete Fourier transform (DFT) and its computation with the fast Fourier transform (FFT) are explained in detail. Students are introduced to MATLAB and get hands-on experience with a series of laboratory assignments that illustrate and apply the theory. Single variable calculus is the only essential background although some knowledge of differential equations, linear algebra, and vector spaces is helpful. The materials covered in this text have

[1] *Fundamentals of Modeling and Analyzing Engineering Systems* by P. D. Cha, J. J. Rosenberg and C. L. Dym, Cambridge University Press, UK, 2000.

grown out of lectures given primarily to sophomores at Harvey Mudd College. These notes have been classroom-tested over a period of six semesters.

Engineers, scientists, and mathematicians are increasingly faced with acquiring, processing, interpreting, and extracting information from data, which are usually provided as a series of discrete samples, independent of whether the original underlying signals and systems were continuous or discrete in nature. Discrete-time techniques are used almost exclusively for simulating both continuous-time and discrete-time systems. Effective use of modern analysis, design, and simulation tools such as MATLAB require a clear understanding of the underlying theory as well as a good bit of practice with applications.

We begin by developing representations of continuous-time signals as functions and discrete-time signals as sequences (Chapter 1). We then explore various transformations such as shifts, reversal, compression, and expansion for continuous-time signals. We also cover upsampling and downsampling for discrete-time signals. Next, the construction of complicated signals from basic building blocks is introduced using the orthogonality principle (Chapter 2). This provides the foundation and context for the later focus on complex exponentials and the Fourier series. By starting with the generally applicable approach of minimizing the integrated squared error through the use of the orthogonality principle, the student is given a much deeper understanding and appreciation for a broad class of applications, including the use of Walsh functions in cellular phones to various series expansions using a variety of building blocks or basis functions. The difference between the orthogonality principle and orthogonal basis functions is carefully explained. An appendix provides a natural development of basis functions starting with the familiar three-dimensional vectors.

Complex exponentials as building blocks provide the foundation for the development of the spectrum of a continuous-time signal. Its importance in characterizing and extracting information from signals leads naturally to the need for numerical techniques to compute the spectrum of complicated signals produced by phenomena such as earthquakes, space photographs or communication systems. This, in turn, leads to the discrete Fourier transform and its efficient computation using the fast Fourier transform algorithm. While we do not derive the fast Fourier transform algorithm, the text describes its use in considerable detail and laboratory exercises provide the opportunity for students to explore practical applications.

Continuous-time and discrete-time signals are connected via the sampling theorem (Chapter 3). The spectrum of a discrete-time signal provides the basis for exploring the phenomena of aliasing and folding, which must be fully understood in order to correctly acquire, process, analyze and interpret data. The behavior of continuous-to-discrete (analog-to-digital) and discrete-to-continuous (digital-to-analog) converters is explored as the culmination of the first part of the text.

The next major division of the text is devoted to the lumped element modeling of mechanical and electrical systems (Chapters 4 and 5). We start with the basic elements (building blocks) for mechanical systems consisting of springs, dampers, and masses. First-order and second-order governing equations are developed and canonical forms are defined for both the translational and rotational systems. Parallel and series combinations of elements as well as the division of force and displacement are covered. A parallel development for electrical systems leads naturally to the force–current and velocity–voltage analogs. Solution of first-order and second-order differential governing equations is introduced, transient response specifications are defined that are used in system design, and a state space approach is formulated as an alternative means to analyze the free and forced responses of a system (Chapter 6).

Frequency response builds on the concept of the complex exponential building blocks that are covered in the first part of the text (Chapter 7). The complex exponentials also serve as the eigenfunctions of linear time-invariant systems, and the concept of frequency response provides the bridge between signals and systems. Bode plots of first-order and second-order factors are developed. The complex exponential building blocks and the concept of frequency response are then used to define impedance and its application in combining various elements of mechanical and electrical systems.

An introduction to the analysis and design of finite-impulse response filters forms the final major part of the text (Chapter 8). The ease with which arbitrary frequency response functions can be implemented is developed as another application of the Fourier series and demonstrated with a whimsical example. A series of applications from a variety of disciplines then follows, providing the student with an appreciation of the power and scope of the concepts, tools, and techniques developed throughout the text (Chapter 9). The text concludes with a short transition section designed to relate the fundamentals to concepts covered in more advanced texts (Chapter 10).

This text offers numerous special features that distinguish itself from other texts on signals and systems, and they are summarized in the following:

- A rigorous development of the construction of signals from building blocks (basis functions) via the orthogonality principle is given.
- The building block approach to develop a clear understanding of the spectra of both continuous-time and discrete-time signals as well as the frequency responses of continuous-time and discrete-time systems all without the use of the continuous-time impulse, Fourier transform, or Laplace transform is used.
- A solid understanding of the use of the FFT in extracting information from signals and determining the response of electrical and mechanical systems to realistic inputs is provided.
- Signal processing and systems modeling are treated on equal footing. Electrical engineering departments usually teach systems with primary emphasis on signal processing while mechanical engineering departments put the primary emphasis on dynamic modeling and control. In order to cover all of the desired topics, this text does not include the control of dynamical systems. This is a conscious decision the authors made in order to present a complete end-to-end analysis from characterizing the input signal, to modeling the physical system, to determining the response or output of the system to arbitrary inputs.
- A thorough treatment of the modeling of complicated mechanical and electrical systems is provided. The analogy between lumped mechanical and electrical systems is introduced in detail.
- Detailed examples of characterizing both simple and realistic (complicated) input signals, modeling physical systems, and determining their response to these inputs are provided. Most modeling texts focus extensively on how to describe the physical system and determine its response to classical inputs such as impulse, step, and sinusoid. In addition to these standard inputs, this text shows how complicated inputs can be represented using simple building blocks, thus allowing the determination of the response of systems to realistic inputs.

Finally, seven MATLAB laboratory exercises are included at the end of the text to allow students to gain a deeper understanding and mastery of the topics covered in the text. We hope that the use of computational software will enhance the learning experience and stimulate the students' interest in signals and systems.

Acknowledgments

The creation of this text was truly a team effort. As part of the teaching team Professor David Harris diligently took notes on his laptop during the authors' lectures and carefully edited them. Without these notes as a basis or building blocks, the writing of the text may never have gotten off the ground. In addition, David wrote the MATLAB-based laboratory exercises and created the review that is part of Chapter 10. Professors Lori Bassman, Mary Cardenas, David Harris, Elizabeth Orwin, Jennifer Rossmann, and Qimin Yang have all taught parts of the course over the past three years and contributed significantly to the final form and content of the text. Professor Anthony Bright, our department chair, gave us unwavering support and encouragement and allowed us to use his introduction to MATLAB that is the basis of Laboratory Exercise 1. Paul Nahin, a former colleague and long-time friend of the second author gave inspiration through his wide-ranging books from science fiction to a truly innovative blend of science technology, and history, as well as encouragement when this text was just an idea. During his senior year, student Warren Katzenstein provided support in developing the initial drafts of the text and figures. We also want to acknowledge the hundreds of students who have contributed through their questions, evaluations, and comments. Finally, we want to acknowledge the moral support of our families, who have all shared the joy and pain of the writing of this text. In particular, the first author would like to thank his Dad, Mom, brother Paul and sister Pauline, the second author his wife Janet, son Tim, and daughter Karen who always encouraged him but wondered if he would ever actually write a book.

Philip D. Cha
John I. Molinder

Harvey Mudd College
Claremont, California

1

Introduction to signals and systems

Welcome to *Introduction to Signals and Systems.* This text will focus on the properties of signals and systems, and the relationship between the inputs and outputs of physical systems. We will develop tools and techniques to help us analyze, design, and simulate signals and systems.

1.1 Signals and systems

A signal is a pattern of variation of a physical quantity: a definition which covers a wide territory. You are processing signals as you read this text. A lecturer creates a signal as he or she talks and your ear processes these signals. Signals are all around us. Examples include acoustical, electrical, and mechanical signals. Signals may depend on one or more independent variables. As the name implies, one-dimensional signals depend on one independent variable. An example is the location of a particle moving in a rectilinear motion, in which case the independent variable is time, t. Two-dimensional signals depend on two independent variables. An example is a picture that varies spatially, in which case the independent variables are the spatial coordinates, x and y. Many of the signals and systems that you have routinely dealt with have interesting properties that this text will explore.

A system processes signals. For example, a compact disc (CD) player is a system that reads a digital signal from a CD and transforms it into an electrical signal. The electrical signal goes to the speaker, which is another system that transforms electrical signals into acoustical signals. Many signals contain information. Other signals are used only to transport energy. For example, the signal from a wall socket is boring in terms of information content, but very useful for carrying energy.

Signals can be categorized as analog, discrete or digital. They are summarized as follows:

- **Analog signals:** signals that vary continuously in amplitude and time. The independent variable is not necessarily time, it could also be a spatial coordinate.
- **Discrete-time signals:** signals that have continuous amplitude but only exist at discrete times. These signals are represented as sequences of numbers.
- **Digital signals:** signals that have discrete amplitude and time. These signals are represented by sequences of numbers with finite precision. They are used when processing information by computer.

We will often be interested in converting signals from one type to another. For example, in a chemistry laboratory, a continuous-time transducer measures the analog value of a physical quantity. We will often use a continuous-to-discrete (analog-to-digital) converter to capture the signal into a computer for processing.

Systems can be categorized as linear or nonlinear. Systems are said to be linear when scaling and superposition hold. For linear systems, if the input to the system is scaled by some constant a, the output of the system will be scaled by the same amount. Thus, if the input to the system is doubled, so will be the output of the system. Linear systems also obey the superposition principle. Thus, for a linear system, the response of the system to a combination of N inputs is simply the sum of the responses to each input considered individually.

Systems can also be classified as time-variant or time-invariant. When the parameters of a system remain constant during operation, the system is said to be time-invariant. When the parameters can vary as a function of time, the system is said to be time-variant. For time-invariant systems, the system responds the same yesterday, today, and tomorrow. Although time-varying systems without nonlinearity are still considered linear, such systems are considerably more difficult to analyze and design.

In this text, we will primarily be interested in linear and time-invariant systems, or LTI systems for short. They are very useful for signal processing and system modeling. While most physical signals and systems are not LTI, surprisingly many can be approximated as LTI over a specified time domain of interest. Nonlinear systems have some very interesting and surprising properties but are usually much more difficult to handle mathematically, and there are limited methods available for solving nonlinear systems.

A familiar example of signals and systems is the recording and playback of audio signals such as music and voice. Over the years there have been major changes in the technology used resulting in dramatic improvements in quality. Milestones in audio technology include:

- Phonograph: invented by Thomas Edison in 1877.
- Gramophone: developed by Emile Berliner in 1887 (70 revolution per minute or rpm).
- 78 rpm record: 1930s.
- AM (amplitude modulation) radio: 1920s
- 33 1/3 LP (long-playing) record: introduced by CBS in 1948 (held about 20 minutes on each side).
- 45 EP (extended-playing) record: introduced by RCA in 1948 (more portable, held 5–6 minutes on each side).
- Stereo: 1956.
- Stereo FM (frequency modulation): 1960s.
- CD (compact disc): developed by Sony and Philips in 1982 (holds about 70 minutes).

The information density has improved over time. On a conventional LP record, each track is about 100 micrometers wide. On a CD, the tracks are 1.6 micrometers wide, so tracks are packed 60 times more densely. On a CD, the information is stored by a series of pits burned with a powerful laser. The information is read by measuring the reflection from another laser to determine where the pits are located. The locations encode a series of binary numbers, so a CD is really just a physical encoding of a long sequence of 1s and 0s.

Signals may be represented as a graph with time on the horizontal axis and amplitude of the signals on the vertical axis. An *oscilloscope* is a system that converts an electrical signal into an optical signal showing such a graph, allowing us to examine how the signal varies as a function of time. A *spectrum analyzer* is a system that converts an electrical signal into an optical signal showing a graph of what frequencies are in the input, called the *spectrum* of a signal. We will have much more to say about the spectrum of a signal. Knowing the frequency content of a signal allows us to characterize the signal.

Fig. 1.1. A 200 Hz sine wave,
$x(t) = \cos[2\pi(200t)]$.

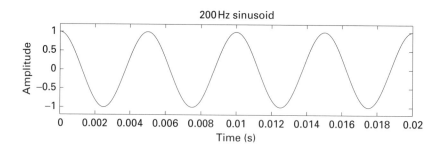

Fig. 1.2. The spectrum of a 200 Hz sine wave.

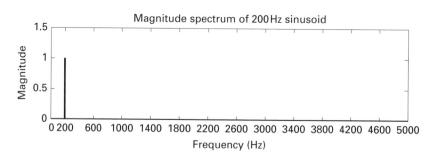

1.2 Examples of signals

Let us now consider some simple examples of signals. Sinusoidal signals will prove to be one of the most useful signals we will encounter. Mathematically, a sinusoidal signal may be represent as

$$x(t) = A\cos(2\pi ft + \phi), \tag{1.1}$$

where A denotes the *amplitude* or magnitude of the signal, f its corresponding *frequency*, and ϕ its *phase*. Figure 1.1 shows a sine wave whose amplitude is 1 and phase is 0, and which has a frequency of 200 Hz (0.2 kHz). Figure 1.2 shows the spectrum of the same sine wave. The spectrum of a signal consists of a graph that shows what frequencies are present in the signal as well as the magnitudes of the frequency components. Thus, as expected, the spectrum of a 200 Hz sine wave (Figure 1.2) shows a single peak centered at 200 Hz with amplitude 1.

Figure 1.3 shows a periodic square wave at 200 Hz, and Figure 1.4 shows its spectrum. The square wave alternates between 1 and -1 with a period of 0.005 s. The spectrum consists of spikes at 200 Hz, 600 Hz, 1000 Hz, 1400 Hz, etc. The largest peak is at the fundamental frequency f_0 of 200 Hz. The next peaks are at the third harmonic $3f_0$, fifth harmonic

Fig. 1.3. A 200 Hz square wave.

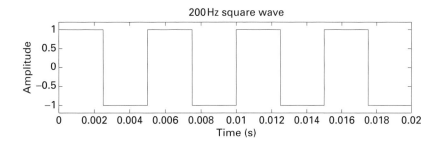

Fig. 1.4. The spectrum of a 200 Hz square wave.

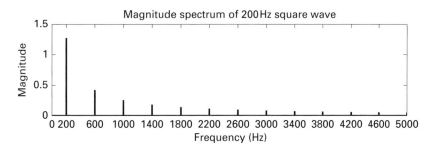

Fig. 1.5. A 200 Hz triangle wave.

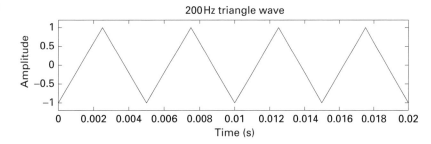

$5 f_0$, seventh harmonic $7 f_0$, and so forth. The amplitudes drop off for higher harmonics. Later we will show how these amplitudes and frequencies can be determined analytically using Fourier series and numerically using the MATLAB command fft.

Figure 1.5 shows a triangle wave at 200 Hz and Figure 1.6 shows its spectrum. Note that the spectrum also contains the odd harmonics, but the amplitudes drop off quickly compared to the amplitudes of the square wave of Figure 1.4.

The previous examples suggest that we could construct any signal by summing sinusoids of different amplitudes and frequencies. This underlying principle forms the basis of *Fourier series*. The spectrum shows the am-

Fig. 1.6. The spectrum of a
200 Hz triangle wave.

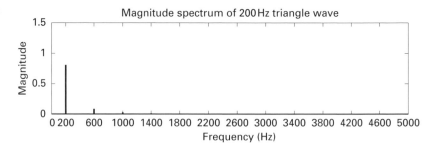

Fig. 1.7. A graph of
digitized signal.

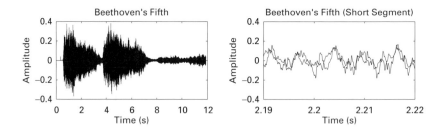

plitude of the coefficients of the Fourier series. We will be interested in both
synthesizing signals from sums of other signals (for example sinusoids) and
analyzing signals by determining the amplitudes of the frequencies within
the signal.

As a more elaborate example, Figure 1.7 shows a digital representation
of a two-channel audio signal sampled at 44 kHz (CD quality). Clearly, the
eye is not trained to interpret the signal, even when zooming in on a short
segment. Transforming the signal into an acoustical signal, however, makes
it clearly recognizable as the beginning of Beethoven's Fifth Symphony.
The above examples serve to remind us that there are many different ways
to represent a signal. We can identify a signal by plotting its variation with
time, analyzing its spectrum, or listening to how the signal changes as a
function of time. These approaches can all be used to characterize the same
signal. Depending on the application, one approach may be more useful
than the others.

Another example of a system that uses signals is the telephone. Dialing
a touch-tone telephone generates a series of tones. These tones are the
superposition or sum of a pair of sine waves, as shown in Table 1.1. You
could build your own dialer by producing these tones with another system
such as your computer. Similarly, you could determine which phone number

Table 1.1. Touch-tone telephone tones

Frequencies	1209 Hz	1336 Hz	1477 Hz
697 Hz	1	2	3
770 Hz	4	5	6
852 Hz	7	8	9
941 Hz	*	0	#

was dialed by looking at the spectrum of the tones and observing what frequencies are present.

Having seen various examples of signals and systems, we are now ready to lay the mathematical foundations to understand them in detail.

1.3 Mathematical foundations

Euler's formula (or identity) was introduced in calculus. It states that a complex exponential[1] can be expressed as the sum of a cosine function and a sine function:

$$e^{j\theta} = \cos\theta + j\sin\theta, \tag{1.2}$$

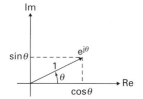

Fig. 1.8. Complex exponential plotted on a complex plane.

where $j = \sqrt{-1}$.[2] Euler's identity can be easily proved by expanding e^x in Taylor series and replacing x by $j\theta$. Figure 1.8 shows Eq. (1.2) plotted as a vector on the complex plane, where the horizontal axis corresponds to the real axis, and the vertical axis corresponds to the imaginary axis. Observe that the length or magnitude of the vector is 1 and the angle is θ, where θ is measured in radians, and is positive in the counterclockwise sense from the horizontal or real axis. Similarly, we can show that

$$e^{-j\theta} = \cos\theta - j\sin\theta. \tag{1.3}$$

Therefore, in terms of complex exponentials, cosine and sine can be expressed as

$$\cos\theta = \frac{1}{2}(e^{j\theta} + e^{-j\theta}) \quad \text{and} \quad \sin\theta = \frac{1}{2j}(e^{j\theta} - e^{-j\theta}). \tag{1.4}$$

---

[1] Because complex arithmetic will be used extensively throughout the text, the reader is encouraged to review Appendix A for a detailed discussion.

[2] Engineers often use j to represent complex unity. The variable i is reserved to denote the current of electrical systems.

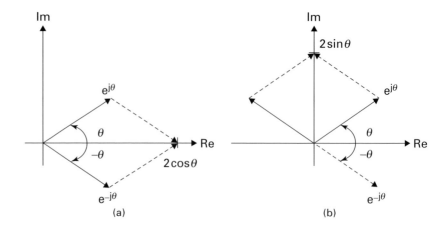

(a)   (b)

These sums are represented graphically in Figure 1.9. Recall that the complex conjugate of a complex number $a + jb$ is simply $a - jb$, where a and b are real constants. Thus, $e^{j\theta}$ and $e^{-j\theta}$ are complex conjugates, because their real parts are identical and their imaginary parts are the negative of one another.

In terms of complex exponentials, we can now write a simple sine wave as follows:

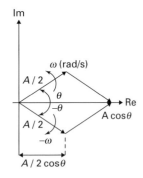

$$x(t) = A\cos(\omega t + \phi) = \frac{A}{2}\left[e^{j(\omega t + \phi)} + e^{-j(\omega t + \phi)}\right], \tag{1.5}$$

where A and ϕ are real constants. The units of ω are radians per second. This is shown graphically in Figure 1.10, where both vectors are now functions of time, t. Note that as t increases, the upper vector rotates counterclockwise at a rate determined by the angular frequency ω and the lower vector rotates clockwise at an angular frequency of $-\omega$. Thus, we define a vector with a negative frequency $(-\omega)$ as one that rotates in the clockwise direction.

There are many different ways of representing a sinusoid. Another representation utilizes complex exponentials and the definition of the real part of a complex number as follows:

$$x(t) = A\cos(\omega t + \phi) = \text{Re}\left\{Ae^{j(\omega t + \phi)}\right\}, \tag{1.6}$$

where A and ϕ are real constants, and Re{ } indicates the real part of an expression. Using Euler's identity, we have:

$$Ae^{j(\omega t + \phi)} = A\cos(\omega t + \phi) + jA\sin(\omega t + \phi). \tag{1.7}$$

By inspection, note that $x(t)$ of Eq. (1.5) is simply the real part of Eq. (1.7).

Finally, using Eq. (1.6), we can also represent a sinusoid as:

$$x(t) = \mathrm{Re}\{Ae^{j\phi}e^{j\omega t}\} = \mathrm{Re}\{Xe^{j\omega t}\}. \tag{1.8}$$

This is called the *phasor* notation. The phasor $X = Ae^{j\phi}$ contains the amplitude (A) and phase (ϕ) information of the sinusoid, corresponding to the length or magnitude and angle of the vector in the complex plane.

We have introduced various ways of representing signals. They will be used extensively throughout the text, and the reader should become very familiar with these different representations and how they are related.

1.4 Phasors

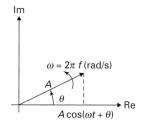

Fig. 1.11. The real part of a rotating phasor.

We have seen that sine waves appear to be important building blocks. Thus, we expect that we can construct more elaborate signals as the sums of sine waves of various frequencies, amplitudes, and phases. We can write any sine wave as

$$x(t) = A\cos(\omega t + \theta). \tag{1.9}$$

This can also be represented with a phasor as follows:

$$x(t) = \mathrm{Re}\{Ae^{j\theta}e^{j\omega t}\}, \tag{1.10}$$

which can be viewed as a vector on the complex plane, as shown in Figure 1.11. The vector rotates at a rate (called the *angular frequency*) of ω in units of radians per second. The real part of the vector corresponds to the value of the sinusoid of Eq. (1.8) at time t. The wave is periodic with a period T between repetitions, where the period is given by $T = 2\pi/\omega$. We often wish to refer to the rate at which the wave repeats. Let us define this rate as $f = \omega/(2\pi) = 1/T$, where f is called the *frequency*, measured in units of hertz (Hz) or cycles/second. Therefore, in terms of frequency, we often rewrite Eq. (1.9) as

$$x(t) = A\cos(2\pi f t + \theta). \tag{1.11}$$

Also observe that the peaks occur at times

$$t = \frac{2\pi n - \theta}{2\pi f}, \quad n = 0, \pm 1, \pm 2, \ldots. \tag{1.12}$$

Signals generally represent physical quantities and should be dimensioned appropriately. For example, the x-axis may be in dimensions of seconds, minutes, or feet. The y-axis may be in volts, meters, degrees Celsius, etc. For example, a facetious set of dimensions for calibrating speedometers is furlongs/fortnight. In this text we will typically use the SI (Système International) metric units.

In analyzing signals, we often will be interested in adding multiple sine waves of the same frequency. Consider two signals $x_1(t)$ and $x_2(t)$ of the same angular frequency, ω:

$$
\begin{aligned}
x_1(t) &= A_1 \cos(\omega t + \theta_1) \\
x_2(t) &= A_2 \cos(\omega t + \theta_2),
\end{aligned}
\tag{1.13}
$$

whose sum is given by

$$
x(t) = x_1(t) + x_2(t) = A \cos(\omega t + \theta).
\tag{1.14}
$$

Not surprisingly, the amplitude (A) and phase (θ) of the new signal $x(t)$ are related to the amplitudes and phases of $x_1(t)$ and $x_2(t)$, and they can be obtained through simple algebraic manipulations. Let us first expand $x_1(t)$ and $x_2(t)$ using the trigonometric identity for the cosine of the sum of two angles,

$$
\begin{aligned}
x_1(t) &= A_1 \cos(\omega t) \cos \theta_1 - A_1 \sin(\omega t) \sin \theta_1 \\
x_2(t) &= A_2 \cos(\omega t) \cos \theta_2 - A_2 \sin(\omega t) \sin \theta_2.
\end{aligned}
\tag{1.15}
$$

Summing the coefficients of the cosine and sine terms, we get

$$
x(t) = \left(\underbrace{A_1 \cos \theta_1 + A_2 \cos \theta_2}_{A_c} \right) \cos(\omega t)
$$
$$
- \left(\underbrace{A_1 \sin \theta_1 + A_2 \sin \theta_2}_{A_s} \right) \sin(\omega t),
\tag{1.16}
$$

which can be rewritten as

$$
\begin{aligned}
x(t) &= A_c \cos(\omega t) - A_s \sin(\omega t) \\
&= A \cos \theta \cos(\omega t) - A \sin \theta \sin(\omega t) \\
&= A \cos(\omega t + \theta).
\end{aligned}
\tag{1.17}
$$

By matching the coefficients of $\cos(\omega t)$ and $\sin(\omega t)$, we note immediately that

$$
\begin{aligned}
A \cos \theta &= A_c \\
A \sin \theta &= A_s,
\end{aligned}
\tag{1.18}
$$

from which we find:

$$A = \sqrt{A_c^2 + A_s^2}$$
$$\theta = \tan^{-1}\left(\frac{A_s}{A_c}\right). \tag{1.19}$$

While conceptually simple, this direct approach of determining the new amplitude and phase is quite lengthy, leaves ample room for error, and does not give us very much insight. The mathematics becomes much simpler and cleaner using the phasor notation. To find the sum of $x_1(t)$ and $x_2(t)$, we first rewrite them in phasor form as illustrated below:

$$x_1(t) = \mathrm{Re}\left\{\underbrace{A_1 e^{j\theta_1}}_{X_1} e^{j\omega t}\right\}$$

$$x_2(t) = \mathrm{Re}\left\{\underbrace{A_2 e^{j\theta_2}}_{X_2} e^{j\omega t}\right\} \tag{1.20}$$

$$x(t) = x_1(t) + x_2(t) = \mathrm{Re}\left\{\underbrace{(X_1 + X_2)}_{X} e^{j\omega t}\right\}.$$

Their sum can also be expressed in phasor form, as shown in Eq. (1.20). To determine the new phasor X, we simply equate it with the sum of X_1 and X_2 as follows:

$$X = X_1 + X_2 = A_1 e^{j\theta_1} + A_2 e^{j\theta_2}$$
$$= [A_1 \cos\theta_1 + A_2 \cos\theta_2] + j[A_1 \sin\theta_1 + A_2 \sin\theta_2] \tag{1.21}$$
$$= A_c + jA_s = A e^{j\theta},$$

where A and θ are given by Eq. (1.19).

The above mathematical operation can be viewed as the addition of two vectors in the complex plane. The magnitude and phase are the length and angle of the resultant vector X, as shown in Figure 1.12. Physically, the constants A_c and A_s correspond to the real and imaginary parts of the right triangle, and the angle θ corresponds to the angle of the resulting vector, measured counterclockwise from the horizontal or the real axis.

Phasors are also convenient for calculating the time derivatives of sinusoids. The conventional approach is to differentiate the cosine, giving a new

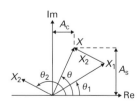

Fig. 1.12. Graphical addition of sinusoids with phasors.

cosine scaled by ω and shifted in phase by $\pi/2$, as shown below:

$$x(t) = A\cos(\omega t + \theta)$$
$$y(t) = \frac{\mathrm{d}x}{\mathrm{d}t} = -A\omega\sin(\omega t + \theta) = A\omega\cos\left(\omega t + \theta + \frac{\pi}{2}\right): \tag{1.22}$$

Because exponentials are trivial to differentiate, taking the derivative is easier using phasors as illustrated below:

$$y(t) = \frac{\mathrm{d}x}{\mathrm{d}t} = \frac{\mathrm{d}}{\mathrm{d}t}\mathrm{Re}\{X\mathrm{e}^{\mathrm{j}\omega t}\} = \mathrm{Re}\left\{\frac{\mathrm{d}}{\mathrm{d}t}(X\mathrm{e}^{\mathrm{j}\omega t})\right\} = \mathrm{Re}\left\{\underbrace{\mathrm{j}\omega X}_{Y}\,\mathrm{e}^{\mathrm{j}\omega t}\right\}. \tag{1.23}$$

This could be simplified by noting the relationship between the phasors X and Y:

$$Y = \mathrm{j}\omega X = \mathrm{j}\omega A\mathrm{e}^{\mathrm{j}\theta} = \omega A\mathrm{e}^{\mathrm{j}\frac{\pi}{2}}\mathrm{e}^{\mathrm{j}\theta} = \omega A\mathrm{e}^{\mathrm{j}(\theta + \frac{\pi}{2})}. \tag{1.24}$$

By inspection, we note that the new phasor Y is scaled by ω and is shifted in phase by $\pi/2$ relative to the phasor X. The second derivative of $x(t)$ can also be expressed with a phasor whose amplitude and phase are scaled by another factor of ω and shifted by another phase of $\pi/2$, respectively. Simply put, taking the nth derivative of the following phasor $X\mathrm{e}^{\mathrm{j}\omega t}$ corresponds to multiplying $X\mathrm{e}^{\mathrm{j}\omega t}$ by $(\mathrm{j}\omega)^n$. Similarly, it can be shown that taking the nth integral of $X\mathrm{e}^{\mathrm{j}\omega t}$ corresponds to dividing $X\mathrm{e}^{\mathrm{j}\omega t}$ by $(\mathrm{j}\omega)^n$. Thus, to compute

$$y(t) = \frac{\mathrm{d}^{121}}{\mathrm{d}t^{121}}[A\cos(\omega t + \theta)] \tag{1.25}$$

we can either differentiate $A\cos(\omega t + \theta)$ 121 times, or differentiate it in one line using the concept of phasors as illustrated below:

$$y(t) = \mathrm{Re}\{(\mathrm{j}\omega)^{121}X\mathrm{e}^{\mathrm{j}\omega t}\}. \tag{1.26}$$

In summary, phasors can often simplify the mathematical operations of adding (of the same frequencies) and differentiating sinusoids.

1.5 Time-varying frequency and instantaneous frequency

Consider a sinusoidal signal with a constant angular frequency ω (in rad/s) given by

$$x(t) = A\cos(\omega t + \theta) = A\cos(\psi(t)), \tag{1.27}$$

where $\psi(t) = \omega t + \theta$. The function $\psi(t)$ is known as the angle or phase of the signal. Taking the derivative of $\psi(t)$, we get

$$\frac{d\psi(t)}{dt} = \omega = 2\pi f. \tag{1.28}$$

We note that the time derivative of $\psi(t)$ yields the constant angular frequency ω (in rad/s) or the constant frequency f (in Hz). We will now extend the above result to analyze a signal whose frequency varies with time.

Consider a signal with a time-varying frequency as follows:

$$x(t) = A\cos(\psi(t)). \tag{1.29}$$

Motivated by the result of Eq. (1.28), we define an *instantaneous frequency* as the time derivative of $\psi(t)$, given by

$$\frac{d\psi(t)}{dt} = \omega_i(t) = 2\pi f_i(t), \tag{1.30}$$

where $\omega_i(t)$ is known as the instantaneous angular frequency in radians per second, and $f_i(t)$ is the instantaneous frequency in hertz. Upon rearranging, we find

$$f_i(t) = \frac{1}{2\pi}\frac{d\psi(t)}{dt}. \tag{1.31}$$

Assume we desire a signal with an instantaneous frequency of $f_i(t) = at + b$, where a and b are real constants. If we wish to sweep the frequency from f_1 at time $t = 0$ to f_2 at time $t = T$, then we can readily determine the unknown constants a and b by solving the following algebraic equations:

$$\begin{aligned}f_i(t = 0) &= f_1 = b\\ f_i(t = T) &= f_2 = a + bT.\end{aligned} \tag{1.32}$$

Equation (1.32) yields two equations with two unknowns, whose solutions are

$$b = f_1 \quad \text{and} \quad a = \frac{f_2 - f_1}{T}. \tag{1.33}$$

Thus, we can solve for $\psi(t)$ as follows:

$$\frac{1}{2\pi}\frac{d\psi(t)}{dt} = f_i(t) = at + b, \tag{1.34}$$

from which we obtain

$$\psi(t) = 2\pi\left(\frac{a}{2}t^2 + bt + k\right), \tag{1.35}$$

where k is a constant of integration. The constant k is uniquely determined from the initial condition of the signal, given by $x(t = 0)$. Hence, a signal

Fig. 1.13. Plot of a sweep.

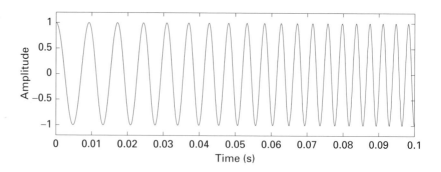

that varies from a frequency of 100 Hz to a frequency of 2000 Hz over a 1 second interval may be written as

$$x(t) = A \cos\{2\pi[(950)t^2 + 100t + k]\}. \tag{1.36}$$

A portion of this signal is plotted in Figure 1.13 (zoomed in for 0.1 seconds to show how the repetition rate increases with time; for simplicity, we arbitrarily set $k = 0$). Such a signal is often called a *sweep*.

Often we are interested in understanding how a system responds to sine waves of different frequencies. One way to do this is to apply a signal of a particular frequency and measure the response, then repeat for each different frequency of interest. Another approach is to apply a single sweep, which will return information about the response of the system to many different frequencies. In this case, the rate of sweep must be significantly lower than the response characteristics of the system in order to give the system sufficient time to respond.

1.6 Transformations

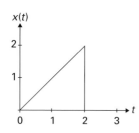

Fig. 1.14. A sawtooth signal.

Frequently we will also be interested in performing various transformations on signals. For example, let us consider a *sawtooth* signal $x(t)$, as shown in Figure 1.14. Mathematically, $x(t)$ is defined as

$$x(t) = \begin{cases} t & 0 < t < 2 \\ 0 & \text{otherwise.} \end{cases} \tag{1.37}$$

Let us define two new signals as follows:

$$\begin{aligned} y_1(t) &= x(t - 2) \\ y_2(t) &= x(t + 1). \end{aligned} \tag{1.38}$$

Fig. 1.15. The signal of Figure 1.14 time shifted by (a) −2 and (b) +1.

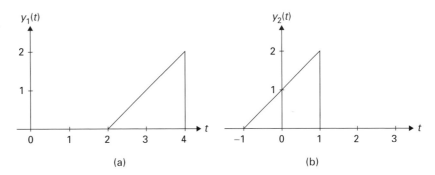

(a)   (b)

Fig. 1.16. The signal of Figure 1.14 vertically shifted by (a) −2 and (b) +1.

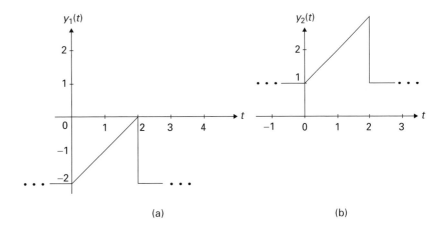

(a)   (b)

These new functions are *shifted* versions of the original, and are shown in Figure 1.15. The first signal $y_1(t)$ is simply $x(t)$ shifted to the right by 2 units, and the second signal $y_2(t)$ is $x(t)$ shifted to the left by 1. In general, $x(t - a)$ is a copy of the signal $x(t)$ shifted to the right by a units (assuming that a is positive). Similarly, $x(t + a)$ is a copy of the signal $x(t)$ shifted to the left by a units, (again where $a > 0$).

In addition to shifting a signal along the t-axis, we can also shift a signal vertically along the y-axis. Let

$$y_1(t) = x(t) - 2$$
$$y_2(t) = x(t) + 1. \tag{1.39}$$

These new functions are shown in Figure 1.16. The first signal $y_1(t)$ is the original shifted downwards by 2 units, and the second signal $y_2(t)$ is the original shifted upwards by 1. In general, $x(t) + a$ is a copy of $x(t)$ shifted vertically by a. Finally, note that the regions that were previously zeros are now a.

Fig. 1.17. The reverse of the signal of Figure 1.14.

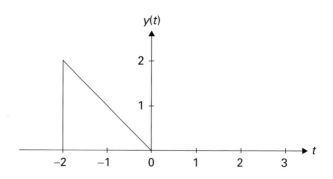

Fig. 1.18. The signal of Figure 1.14 (a) compressed by a factor of 2 and (b) expanded by a factor of 2.

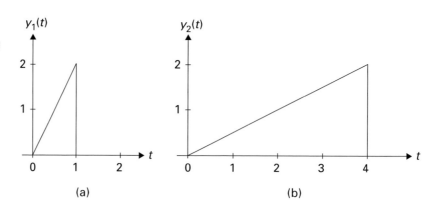

Another commonly encountered transformation is the *reverse*, given by

$$y(t) = x(-t) \tag{1.40}$$

and shown in Figure 1.17. For example, if $x(t)$ is a signal recorded on a magnetic tape, then the new signal $y(t)$ is the signal obtained by playing the tape backwards.

We can also consider multiplying the rate of a signal by a positive constant, as given by

$$y_1(t) = x(2t)$$
$$y_2(t) = x\left(\frac{1}{2}t\right). \tag{1.41}$$

In Figure 1.18, the first signal $y_1(t)$ is the original *compressed* by a factor of two, and the second signal $y_2(t)$ is the original *expanded* by a factor of 2. In general, $x(at)$ is a copy of $x(t)$ compressed or expanded by a factor of a (assuming $a > 0$), depending on whether $a > 1$ or $a < 1$.

Fig. 1.19. The signal of
Figure 1.14 (a) amplified
by a factor of 2 and
(b) attenuated by a factor
of 2.

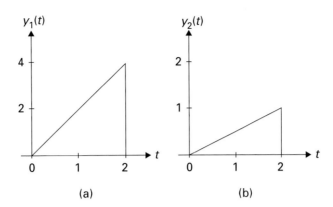

Fig. 1.19. The signal of Figure 1.14 (a) amplified by a factor of 2 and (b) attenuated by a factor of 2.

We can also consider multiplying the amplitude of a signal by a positive constant a, as shown below:

$$y_1(t) = 2x(t)$$
$$y_2(t) = \frac{1}{2}x(t). \tag{1.42}$$

In Figure 1.19, the first signal $y_1(t)$ is the original *amplified* by a factor of two, and in the second signal $y_2(t)$ is the original *attenuated* by a factor of two. In general, $ax(t)$ is a copy of the signal $x(t)$ amplified or attenuated by a factor of a (assuming $a > 0$), depending on whether $a > 1$ or $a < 1$.

More complicated transformations may be viewed as a combination of shift, reverse, and compress operations. Consider the following signal:

$$y(t) = x(2t + 3) = x\left[2\left(t + \frac{3}{2}\right)\right]. \tag{1.43}$$

One way to analyze the new signal $y(t)$ is to consider successive transformations as follows. Let us first consider a transformation to a new signal $y_1(t)$, followed by another transformation to yield $y(t)$, as shown below:

$$y_1(t) = x(t + 3)$$
$$y(t) = y_1(2t). \tag{1.44}$$

These signals are illustrated in Figure 1.20.

Alternatively, we can analyze the transformation in a single step by considering the second representation of $y(t)$ as shown in Eq. (1.43), namely $y(t) = x(2(t + 3/2))$. We can immediately see that this transformation is merely a compression by a factor of 2 followed by a leftward shift of $3/2$. Finally, after the new signal is obtained, it is always helpful to substitute

Fig. 1.20. The successive
transformations of
Eq. (1.44).

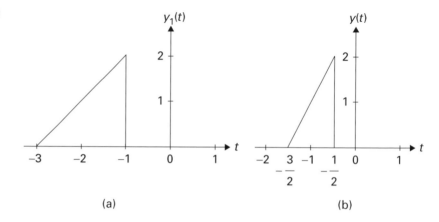

Fig. 1.20. The successive transformations of Eq. (1.44).

(a)   (b)

values at the endpoints or other representative points to check for the correctness of the transformation.

1.7 Discrete-time signals

The signals shown so far have been continuous-time signals, represented with continuous graphs and the signal $x(t)$. Now we focus our attention on *discrete-time signals* or signals that are only defined at isolated time intervals. Discrete-time signals are represented with a sequence of numbers and as a convention are denoted by $x[n]$. For example, consider the following sequence:

$$x[n] = \{-0.7, 0.6, 1.0, 0.3, 0.8, -0.3\} \quad \text{for}$$
$$n = -2, -1, 0, 1, 2, 3. \tag{1.45}$$

Figure 1.21 shows how such a signal may be graphed. This is sometimes called a *lollipop diagram* or *stem diagram*. Note that this signal is only defined for integer values of n. For example, $x[1/2]$ is meaningless. It is not zero.

A discrete-time signal may be represented by a sequence of numbers or graphed on a lollipop diagram. It may also be given in a table, such as Table 1.2. The table may be easily stored in a computer and processed with software such as MATLAB.

There are a number of special discrete-time signals that we will often see. They consist of basic building blocks with which other complicated signals can be constructed. These include a unit step sequence and a unit impulse

Table 1.2. Discrete-time signal values

n	$x[n]$
< -2	0
-2	-0.7
-1	0.6
0	1
1	0.3
2	0.8
3	-0.3
> 3	0

Fig. 1.21. The lollipop diagram for the discrete-time signal of Eq. (1.45).

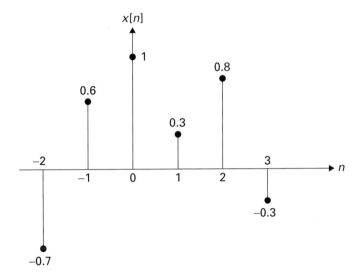

sequence. They are defined as follows and can be seen schematically in Figure 1.22 and Figure 1.23, respectively. A discrete-time unit step sequence is given by

$$\text{Unit step}: u[n] = \begin{cases} 1 & n \geq 0 \\ 0 & n < 0, \end{cases} \tag{1.46}$$

and a discrete-time unit impulse sequence is defined as

$$\text{Unit impulse}: \delta[n] = \begin{cases} 1 & n = 0 \\ 0 & n \neq 0. \end{cases} \tag{1.47}$$

A unit step sequence consists of a sequence of ones for $n \geq 0$ and zeros otherwise. A unit impulse sequence consists of a one for $n = 0$ and zeros

Fig. 1.22. The
discrete-time unit step
function.

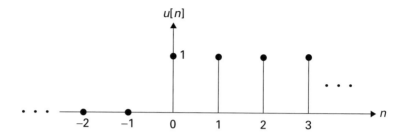

Fig. 1.23. The
discrete-time unit impulse
function.

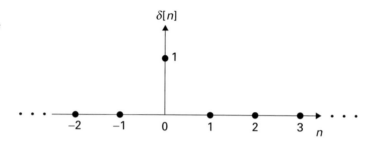

Fig. 1.24. A
continuous-time unit step
function.

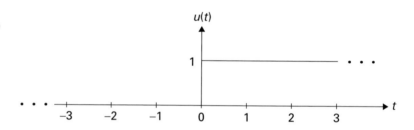

otherwise. These sequences can also be shifted by an integer either to the right or to the left. Thus, $\delta[n - m]$ is defined to be one for $n = m$, and zeros elsewhere.

Note that there are similar signals defined for continuous-time. However, for continuous-time, their definitions may depend on the specific application and may appear abstract. For example, one possible definition for the continuous-time unit step function is

$$\text{Unit step} : u(t) = \begin{cases} 1 & t > 0 \\ 1/2 & t = 0 \\ 0 & t < 0 \end{cases} \tag{1.48}$$

and is shown in Figure 1.24. Note that $u(t)$ is defined to be exactly $1/2$ at $t = 0$. Different texts may define $u(t)$ differently for $t = 0$. Other possible definitions for a unit step function include

$$\text{Unit step}: u(t) = \begin{cases} 1 & t \geq 0 \\ 0 & t < 0 \end{cases} \tag{1.49}$$

or

$$\text{Unit step}: u(t) = \begin{cases} 1 & t > 0 \\ 0 & t \leq 0 \end{cases} \tag{1.50}$$

or

$$\text{Unit step}: u(t) = \begin{cases} 1 & t > 0 \\ 0 & t < 0. \end{cases} \tag{1.51}$$

A continuous-time unit impulse function is defined as follows:

$$\delta(t) = \begin{cases} \infty & t = 0 \\ 0 & t \neq 0 \end{cases} \tag{1.52}$$

such that

$$\int_{-\infty}^{\infty} \delta(t)\mathrm{d}t = 1. \tag{1.53}$$

Thus, for continuous-time, a unit impulse function is defined to have zero width, infinite height, and unity area. Continuous-time unit step and unit impulse functions will not be treated in this text. Interested readers can refer to *Signals and Systems* by Oppenheim, Willsky, and Nawab.[3]

Any arbitrary signal may be constructed from a sum of these special signals. For example, $x[n]$ from Table 1.2 may be constructed using scaled and shifted unit impulses as follows:

$$x[n] = -0.7\delta[n+2] + 0.6\delta[n+1] + \delta[n] + 0.3\delta[n-1]$$
$$+ 0.8\delta[n-2] - 0.3\delta[n-3]. \tag{1.54}$$

Similarly, the unit impulse may be expressed as the difference between two unit step functions as follows:

$$\delta[n] = u[n] - u[n-1]. \tag{1.55}$$

In general, any discrete-time signal may be written as a sum of scaled and shifted unit impulses as shown below:

$$x[n] = \sum_{k=-\infty}^{\infty} x[k]\,\delta[n-k], \tag{1.56}$$

---

[3] *Signals and Systems* by A. V. Oppenheim, A. S. Willsky and S. H. Nawab, Prentice Hall, New Jersey, 1997.

Fig. 1.25. Sampling of an
arbitrary signal.

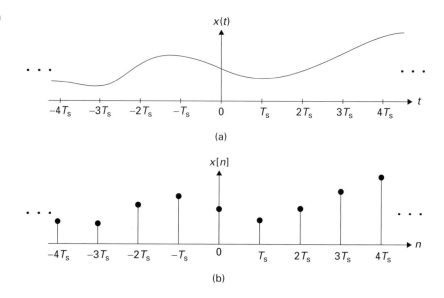

where $x[k]$ represents the signal at index k, and $\delta[n-k]$ represents the shifted unit impulse function. Using this notation, we note that the unit step function can be expressed as the sum of an infinite number of unit impulses:

$$u[n] = \sum_{k=-\infty}^{\infty} u[k]\delta[n-k] = \sum_{k=0}^{\infty} \delta[n-k]. \tag{1.57}$$

1.8 Sampling

Consider a continuous-time signal $x(t)$. We can define an associated discrete-time signal as follows:

$$x[n] = x(nT_\mathrm{s}), \tag{1.58}$$

where n is some integer and T_s is the sampling period, which measures how fast the signal is acquired or sampled. For notation purposes, the continuous-time signal is indicated with ( ) and the discrete-time signal is indicated with [ ]. For example, Figure 1.25(a) shows a continuous-time signal, and Figure 1.25(b) shows the discrete-time signal, obtained from the continuous-time signal by *sampling* at some interval T_s. It is impossible to store a continuous-time signal in a computer because it requires an infinite amount of data. However, we could store the associated sampled discrete-time signal, then attempt to recreate the continuous-time signal from the sampled data. We will discover that under certain practical conditions it is possible to exactly reconstruct the continuous-time signal from the samples. Of course we cannot store the *exact* values of the samples in the computer, but we will ignore the effects of finite precision in this text.

We will often sample a signal $x(t)$ at intervals T_s to get $x[n]$. This can be easily achieved by using Eq. (1.58). Consider a simple sinusoidal signal $x(t)$, defined as

$$x(t) = A\cos(\omega t + \theta) = A\cos(2\pi f t + \theta). \tag{1.59}$$

To find the corresponding discrete-time signal, we simply replace t by nT_s as follows:

$$x[n] = x(nT_s) = A\cos\left(n\underbrace{\omega T_s}_{\hat{\omega}} + \theta\right) = A\cos\left(2\pi n\underbrace{f T_s}_{\hat{f}} + \theta\right). \tag{1.60}$$

In this notation, $\hat{\omega}$ is called the *digital angular frequency* and is measured in units of radians. Similarly, $\hat{f}$ is known as the *digital frequency* and is measured in units of cycles.

1.9 Downsampling and upsampling

In many different discrete-time sampling applications such as filter design and implementation or communication applications, it is very inefficient to transmit or store the entire sequence of sampled data, because there may be many zeros in the sampled data. Thus, the sampled sequence is typically replaced by a new sequence, $x_{(N)}[n]$, which is sampled every Nth value, that is,

$$x_{(N)}[n] = x[nN]. \tag{1.61}$$

The operation of extracting every Nth sample and discarding the rest is defined as *downsampling* or *decimation*. For example, for the $x[n]$ shown in Figure 1.26(a), $x_{(2)}[n]$ is given by

$$x_{(2)}[n] = 2.5, 2.0, \quad \text{where } n = 0, 1. \tag{1.62}$$

It should be noted that the original signal cannot be directly reconstructed after downsampling because some information would be lost.

Fig. 1.26. (a)
Downsampling and (b)
upsampling of an arbitrary
discrete-time signal.

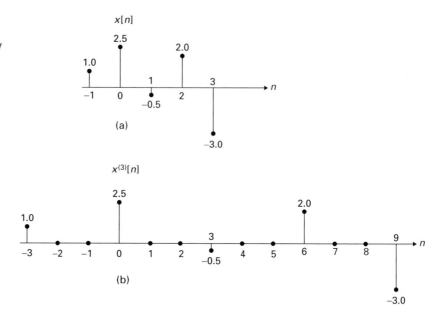

(a)

(b)

In some applications it is useful to downsample, but in other cases it may be useful to convert a sequence to a higher equivalent sampling rate, a process known as *upsampling* or *zero-stuffing*, defined as

$$x^{(N)}[n] = \begin{cases} x[n/N] & \text{for } n = 0, \pm N, \pm 2N, \ldots \\ 0 & \text{otherwise.} \end{cases} \tag{1.63}$$

It is often part of a process used to interpolate between samples in applications such as a CD player. For example, if $x[n]$ is given in Figure 1.26(a), then $x^{(3)}[n]$ is shown in Figure 1.26(b). Unlike downsampling, the original signal can be successfully reconstructed because no information is lost.

1.10 Problems

1-1 Express the following complex-valued expressions in both Cartesian and polar forms. Also, sketch the results in a complex plane.

(a) $3e^{j\pi/3} + 4e^{-j\pi/6}$

(b) $(\sqrt{3} - 3j)^{10}$

(c) $(\sqrt{3} - 3j)^{-1}$

(d) $\text{Re}\{je^{-j\pi/3}\}$

1-2 Express the following functions in polar form. What are the corresponding expressions for the magnitude and phase? Other than s, all the parameters are real constants.

(a) $T(s) = \dfrac{a}{s + 1/\tau}$, $s = j\omega$

(b) $T(s) = \dfrac{s}{s^2 + 2\zeta\omega_n s + \omega_n^2}$, $s = j\omega$

(c) $T(s) = \dfrac{3s + 4}{s^2 + 3s + 4}$, $s = 7j$

1-3 Use phasors to derive the following trigonometric identities:

$$\cos(\theta_1 \pm \theta_2) = \cos\theta_1 \cos\theta_2 \mp \sin\theta_1 \sin\theta_2$$
$$\sin(\theta_1 \pm \theta_2) = \sin\theta_1 \cos\theta_2 \pm \cos\theta_1 \sin\theta_2.$$

Use the above results to show that:

$$\cos(\theta_1)\cos(\theta_2) = \frac{1}{2}\cos(\theta_1 + \theta_2) + \frac{1}{2}\cos(\theta_1 - \theta_2).$$

1-4 Determine A and θ in the expression below:

$$x(t) = A\cos(\omega t + \theta) = \cos(\omega t) + 2\cos(\omega t + \pi/4)$$
$$+ 3\cos(\omega t + \pi/2).$$

1-5 Determine the values of $z = j^j$.

1-6 In his book *The Science of Radio*,[4] Paul Nahin describes an interesting walk in the complex plane. Start at the origin and walk forward along the positive real axis a distance of one unit. Then pivot on your heel counterclockwise (CCW) by an angle θ and walk forward a distance of one-half unit. Again pivot on your heel CCW by an angle θ and walk forward a distance of one-quarter unit. Continue doing this for an infinity of equal rotations and ever-decreasing distances (each one-half the previous distance). Where do you end up in the complex plane and for what angle θ are you farthest away from the real axis?

1-7 Solve the following simultaneous equations using phasors, that is determine the values of A_1, A_2, ϕ_1 and ϕ_2. Sketch your results using a phasor diagram to explain your solution.

$$\cos(\omega t) = A_1 \cos(\omega t + \phi_1) + A_2 \cos(\omega t + \phi_2)$$
$$\sin(\omega t) = A_1 \cos(\omega t + \phi_1) - A_2 \cos(\omega t + \phi_2)$$

---

[4] *The Science of Radio* by P. J. Nahin, Springer-Verlag, New York, 2001.

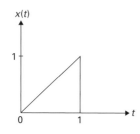

$x(t)$

1

0   1 t

Fig. 1.27. Figure for Problem 1-8.

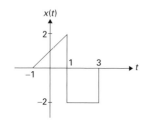

$x(t)$

2

−1

1   3 t

−2

Fig. 1.28. Figure for Problem 1-10.

1-8 Consider the function $x(t)$ shown in Figure 1.27. Sketch and dimension the following transformations:

(a) $y_1(t) = 2x(t/2) - 1$

(b) $y_2(t) = -x(-2t + 4) + 1$

(c) $y_3(t) = y_1(t) + y_2(t)$

1-9 If $x[n] = \{1, 2, 3, 4, 5, 6, 7, 8, 9, 10\}$, for $n = 0, 1, \ldots, 9$, respectively, and is 0 otherwise, sketch and dimension the following transformations:

(a) $x[n]$

(b) $y_1[n] = 2x[4n] - 1$

(c) $y_2[n] = \begin{cases} x[n/3], & \text{where } n \text{ is a multiple of 3} \\ 0 & \text{otherwise} \end{cases}$

(d) $y_3[n] = y_1[n] + y_2[n]$

1-10 Consider the continuous-time signal $x(t)$ shown in Figure 1.28. Sketch and label each of the following transformed signals:

(a) $y(t) = x(t - 2) + 1$

(b) $y(t) = x(5 - t)$

(c) $y(t) = x(4t - 8)$

(d) $y(t) = x(t/2 + 2)$

1-11 The continuous-time output $x(t)$ of a sensor used to monitor temperature is shown in Figure 1.29.

(a) If the sensor output is sampled every 0.1 minutes, determine and sketch the resulting discrete-time signal $x[n]$.

(b) A discrete-time system is defined by the following difference equation:

$$y[n] = x[n] + x[n - 1] + x[n - 2],$$

where $x[n]$ is the input and $y[n]$ is the output. Determine and sketch the response of the system when the input is an impulse, that is $x[n] = \delta[n]$.

(c) Sketch and label the output $y[n]$ of the system described in part (**b**) when the input is the signal $x[n]$ determined in part (**a**). Assume $x(t)$ is zero outside the 1 minute time interval shown in Figure 1.29.

Fig. 1.29. Figure for Problem 1-11.

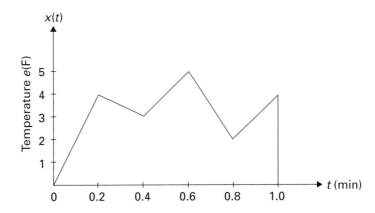

Fig. 1.30. Figure for Problem 1-12.

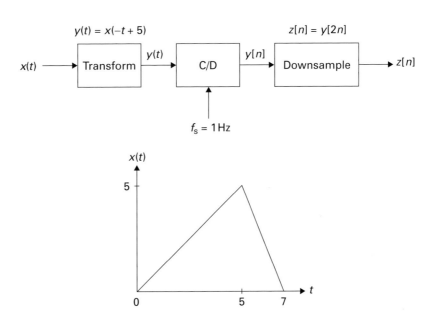

1-12 A system and its inputs are shown in Figure 1.30. Sketch $y(t)$, $y[n]$ and $z[n]$. Label all relevant amplitudes, times and indices.

1-13 Consider the following continuous-time signal:

$$x(t) = 12\cos(200\sqrt{2}\,t) - 8\sin(140\sqrt{2}\,t)$$
$$+ 19\cos\left(90\sqrt{2}\,t + \frac{\pi}{4}\right) + 16.$$

(a) Determine the fundamental frequency ω_0 and period T_0 of $x(t)$.

(b) Express $x(t)$ in the phasor form

$$x(t) = X_0 + \mathrm{Re}\left\{\sum_{k=1}^{K} X_k e^{\mathrm{j}k\omega_0 t}\right\}.$$

1-14 Evaluate:

$$z = e^{\mathrm{j}\pi/3} + 2e^{\mathrm{j}2\pi/3}.$$

Express the result in both rectangular and polar forms.

1-15 Write

$$x(t) = \mathrm{Re}\left\{e^{\mathrm{j}(\pi/3+\omega t)} + 2e^{\mathrm{j}(2\pi/3+\omega t)}\right\}$$

in the form

$$x(t) = A\cos(\omega t + \theta).$$

2

Constructing signals from building blocks

In this chapter we focus on constructing signals from basic building blocks (that is, a set of simpler signals also referred to as basis functions). We then introduce the important concept of integrated squared error and the orthogonality principle. The advantage of using orthogonal basis functions is developed. Complex exponentials (the most widely used basis functions) are then used to define the spectrum of a signal that often characterizes the relevant information that it contains. Finally, we introduce analytical and numerical techniques to compute the weights of the complex exponentials that are frequently used to describe a signal.

2.1 Basic building blocks

We start by approximating a signal as a weighted sum of basic building blocks. In general, we may approximate a signal $x(t)$ by $\hat{x}(t)$, which in turn can be expressed as the following sum:

$$x(t) \approx \hat{x}(t) = \sum_k c_k \phi_k(t), \tag{2.1}$$

where k takes on particular integer values (to be defined later) and $\phi_k(t)$ are the basic building blocks, also known as the *basis functions*. We have already seen an example of this in discrete-time when we constructed a signal from an infinite sum of weighted shifted unit impulses (see Eq. (1.56)). The infinite set of unit impulses formed the set of basis functions. In general, we may have a finite or infinite number of basis functions.

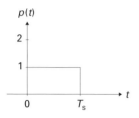

Fig. 2.1. A rectangular pulse.

For another example, let $p(t)$ be the *rectangular pulse* of width T_s, as shown in Figure 2.1. We can approximate the continuous-time signal $x(t)$ with $\hat{x}(t)$, which consists of a series of steps as shown in Figure 2.2. Let us make each step T_s wide. We can do this using basis functions made of shifted pulses, where $\phi_k(t)$ is the rectangular pulse shifted to the right by k pulse widths:

$$\phi_k(t) = p(t - kT_s). \tag{2.2}$$

What should we use for the weights c_k? From Figure 2.2, we see that we can simply use the value of the function at the beginning of the pulse time, so that an approximation to $x(t)$ is given by

$$\hat{x}(t) = \sum_{k=-\infty}^{\infty} \underbrace{x(kT_s)}_{x[n]} p(t - kT_s). \tag{2.3}$$

This is part of the process performed in converting music to a sequence of numbers that can be stored on a compact disc.

Visually, using rectangular pulses does not appear to lead to such a good approximation. Could we somehow do better? To answer this question, we need to define what "better" means. Figure 2.3 shows one approach in which

Fig. 2.2. Approximating a continuous-time signal with a series of pulses.

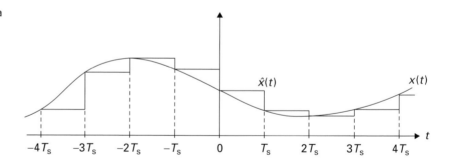

Fig. 2.3. A better approximation for the initial part of a continuous-time signal.

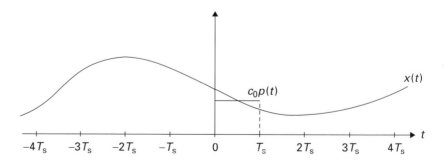

the pulse height is varied to make a "better" fit, where "better" means to minimize the *integrated squared error* (ISE), defined as:

$$\text{ISE} = \int_0^{T_s} [x(t) - \hat{x}(t)]^2 \, dt, \tag{2.4}$$

where $x(t) - \hat{x}(t)$ represents the error between the actual signal and its approximation. Why square the error? If the error were taken to the first power, large positive and negative power errors could cancel, giving a deceptively good result for poor fits. But taking the absolute value or the fourth power would also give meaningful results and are perfectly legitimate approaches. The squaring is in a sense arbitrary, but it leads to tractable mathematics and useful results. This is related to the method of *least squares* that is often used in statistics to fit a set of data to a curve.

In Figure 2.3, we wish to solve for the coefficient c_0 that minimizes the integrated squared error over the area it is approximating, defined as

$$\text{ISE} = \int_0^{T_s} [x(t) - c_0 p(t)]^2 \, dt = f(c_0). \tag{2.5}$$

Note that the integrated squared error is a function of c_0, shown mathematically as $f(c_0)$. To minimize the ISE, we take its derivative with respect to c_0 and set the resulting expression to zero. Differentiating integrals takes some care, but in this case it is straightforward and leads to

$$\frac{d}{dc_0}(\text{ISE}) = \int_0^{T_s} 2\,[x(t) - c_0 p(t)]\,[-p(t)]\, dt = 0, \tag{2.6}$$

which can be rearranged to give

$$-\int_0^{T_s} x(t)p(t)dt + c_0 \int_0^{T_s} p^2(t)\, dt = 0 \tag{2.7}$$

from which we obtain

$$c_0 = \frac{\displaystyle\int_0^{T_s} x(t)p(t)\, dt}{\displaystyle\int_0^{T_s} p^2(t)\, dt}. \tag{2.8}$$

It would be wise to check that we have really found a minimum instead of a maximum. Taking the second derivative of Eq. (2.5) yields

$$\frac{d^2}{dc_0^2}(\text{ISE}) = \int_0^{T_s} p^2(t)\, dt > 0, \tag{2.9}$$

which is always positive, verifying that we have indeed obtained a minimum.

Fig. 2.4. Approximating a
sawtooth wave with a
single pulse.

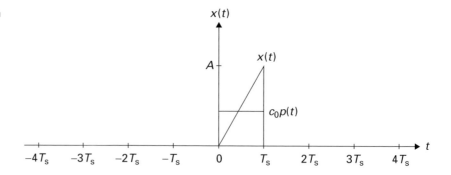

For example, suppose $x(t)$ is the sawtooth wave given in Figure 2.4. Because the area under the graph of a nonnegative continuous function is simply the definite integral of the function over the same interval, the numerator integral in Eq. (2.8) corresponds to the area of the sawtooth function from 0 to T_s (over this interval, the rectangular pulse has a magnitude of 1), and the denominator integral is simply T_s. Solving for c_0 using Eq. (2.8), we obtain

$$c_0 = \frac{AT_s/2}{T_s} = \frac{A}{2}. \tag{2.10}$$

The result appears reasonable, because without solving the problem mathematically, we expect the best approximation to the sawtooth to be a single pulse with half the height of the continuous signal. Later we will examine ways to improve the approximations using multiple pulses.

In general, we would like to construct or approximate $x(t)$ over some time interval T with a weighted sum of known building blocks. How do we calculate these *weights* or *coefficients*? Mathematically, the problem is how to choose the coefficients c_k such that the approximated signal

$$x(t) \approx \hat{x}(t) = \sum_k c_k \phi_k(t) \tag{2.11}$$

minimizes the integrated squared error

$$\text{ISE} = \int_T |x(t) - \hat{x}(t)|^2 \, dt \tag{2.12}$$

for a given function $x(t)$, the time interval of interest T, and the basis functions $\phi_k(t)$. The coefficients c_k are constants and may be either real or complex, so we take the magnitude squared to get the proper behavior for complex values. The function $x(t)$ is usually real but our results will be valid for both real and complex functions. The basis functions may be either real or complex. If the signal and basis functions are real, Eq. (2.12) reduces to Eq. (2.4) integrated over the time interval of interest T.

The mathematics to obtain the coefficients c_k such that the integrated squared error is minimized is interesting but a bit abstract. The results that we obtain, however, will be extremely useful. Combining Eq. (2.11) and Eq. (2.12), we wish to minimize

$$\text{ISE} = \int_T \left| x(t) - \sum_k c_k \phi_k(t) \right|^2 dt. \tag{2.13}$$

We could minimize the ISE by taking derivatives with respect to each of the coefficients c_k and setting them to zero. This gives us a set of simultaneous equations (one equation for each basis function) that can be solved to give the desired coefficients. The mathematics is straightforward but a bit tedious and will not be provided here.

2.2 The orthogonality principle

A more elegant approach is to use the *orthogonality principle* (OP) (whose proof is given in Section 2.12), which states that the coefficients c_k that minimize the ISE must satisfy the following relationship:

$$\int_T [x(t) - \hat{x}(t)] \, \phi_k^*(t) \, dt = 0, \quad \text{for all } k, \tag{2.14}$$

where $\phi_k^*(t)$ denotes the complex conjugate of the basis function $\phi_k(t)$. For a real function, its complex conjugate is just itself. Substituting the expression of $\hat{x}(t)$ into Eq. (2.14), we obtain

$$\int_T \left[x(t) - \sum_k c_k \phi_k(t) \right] \phi_k^*(t) \, dt = 0, \quad \text{for all } k. \tag{2.15}$$

One confusing aspect of this particular equation is that index k is used in two ways. Outside of the brackets, k is a specific number. Inside the brackets, k is used again as a dummy index for the summation. To distinguish between the two indices, let us rewrite the OP criterion with a different dummy index variable n for the summation. Thus, Eq. (2.15) becomes

$$\int_T \left[x(t) - \sum_n c_n \phi_n(t) \right] \phi_k^*(t)\, dt = 0, \quad \text{for all } k, \tag{2.16}$$

which yields a set of simultaneous equations that can be used to solve for all the coefficients c_k.

To illustrate how Eq. (2.16) can be used to determine the coefficients, consider the following example. Suppose $x(t)$ is a sawtooth signal with a maximum height A, as illustrated in Figure 2.5, and we are given two basis functions, shown in Figure 2.6. Our objective is to find $\hat{x}$, which consists of the linear combination of the basis functions, that best approximates $x(t)$ over the interval $[0, T]$. While for this simple problem the coefficients may be found by inspection, let's solve for the coefficients by applying the orthogonality principle to see how it works. Utilizing the OP, we obtain the following two equations, one for each coefficient:

$$\int_0^T [x(t) - c_0\phi_0(t) - c_1\phi_1(t)]\, \phi_0^*(t)\, dt = 0$$

$$\int_0^T [x(t) - c_0\phi_0(t) - c_1\phi_1(t)]\, \phi_1^*(t)\, dt = 0. \tag{2.17}$$

Because the basis functions are real, their conjugates are the basis functions themselves, that is, $\phi^*(t) = \phi(t)$. When rearranged, we get

$$\int_0^T x(t)\phi_0(t)\, dt - c_0\int_0^T \phi_0^2(t)\, dt - c_1\int_0^T \phi_0(t)\phi_1(t)\, dt = 0$$

$$\int_0^T x(t)\phi_1(t)\, dt - c_0\int_0^T \phi_1(t)\phi_0(t)\, dt - c_1\int_0^T \phi_1^2(t)\, dt = 0. \tag{2.18}$$

Evaluating these integrals, we obtain the following two algebraic equations

$$\frac{AT}{8} - c_0\frac{T}{2} - c_1\frac{T}{2} = 0$$

$$\frac{AT}{2} - c_0\frac{T}{2} - c_1 T = 0. \tag{2.19}$$

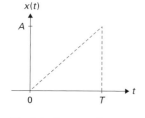

Fig. 2.5. A sawtooth signal.

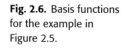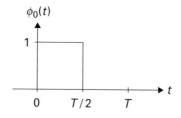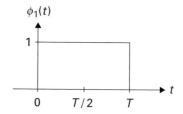

Fig. 2.6. Basis functions for the example in Figure 2.5.

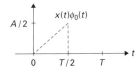

Fig. 2.7. Product of $x(t)$ and $\phi_0(t)$.

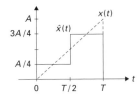

Fig. 2.8. Approximation to the sawtooth signal of Figure 2.5 using the basis functions of Figure 2.6.

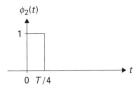

Fig. 2.9. Third basis function for the example in Figure 2.5.

To clarify how each integral was evaluated, let us consider the first term of the top equation in Eq. (2.18). The product of the given signal and the first basis function, $x(t)\phi_0(t)$, is sketched in Figure 2.7. The integral of their product over the interval of interest is simply the area of the triangle, which can be easily determined from Figure 2.7 to be $1/2(AT/4) = AT/8$.

Equation (2.19) consists of two equations in terms of the two unknown coefficients, c_0 and c_1. Canceling the Ts and simplifying, we find

$$\frac{c_0}{2} + \frac{c_1}{2} = \frac{A}{8}$$
$$\frac{c_0}{2} + c_1 = \frac{A}{2}.$$

(2.20)

Solving these two equations simultaneously, we get $c_0 = -\frac{1}{2}A$ and $c_1 = \frac{3}{4}A$. This is plotted in Figure 2.8, and the resulting $\hat{x}(t)$ is the approximation to $x(t)$ that minimizes the ISE.

Suppose we wish to repeat the previous example with a third basis function shown in Figure 2.9. Now we need to evaluate all the integrals again, which is tedious and highly inefficient. It would be very convenient if we could add another basis function without having to re-evaluate the coefficients corresponding to the other basis functions. Fortunately, this is possible if we carefully select the building blocks that are used to approximate the given signal. Such building blocks are themselves called *orthogonal*.

2.3 Orthogonal basis functions

The orthogonality principle remains valid for any set of basis functions, but is easier to use for a special class of basis functions called *orthogonal functions*. By definition, a pair of functions $\phi_l(t)$ and $\phi_k(t)$ are said to be orthogonal over the interval T if they satisfy the following condition:

$$\int_T \phi_l(t)\phi_k^*(t)dt = \begin{cases} K_k, & k = l \\ 0, & k \neq l, \end{cases}$$

(2.21)

where K_k is some non-zero constant. Now when we apply the OP, we find that by virtue of the orthogonality conditions of these basis functions, all of the products of basis functions integrate to 0, except for the product in which $k = l$. Mathematically, Eq. (2.21) leads to a set of *decoupled* equations that can be used to solve for each coefficient individually:

$$\int_T \left[x(t) - \sum_l c_l \phi_l(t) \right] \phi_k^*(t) dt = 0, \quad \text{for all } k. \tag{2.22}$$

Upon rearranging, Eq. (2.22) becomes

$$\int_T x(t)\phi_k^*(t) dt - \sum_l c_l \int_T \phi_l(t)\phi_k^*(t) dt = 0. \tag{2.23}$$

Because of the orthogonal conditions of Eq. (2.21), all terms within the summation vanish except for the term $l = k$: thus Eq. (2.23) reduces to

$$\int_T x(t)\phi_k^*(t) dt - c_k \int_T \phi_k(t)\phi_k^*(t) dt = 0 \tag{2.24}$$

or

$$\int_T x(t)\phi_k^*(t) dt - c_k K_k = 0, \tag{2.25}$$

from which the coefficient c_k is found to be

$$c_k = \frac{1}{K_k} \int_T x(t)\phi_k^*(t) dt. \tag{2.26}$$

This is substantially more efficient, because we can now solve for each coefficient separately instead of solving a set of simultaneous equations. For this reason, we will find that orthogonal basis functions are particularly convenient. Examples of orthogonal basis functions include sinusoidal functions, complex exponentials, Bessel functions, Walsh functions and Haar functions.

The two building blocks of Figure 2.6 are not orthogonal. We can easily verify this by integrating their product. These basis functions are real, so we can drop the conjugate. Thus, the integral of their product is

$$\int_T \phi_0(t)\phi_1(t) dt = \int_T \phi_0(t) dt = \frac{T_0}{2}, \tag{2.27}$$

which is non-zero, implying that the basis functions are not orthogonal. However, the building blocks shown in Figure 2.10 are orthogonal. Their product is given by $\phi_0(t)$, and its integral is

$$\int_T \phi_0(t)\phi_1(t) dt = 0. \tag{2.28}$$

As an exercise, the reader should try finding the weights that best fit the sawtooth signal of Figure 2.5 using the orthogonal basis functions of Figure 2.10. Figure 2.11 shows yet a third basis function that is orthogonal

Fig. 2.10. Example of a pair of orthogonal basis functions.

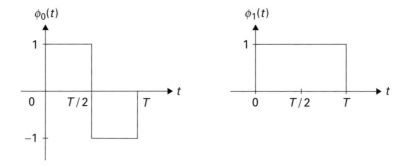

Fig. 2.11. A third orthogonal basis function.

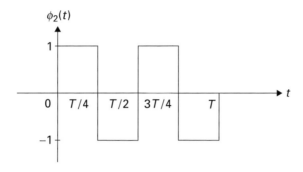

Fig. 2.12. Multiplication of $\phi_0(t)$ and $\phi_2(t)$.

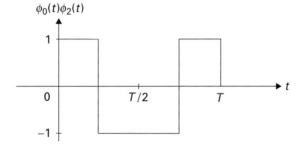

to the two basis functions shown in Figure 2.10. We will ask the reader to verify this as an exercise.

As a side remark, it should be noted that multiplication of the functions is accomplished by multiplying the values of the functions at each point. For example, $\phi_0(t)\phi_2(t)$ is shown in Figure 2.12. Thus, the product of two distinct orthogonal functions has the same area above and below the x-axis.

2.3.1 Examples

Suppose we wish to approximate the pulse shown in Figure 2.13 with a sum of the sinusoidal basis functions that are given in Figure 2.14. The basis

Fig. 2.13. Approximation of a pulse function using sinusoids.

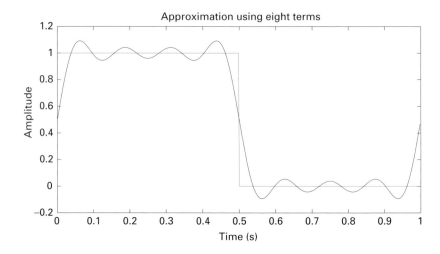

Fig. 2.14. Sinusoidal basis functions (Fourier series).

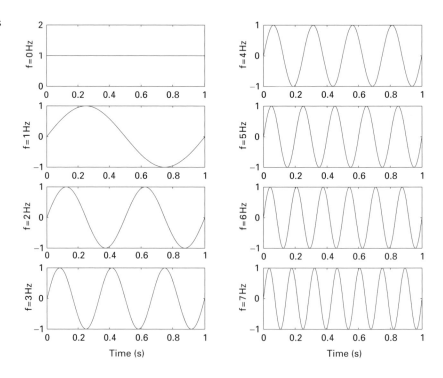

functions are all orthogonal (you are encouraged to check this), each with a distinct frequency f. Table 2.1 shows the coefficients that give the best fit and the resulting approximation is plotted overlaying the original pulse. For more complicated signals computing these coefficients directly using Eq. (2.26) is tedious by hand and would be best done on a computer. This example illustrates the important concept of *Fourier series*.

Table 2.1. Coefficients of
the pulse function of Figure
2.13 using sinusoids

k	c_k
0	$1/2$
1	$2/\pi$
2	0
3	$2/(3\pi)$
4	0
5	$2/(5\pi)$
6	0
7	$2/(7\pi)$

Fig. 2.15. Approximation
of sinusoid using Walsh
functions.

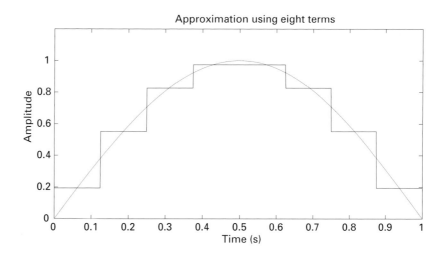

As another example, let us approximate a sinusoidal signal as shown in Figure 2.15 using the Walsh functions of Figure 2.16. The Walsh functions consist of trains of rectangular pulses, with the allowed states being -1 and 1. The Walsh functions may look a bit odd, but they are all orthogonal and constitute an important set of basis functions. The coefficients for the approximation are shown in Table 2.2.

The previous examples are a bit contrived. In the first example we tried to approximate a square pulse with sinusoids, and in the second example we tried to approximate a sinusoid with square pulses. It might be more sensible to approximate pulses with pulses and sinusoids with sinusoids. The above examples illustrate how different basis functions can be used to

Table 2.2. Coefficients of the sinusoid of Figure 2.15 using Walsh functions

k	c_k
0	0.636
1	0.000
2	0.000
3	−0.053
4	0.000
5	−0.127
6	−0.264
7	0.000

Fig. 2.16. Walsh basis functions.

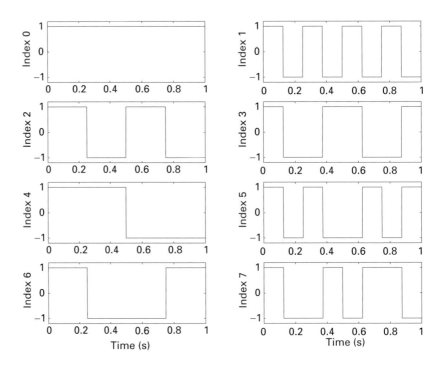

approximate a given signal. They also show the importance of using proper judgment in choosing the basis functions.

In a final example of orthogonal functions, let us approximate the sinusoidal signal of Figure 2.17 using a set of *Harr Wavelet* basis functions (discovered in 1910) as shown in Figure 2.18. The coefficients for the approximation are shown in Table 2.3. Interestingly, the wavelets provide

Fig. 2.17. Approximation of sinusoid using Haar Wavelet basis functions.

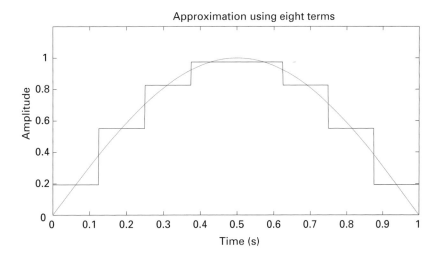

Fig. 2.18. Haar Wavelet basis functions.

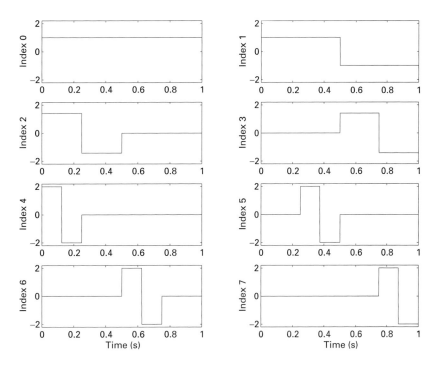

information about how the signal changes not only in frequency (like Fourier series) but also in time. Wavelets[1] became very popular in the 1980s and the FBI uses another set of wavelet basis functions to compress information about fingerprints.

---

[1] See *Wavelets and Filter Banks* by G. Strang and T. Nguyen, Wellesley-Cambridge Press, Wellesley, MA, 1996.

Table 2.3. Coefficients of the
sinusoid of Figure 2.17 using
Haar Wavelet basis functions

k	c_k
0	0.636
1	0.000
2	−0.187
3	0.187
4	−0.090
5	−0.037
6	0.037
7	0.090

2.4 Fourier series

The most widely used approximation of periodic functions is the Fourier
series. Let $x(t)$ be a periodic signal that repeats itself after some given time
interval. Mathematically, $x(t)$ is defined as

$$x(t) = x(t + T_0), \quad \text{for all } t, \tag{2.29}$$

where T_0 is called the *period* of the signal. Let us approximate $x(t)$ with the
following building blocks:

$$\phi_k(t) = e^{jk\left(\frac{2\pi}{T_0}\right)t} = \cos\left(k\frac{2\pi}{T_0}t\right) + j\sin\left(k\frac{2\pi}{T_0}t\right), \tag{2.30}$$

where the integer $k = -\infty$ to ∞. Mathematically this makes sense, because
we would want to approximate a periodic signal with sinusoids at multiples
of the signal frequency, defined as $2\pi/T_0$. These complex exponential basis
functions are orthogonal, as they satisfy

$$\int_{T_0} \phi_l(t)\phi_k^*(t)dt = \begin{cases} T_0, & k = l \\ 0, & k \neq l. \end{cases} \tag{2.31}$$

Let us use an infinite sum of these basis functions to approximate $x(t)$ with
zero ISE:

$$x(t) = \sum_{k=-\infty}^{\infty} c_k\phi_k(t) = \sum_{k=-\infty}^{\infty} c_k e^{jk\omega_0 t}, \tag{2.32}$$

where $\omega_0 = 2\pi/T_0$. Because the basis functions are orthogonal, we can use Eq. (2.26) directly to find these coefficients. Upon substitution, we see that the best weights to approximate $x(t)$ are given by

$$c_k = \frac{1}{T_0} \int_{T_0} x(t) e^{-jk\omega_0 t} dt = \frac{1}{T_0} \int_{T_0} x(t) e^{-j2\pi k f_0 t} dt. \qquad (2.33)$$

Equation (2.33) is known as the complex form of the *Fourier series*. We can also show that the complex Fourier series may be written in terms of real trigonometric functions as follows:

$$x(t) = a_0 + 2 \sum_{k=1}^{\infty} [a_k \cos(k\omega_0 t) + b_k \sin(k\omega_0 t)], \qquad (2.34)$$

where the coefficients a_k and b_k are given by

$$\begin{aligned} a_k &= \frac{1}{T_0} \int_{T_0} x(t) \cos(k\omega_0 t) dt \\ b_k &= \frac{1}{T_0} \int_{T_0} x(t) \sin(k\omega_0 t) dt. \end{aligned} \qquad (2.35)$$

The results of Eq. (2.35) will be derived in the next section.

2.5 Alternative forms of the Fourier series

If the signal $x(t)$ is real (which will be the case for almost all measurable signals), the coefficients will have an interesting symmetry property. Consider the coefficients for the negative frequencies, given by

$$\begin{aligned} c_k^* &= \left[\frac{1}{T_0} \int_{T_0} x(t) e^{-jk\omega_0 t} dt \right]^* \\ &= \frac{1}{T_0} \int_{T_0} [x(t) e^{-jk\omega_0 t}]^* dt = \frac{1}{T_0} \int_{T_0} x^*(t) e^{jk\omega_0 t} dt. \end{aligned} \qquad (2.36)$$

Because $x(t)$ is real, Eq. (2.36) simplifies to

$$c_k^* = \frac{1}{T_0} \int_{T_0} x(t) e^{jk\omega_0 t} dt = \frac{1}{T_0} \int_{T_0} x(t) e^{-j(-k)\omega_0 t} dt = c_{-k}. \qquad (2.37)$$

Thus, if $x(t)$ is real, the coefficients of its Fourier series have a symmetry of $c_{-k} = c_k^*$. It is therefore sufficient to compute only the coefficients for the positive k; the coefficients for the negative k are simply the conjugates.

Another useful result for real $x(t)$ can be derived by rewriting the complex exponential in Eq. (2.36) in terms of sine and cosine functions as follows:

$$c_k = \frac{1}{T_0} \int_{T_0} x(t) e^{-jk\omega_0 t} \, dt = \frac{1}{T_0} \int_{T_0} x(t) [\cos(k\omega_0 t) - j \sin(k\omega_0 t)] \, dt, \quad (2.38)$$

where Euler's formula was used. Thus, we can express c_k in the Cartesian form instead of the complex exponential form:

$$c_k = a_k - jb_k. \tag{2.39}$$

Matching the coefficients of Eqs. (2.38) and (2.39), we can readily determine the coefficients a_k and b_k:

$$
\begin{aligned}
a_k &= \frac{1}{T_0} \int_{T_0} x(t) \cos(k\omega_0 t) \, dt \\
b_k &= \frac{1}{T_0} \int_{T_0} x(t) \sin(k\omega_0 t) \, dt.
\end{aligned}
\tag{2.40}
$$

For real $x(t)$, this shows that a_k and b_k are real, $a_0 = c_0$, and $b_0 = 0$. We can thus rewrite $x(t)$ as a weighted sum of sines and cosines rather than complex exponentials, as shown below:

$$
\begin{aligned}
x(t) &= \sum_{k=-\infty}^{\infty} c_k e^{jk\omega_0 t} \\
&= a_0 + \sum_{k=1}^{\infty} [c_k e^{jk\omega_0 t} + c_{-k} e^{-jk\omega_0 t}] \\
&= a_0 + \sum_{k=1}^{\infty} [c_k e^{jk\omega_0 t} + c_k^* e^{-jk\omega_0 t}] \\
&= a_0 + \sum_{k=1}^{\infty} [(a_k - jb_k) e^{jk\omega_0 t} + (a_k + jb_k) e^{-jk\omega_0 t}] \\
&= a_0 + \sum_{k=1}^{\infty} [a_k(e^{jk\omega_0 t} + e^{-jk\omega_0 t}) - jb_k(e^{jk\omega_0 t} - e^{-jk\omega_0 t})] \\
&= a_0 + 2 \sum_{k=1}^{\infty} \left[\frac{a_k}{2}(e^{jk\omega_0 t} + e^{-jk\omega_0 t}) + \frac{b_k}{2j}(e^{jk\omega_0 t} - e^{-jk\omega_0 t}) \right] \\
&= a_0 + 2 \sum_{k=1}^{\infty} [a_k \cos(k\omega_0 t) + b_k \sin(k\omega_0 t)].
\end{aligned}
\tag{2.41}
$$

Expressing the periodic function $x(t)$ by a trigonometric series may be more intuitive, although determining the Fourier coefficients using sines and cosines is more tedious. To distinguish between the two different representations of the same signal, Eq. (2.32) and Eq. (2.41) are often known

Fig. 2.19. A periodic pulse train.

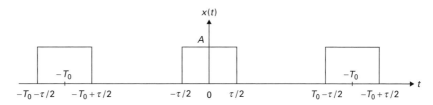

as the complex form and the real form of the Fourier series, respectively. Similarly, the c_k of Eq. (2.41) are called the *complex Fourier coefficients*, and the a_k and b_k of Eq. (2.41) are called the *real Fourier coefficients*.

There is yet a third representation for real $x(t)$ (given without proof) as follows

$$x(t) = c_0 + 2 \sum_{k=1}^{\infty} |c_k| \cos(k\omega_0 t + \theta_k), \tag{2.42}$$

where

$$c_k = |c_k| e^{j\theta_k}. \tag{2.43}$$

In this form, the amplitude and phase of each harmonic can be easily extracted. Finally, a fourth form for real $x(t)$ is given by

$$x(t) = X_0 + \text{Re} \left\{ \sum_{k=1}^{\infty} X_k e^{jk\omega_0 t} \right\}, \tag{2.44}$$

where $X_0 = c_0$ and $X_k = 2c_k$ for $k > 0$. This last form is often more convenient for computational purposes.

In summary, we have presented four different representations of the Fourier series that are all equivalent for real $x(t)$. For mathematical manipulations, we will most often use the first form (the complex Fourier series) with complex exponentials. When to use which form will become more intuitive with experience. Having derived various representations of Fourier series, we now consider some examples.

2.5.1 Examples: periodic pulse trains and square waves

Consider the periodic pulse train of Figure 2.19. Using Fourier series, we can write the pulse train as an infinite sum of complex exponentials. Evaluating

Fig. 2.20. A periodic
square wave.

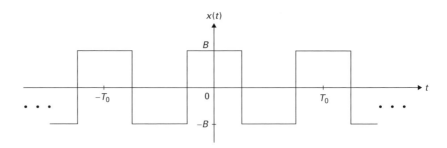

the integrals in Eq. (2.33), we find the Fourier coefficients are given by

$$c_0 = A\,\frac{\tau}{T_0}$$

$$c_k = A\,\frac{\tau}{T_0}\,\frac{\sin\left(k\pi\,\frac{\tau}{T_0}\right)}{k\pi\,\frac{\tau}{T_0}}. \tag{2.45}$$

We often call $\tau/T_0 = d$ the *duty cycle* (ratio of on to off time). Making this substitution in Eq. (2.45), we can show that the coefficients can be expressed more compactly in terms of the duty cycle only as shown:

$$c_0 = Ad$$

$$c_k = Ad\,\frac{\sin(k\pi d)}{k\pi d}. \tag{2.46}$$

Figure 2.20 shows a square wave with a period of T_0. Clearly we can express this periodic square wave in terms of complex exponentials, whose Fourier coefficients can be determined by the direct application of Eq. (2.36). However, we would like to find these coefficients without performing all the laborious integrations. By inspection, note that the square wave Figure 2.20 is simply the pulse train of Figure 2.19 with a duty cycle of $d = 1/2$ and shifted down by $B = A/2$. Recall that c_0 is the average value of the signal, which is 0 for the periodic square wave of Figure 2.20. Thus, shifting signals up or down only changes the coefficient c_0: the other coefficients c_k remain the same. Therefore, all of the coefficients associated with the square wave are identical to those for a pulse train of amplitude $A = 2B$ and duty cycle $d = 1/2$, except that $c_0 = 0$:

$$c_0 = 0$$

$$c_k = B\,\frac{\sin\left(k\frac{\pi}{2}\right)}{k\frac{\pi}{2}}. \tag{2.47}$$

Table 2.4. Periodic square wave
coefficients

k	c_k
0	0
±1	$\dfrac{2}{\pi}B$
±2	0
±3	$-\dfrac{2}{3\pi}B$
±4	0
±5	$\dfrac{2}{5\pi}B$
±6	0
±7	$-\dfrac{2}{7\pi}B$

These coefficients are listed numerically in Table 2.4. The even-numbered coefficients are all 0. The odd-numbered coefficients decrease with k. The frequency associated with $k = 1$ is known as the *fundamental frequency*, while the frequencies associated with the higher (odd) coefficients are called the *odd harmonics*.

2.6 Approximating signals numerically

The Fourier series consists of an infinite summation, which of course cannot be coded numerically. However, we may approximate the signal quite accurately with a finite number of terms. Let us use $2K + 1$ terms in the approximation

$$x(t) = x(t + T_0) \approx \hat{x}(t) = \sum_{k=-K}^{K} c_k \phi_k(t) = \sum_{k=-K}^{K} c_k e^{jk\omega_0 t}. \qquad (2.48)$$

Note that index k in the summation now ranges from $-K$ to K instead of from $-\infty$ to ∞ as shown in Eq. (2.32). The coefficients c_k are still obtained with the same formula, but now we have only a finite number of them.

We can check the quality of these approximations by evaluating them in MATLAB. Consider a periodic square wave. With $K = 8$, amplitude $A = 1$, and period $T_0 = 1$, we obtain the results shown in Figure 2.21(a),

Fig. 2.21. Square wave
approximation showing
Gibbs' phenomenon.

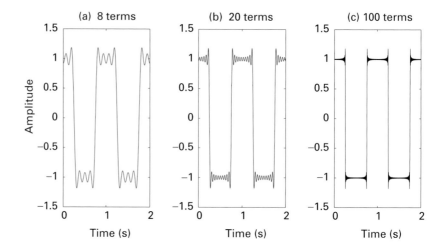

Fig. 2.21. Square wave
approximation showing
Gibbs' phenomenon.

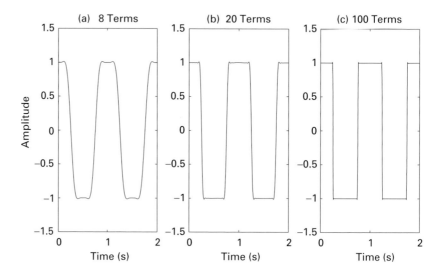

Fig. 2.22. Square wave
approximation using
windowed coefficients.

which captures the general shape but is not such a great approximation. If
we use more terms ($K = 20$) in the approximation, we find the results of
Figure 2.21(b), which better resembles the desired pulse train. If a few terms
are good, then more must be better. With $K = 100$, we find Figure 2.21(c).

Observe that the approximation still has the overshoots (called *ears* or
the *Gibbs phenomenon*), which will never vanish with more terms. The
ISE, however, goes to 0 because the overshoots become narrower (and thus
enclose less area), but not shorter. This suggests that minimizing the ISE
may not always be the criterion we want for "best" fit. Figure 2.22 shows the

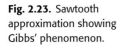

Fig. 2.23. Sawtooth approximation showing Gibbs' phenomenon.

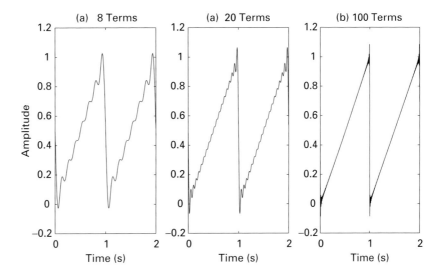

same square wave generated with a "window" that modifies the coefficients. The windowing smooths the function and eliminates the ears but actually does not minimize the ISE. We will return to the concept of windowing later in the text.

Figure 2.23 shows the approximation to a sawtooth signal with $K = 8$, $K = 20$, and $K = 100$. We still observe Gibbs' phenomenon near the discontinuities.

Thus, there is a trade-off between accuracy and computational intensity. Using more terms leads to a better approximation but at the expense of more computational time and effort.

We have been focusing on a particular set of building blocks for constructing signals, namely the complex exponentials. We will see that these are the real workhorses for extracting information from signals and for understanding how systems behave. In the previous section we saw that there are multiple forms in which we could express the Fourier series. In different forms, the Fourier coefficients are given by c_k, X_k, or a_k and b_k.

2.7 The spectrum of a signal

We will now consider the *spectrum* of a signal. The spectrum is a convenient way to display the coefficients of the complex form of the Fourier series. For practical applications, we are interested in approximating periodic signals

Fig. 2.24. Spectrum of a
sinusoid.

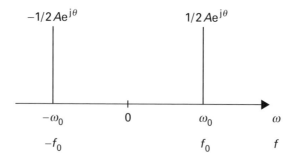

(with a period T_0 and an angular frequency of $\omega_0 = 2\pi/T_0$) using a finite
number of terms

$$x(t) = x(t + T_0) \approx \hat{x}(t) = \sum_{k=-K}^{K} c_k e^{jk\omega_0 t}, \tag{2.49}$$

where the coefficients are determined using

$$c_k = \frac{1}{T_0} \int_{T_0} x(t) e^{-jk\omega_0 t} \, dt. \tag{2.50}$$

This approximation uses $2K + 1$ building blocks. As K approaches infinity,
the integrated squared error between the signal and its approximation goes
to 0.

Consider the following sinusoid signal with $A > 0$:

$$x(t) = A \cos(\omega_0 t + \theta). \tag{2.51}$$

One way to calculate its Fourier coefficients, c_k, is by evaluating the
integral of Eq. (2.50) directly. In this special case, however, we can
bypass this integration and just use Euler's formula to expand the cosine as
follows:

$$\begin{aligned} x(t) &= \frac{A}{2} \left[e^{j(\omega_0 t + \theta)} + e^{-j(\omega_0 t + \theta)} \right] \\ &= \frac{1}{2} A e^{j\theta} e^{j\omega_0 t} + \frac{1}{2} A e^{-j\theta} e^{-j\omega_0 t}. \end{aligned} \tag{2.52}$$

By equating Eq. (2.52) to Eq. (2.49), we see immediately that $c_0 = 0$,
$c_1 = A e^{j\theta}/2$, $c_{-1} = A e^{-j\theta}/2$, and all the other coefficients are 0. Also, note
that because $x(t)$ is real, the coefficients c_k and c_{-k} are merely complex
conjugates, that is $c_{-k} = c_k^*$, as we have determined previously.

The above example leads us to a very important concept: the *spectrum* of
a signal. The spectrum is merely a graph of the coefficients c_k plotted as a
function of frequency. The spectrum of Eq. (2.52) is shown in Figure 2.24.

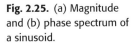

Fig. 2.25. (a) Magnitude and (b) phase spectrum of a sinusoid.

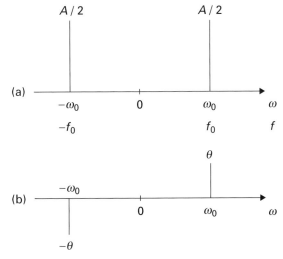

The height of the lines is $A/2$, equal to the magnitude of the coefficient corresponding to the frequency of interest. We have frequencies at ω_0 and $-\omega_0$, where a positive frequency corresponds to a counterclockwise rotation and a negative frequency corresponds to a clockwise rotation. Each line is annotated with the complex amplitude of the signal. Note that the frequency can also be expressed in terms of f with units of hertz.

Representing the spectrum of a signal is not unique. There is an alternative representation of the spectrum that consists of two graphs, one showing how the *magnitude* of the coefficient varies with frequency, and the other showing the *phase* plotted as a function of frequency, as shown in Figure 2.25. Both Figure 2.24 and Figure 2.25 can be used to represent the spectrum of a signal. In this text, we will use the spectrum representation of Figure 2.24 exclusively, where a single plot contains both the amplitude and the phase information.

In another example, let us now consider the periodic pulse train of Figure 2.26. We had determined previously that the coefficients of its Fourier coefficients are given by

$$c_0 = Ad$$
$$c_k = Ad\,\frac{\sin(k\pi d)}{k\pi d},$$

(2.53)

where $d = \tau/T_0$, which is known as the *duty cycle*. The coefficient c_0 does not match the general form for c_k with $k = 0$ because of the division

Fig. 2.26. A periodic pulse train.

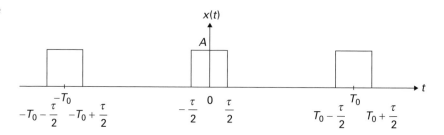

Fig. 2.27. The spectrum of a periodic pulse train.

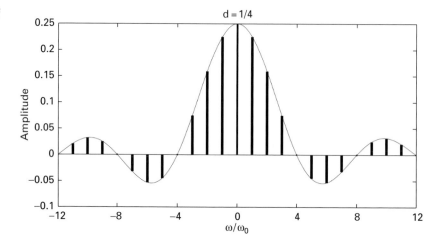

by 0. Instead, c_0 must be calculated as the average value of $x(t)$ over one period:

$$c_0 = \frac{1}{T_0} \int_{T_0} x(t) \mathrm{d}t. \tag{2.54}$$

Figure 2.27 plots the spectrum of the pulse train for $d = 1/4$. In this particular case the coefficients are real.[2] Also notice that the coefficients c_k whose indices are integer multiples of 4 are all 0. This is related to the fact that $d = 1/4$. Finally, observe that if the tips of the vertical bars are connected with a dotted line, we obtain a decaying sinusoidal *envelope*, which can be conveniently described by a function known as the *sinc* function, defined as

$$\mathrm{sinc}\,(x) = \frac{\sin(\pi x)}{\pi x}. \tag{2.55}$$

---

[2] The Fourier coefficients are always real for signals that are real and even functions.

Fig. 2.28. A periodic square wave.

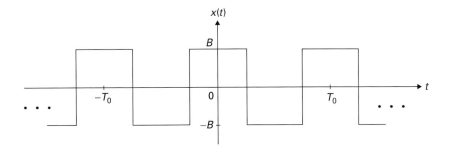

Fig. 2.29. The spectrum of a periodic square wave.

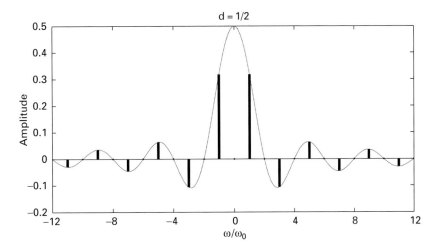

Consider now the periodic square wave of Figure 2.28, whose coefficients were determined previously to be

$$c_0 = 0$$
$$c_k = B \, \frac{\sin \left(k \frac{\pi}{2} \right)}{k \frac{\pi}{2}}.$$

(2.56)

The spectrum of the periodic square wave is given in Figure 2.29. It is quite similar to the periodic pulse train, but has zeros at the origin and for all the even-numbered coefficients. The sinc envelope is plotted again in dashed lines, but this time c_0 is not on the envelope.

In the last two examples, the coefficients are all real. Thus, for simplicity, the coefficients that are negative can be plotted below the x- or ω-axis. Figure 2.27 and Figure 2.29 constitute yet a third representation of the spectrum of a signal. In general, we often plot only the magnitude (which is positive), then annotate the graph with the phase, as in Figure 2.24.

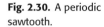

Fig. 2.30. A periodic sawtooth.

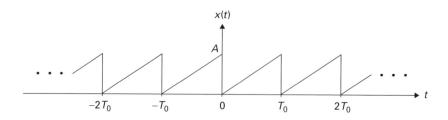

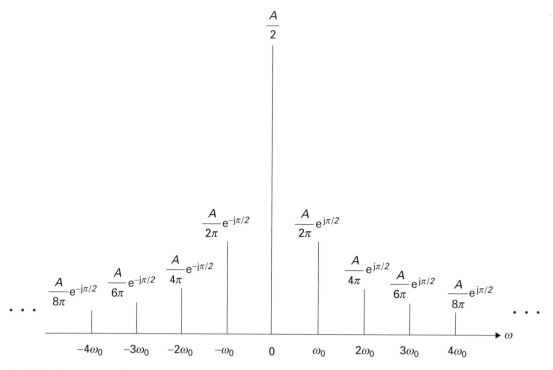

Fig. 2.31. The spectrum of a periodic sawtooth.

Finally, let us consider the periodic sawtooth signal of Figure 2.30. The coefficients are found by direct integration to be

$$c_0 = \frac{A}{2}$$
$$c_k = \frac{jA}{2\pi k}. \tag{2.57}$$

Note that this is another case in which the general form of c_k does not work for $k = 0$ (it blows up to infinity). Instead, c_0 should be obtained from the average value of the function over a period (see Eq. (2.54)). The spectrum is shown in Figure 2.31. Observe that the height of each line is the magnitude of the corresponding coefficient and that each line is annotated with its respective phase.

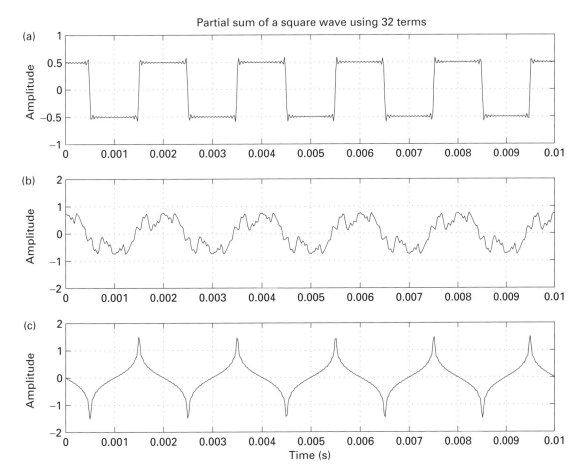

Fig. 2.32. Curious signals.

Finally, Figure 2.32(a) shows a square wave at a fundamental frequency of 500 Hz. Figure 2.32(b) and Figure 2.32(c) show curious waveforms with the same fundamental frequency. The three signals were constructed using coefficients of the same amplitude but with different phases (zero, random, and $\pi/2$). Figure 2.33 shows the magnitude spectrum for each signal. Interestingly, if these different waveforms were played through a speaker, they would all sound identical. Clearly, the eye is sensitive to the phase information that the ear does not detect. The eye senses the differences in phase, while the ear ignores it. Thus, we can deduce that audio amplifiers can distort the phase of a signal without affecting what we hear, but video amplifiers must preserve the phase of a signal in order to maintain the signal's integrity.

We can measure these signals and the spectra in the real world. An oscilloscope plots the signal as a function of time, while a spectrum analyzer

Constructing signals from building blocks

Fig. 2.33. Magnitude
spectra for the signals of
Figure 2.32.

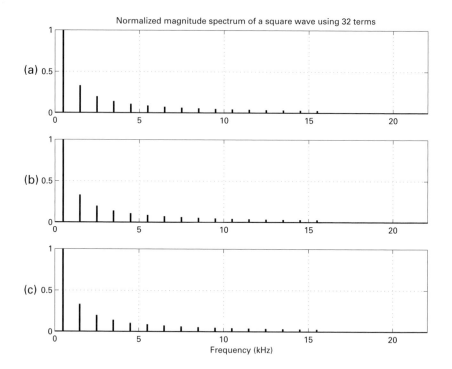

plots the magnitude of the Fourier coefficients as a function of frequency. We have been focusing on constructing signals using building blocks, particularly complex exponential building blocks. Determining the weights on these building blocks (or their sinusoidal cousins) tells us quite a bit of information about the frequencies that characterize a signal. This will be the cornerstone to understanding how systems respond to signals.

In general, it is not practical to write a function for nontrivial signals such as earthquakes or global positioning system (GPS) transmissions. The functions that describe these signals would be too complex or impossible to formulate analytically. Therefore, evaluating the coefficients of the Fourier series with integrals is often not realistic. Instead, we record the signals as a series of data points, then we determine the Fourier coefficients numerically with a computer.

2.8 The discrete Fourier transform

Recall that we can approximate periodic signals $x(t)$ with $\hat{x}(t)$ as follows:

$$x(t) = x(t + T_0) \approx \hat{x}(t) = \sum_{k=-K}^{K} c_k e^{jk\omega_0 t}. \tag{2.58}$$

Fig. 2.34. An arbitrary signal $x(t)$.

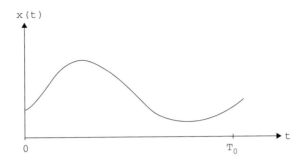

Fig. 2.35. Approximating the integral with a series of rectangular pulses.

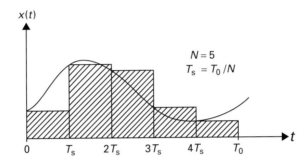

The coefficients c_k, also called the *weights*, are found by direct integration:

$$c_k = \frac{1}{T_0} \int_{T_0} x(t) e^{-jk\omega_0 t} \, dt. \qquad (2.59)$$

Consider an arbitrary signal $x(t)$ over some duration of interest from 0 to T_0, as shown in Figure 2.34. The signal $x(t)$ is not necessarily periodic. We have only recorded the data for the time interval shown, that is over T_0. For the purpose of analysis, however, we will assume $x(t)$ to be periodic with a period of T_0. Because we are only interested in the signal over the specific duration and we do not really care about the behavior of the signal outside that time, this assumption is perfectly acceptable and it will let us accurately approximate the signal over the given time interval.

We would like to find the weights that let us best approximate the signal of Figure 2.34. However, there is no reasonable way to integrate the signal analytically because we only have a graphical representation. Instead, we can sample the signal at times $0, T_s, 2T_s, 3T_s, \dots$. If we take N samples over the period, then each sample interval is $T_s = T_0/N$. The area under the curve is approximately the sum of the areas of the inscribed rectangular pulses, as shown in Figure 2.35. Using this approximation for the integral, we convert the integral of Eq. (2.59) to the following sum:

$$c_k \approx \frac{1}{T_0} \sum_{n=0}^{N-1} x(nT_s) e^{-jk\omega_0 nT_s} T_s. \tag{2.60}$$

This is because the nth rectangular pulse has a height of $x(nT_s)$ and a width of T_s. Substituting $T_s = T_0/N$, we simplify Eq. (2.60) to

$$c_k \approx \frac{1}{N} \sum_{n=0}^{N-1} x(nT_s) e^{-jk\omega_0 nT_s}. \tag{2.61}$$

Noting that $\omega_0 = 2\pi/T_0$, we rewrite Eq. (2.61) as

$$c_k \approx \frac{1}{N} \sum_{n=0}^{N-1} x(nT_s) e^{-jk\frac{2\pi}{T_0} nT_s}. \tag{2.62}$$

Finally, using $T_s = T_0/N$, we get

$$c_k \approx \frac{1}{N} \sum_{n=0}^{N-1} x(nT_s) e^{-jk\frac{2\pi}{N} n}. \tag{2.63}$$

Let us define this approximate coefficient as

$$c_k \approx X[k] \tag{2.64}$$

and the samples

$$x(nT_s) = x[n] \tag{2.65}$$

to rewrite Eq. (2.63) as

$$c_k \approx X[k] = \frac{1}{N} \sum_{n=0}^{N-1} x[n] e^{-jk\frac{2\pi}{N} n}. \tag{2.66}$$

Equation (2.66) is formally known as the *discrete Fourier transform* or DFT for short. Now we have a way to calculate the coefficients numerically. All we need to do is take N samples of a signal at a spacing of T_s, then feed them to a computer that evaluates Eq. (2.66) numerically. Not surprisingly, the approximation gets better as the number of samples N gets larger (or as the sampling interval T_s gets smaller). However, as N increases, it becomes more computationally intensive to determine the DFT numerically. Thus, there is a trade-off between accuracy and computational efficiency.

It can be shown that the samples $x[n]$ are given by

$$x[n] = \sum_{k=0}^{N-1} X[k] e^{jk\frac{2\pi}{N} n}. \tag{2.67}$$

Equation (2.67) is called the *inverse discrete Fourier transform* or IDFT. Notice that the DFT and IDFT are just formulas. The DFT converts a vector of N numbers $x[n]$ to a vector $X[k]$. The IDFT converts the vector $X[k]$ back to $x[n]$. They are used so often that MATLAB has the formula pre-programmed. In fact, MATLAB uses a clever way of evaluating the sum quickly and efficiently. MATLAB utilizes the *fast Fourier transform* or FFT and the *inverse Fast fourier transform* or IFFT to compute the DFT and IDFT, respectively. The two transforms are simply inverses of one another. Specifically, using the MATLAB commands `fft` and `ifft`, we find that `ifft(fft(x))=x`.

2.9 Variations on the DFT and IDFT

The DFT and IDFT are very similarly defined. They differ only in the negative sign in the exponential and the factor of $1/N$. In our derivation (see Eq. (2.66)), the factor $1/N$ was placed in the DFT so that $x[n]$ is approximated by the weighted sum of complex exponentials with the $X[k]$s as weights. Other authors occasionally use different notations and it is imperative that the readers recognize the similarities and differences between various representations.

Unfortunately MATLAB defines the FFT and IFFT with the $1/N$ factor in the IFFT instead of the FFT. Thus, MATLAB uses the following definitions:

$$X[k] = \sum_{n=0}^{N-1} x[n] e^{-jk\frac{2\pi}{N}n} \tag{2.68}$$

and

$$x[n] = \frac{1}{N} \sum_{k=0}^{N-1} X[k] e^{jk\frac{2\pi}{N}n}. \tag{2.69}$$

The above implies that given a vector of samples x, the proper weights (as defined in Eq. (2.66)) are computed by dividing the output of MATLAB function `fft` by N as follows:

$$\text{X=fft(x)/N} \tag{2.70}$$

because Eqs. (2.66) and (2.68) differ by a factor of N.

Other authors seek symmetry between the DFT and IDFT by weighting with the square root in each equation as follows:

$$X[k] = \sqrt{\frac{1}{N}} \sum_{n=0}^{N-1} x[n] e^{-jk\frac{2\pi}{N}n} \tag{2.71}$$

and

$$x[n] = \sqrt{\frac{1}{N}} \sum_{k=0}^{N-1} X[k] e^{jk\frac{2\pi}{N}n}. \tag{2.72}$$

Clearly, the definitions for DFT and IDFT are not unique, and the reader must pay special attention to how they are defined when using any prepackaged software.

2.10 Relationship between $X[k]$ and c_k

Before we proceed and calculate the coefficients, let us make a few observations. First, note that the samples and the coefficients are both periodic and repeat after N samples:

$$x[n] = x[n+N]$$

$$X[k] = X[k+N]. \tag{2.73}$$

We should expect this for $x[n]$ because the original signal is periodic or made periodic with period $T_0 = NT_s$. We can easily check this property for the coefficients by comparing $X[k]$ and $X[k+N]$:

$$
\begin{aligned}
X[k+N] &= \frac{1}{N} \sum_{n=0}^{N-1} x[n] e^{-j(k+N)\frac{2\pi}{N}n} \\
&= \frac{1}{N} \sum_{n=0}^{N-1} x[n] e^{-jk\frac{2\pi}{N}n} e^{-jN\frac{2\pi}{N}n} \\
&= \frac{1}{N} \sum_{n=0}^{N-1} x[n] e^{-jk\frac{2\pi}{N}n} e^{-j2\pi n} \\
&= \frac{1}{N} \sum_{n=0}^{N-1} x[n] e^{-jk\frac{2\pi}{N}n} = X[k].
\end{aligned}
\tag{2.74}
$$

Thus, Eq. (2.74) implies that there are only N distinct coefficients that we can find. Any program that does the calculation will return the N coefficients. For example, the MATLAB `fft` command computes the $X[k]$ for $k =$

Table 2.5. Relation of $X[k]$ to c_k for $N = 4$

c_k	c_{-5}	c_{-4}	c_{-3}	c_{-2}	c_{-1}	c_0	c_1	c_2	c_3	c_4	c_5
$X[k]$	$X[3]$	$X[0]$	$X[1]$	$X[2]$	$X[3]$	$X[0]$	$X[1]$	$X[2]$	$X[3]$	$X[0]$	$X[1]$
f_k	$-5f_0$	$-4f_0$	$-3f_0$	$-2f_0$	$-1f_0$	0	$1f_0$	$2f_0$	$3f_0$	$4f_0$	$5f_0$

Table 2.6. General relation of $X[k]$ to c_k

$X[0]$	$X[1]$	$X[2]$	$X[3]$	...	$X[k]$	...	$X[N-k]$	...	$X[N-2]$	$X[N-1]$
c_0	c_1	c_2	c_3	...	c_k	...	c_{-k}		c_{-2}	c_{-1}
0	$1f_0$	$2f_0$	$3f_0$	...	kf_0	...	$-kf_0$	...	$-2f_0$	$-1f_0$

$0, 1, \ldots, N - 1$, and the MATLAB `ifft` command returns $x[n]$ for $n = 0, 1, \ldots, N - 1$. Finally, having found these coefficients $X[k]$, we need to be able to relate them to the actual weights contained in the original signal. This is demonstrated in Table 2.5 for the case of $N = 4$. The general relation of $X[k]$ to c_k is shown in Table 2.6.

Clearly, we cannot obtain more than four meaningful coefficients. The computer returns the four coefficients $X[0]$, $X[1]$, $X[2]$, $X[3]$. We choose to treat these coefficients to represent c_0, c_1, c_{-2}, c_{-1} because these coefficients correspond to the lowest frequencies, which will give us the best numerical approximation. After we introduce the *sampling theorem* later in the text, we will gain a much clearer understanding of the periodic nature of the DFT.

This is a convention that is convenient to MATLAB. However, there is no reason we could not use the coefficients for c_0, c_1, c_2, c_{-1}. In general if we take an even number N samples, we can obtain coefficients from $c_{-N/2}$ to $c_{(N/2)-1}$. This is illustrated in Table 2.5. For real signals, if N is even $c_{-N/2} = c_{N/2}$ and these coefficients are real. If we take an odd number N samples, we can obtain coefficients from $c_{-(N-1)/2}$ to $c_{(N-1)/2}$.

Reordering the $X[k]$'s as $\{X[2], X[3], X[0], X[1]\}$ corresponding to coefficients $\{c_{-2}, c_{-1}, c_0, c_1\}$ and frequencies $\{-\alpha f_0, -f_0, 0, f_0\}$ allows us to readily identify the two-sided spectrum. MATLAB has a command `fftshift` that does this for us.

There is one more complication with using MATLAB. MATLAB indexes vectors starting at 1 instead of 0. Therefore, we must rewrite the

Table 2.7. MATLAB relation of $X[k]$ to c_k

$X[1]$	$X[2]$	$X[3]$	$X[4]$	...	$X[k+1]$	...	$X[N-k+1]$	...	$X[N-1]$	$X[N]$
c_0	c_1	c_2	c_3	...	c_k	...	c_{-k}		c_{-2}	c_{-1}
0	$1f_0$	$2f_0$	$3f_0$	...	kf_0	...	$-kf_0$	...	$-2f_0$	$-1f_0$

Table 2.8. Impact of the number of samples N on the quality of the approximation (c_k and $X[k]$ are given by Eqs. (2.59) and (2.66), respectively)

k	c_k	$X[k], N = 50$	$X[k], N = 1000$
-5	-0.0450	-0.0524	-0.0443
-4	0.0000	-0.0101	0.0010
-3	0.0750	0.0680	0.0757
-2	0.1592	0.1593	0.1592
-1	0.2251	0.2322	0.2244
0	0.2500	0.2600	0.2490
1	0.2251	0.2322	0.2244
2	0.1592	0.1593	0.1592
3	0.0750	0.0680	0.0757
4	0.0000	-0.0101	0.0010
5	-0.0450	-0.0524	-0.0443

DFT in MATLAB as

$$X[k] = \sum_{n=1}^{N} x[n-1] e^{-jk\frac{2\pi}{N}(n-1)}, \tag{2.75}$$

where the index for the output vector spans from $k = 1$ to $k = N$. This simply corresponds to a change of index, and does not affect the results otherwise. This is illustrated in Table 2.7.

2.11 Examples

Example 1: Consider approximating a pulse train with a duty cycle $d = 1/4$. Some of the weights c_k are shown in Table 2.8. The first column is the exact result obtained by integrating. The later columns are approximations found with the FFT for different numbers of samples N. We see that even with $N = 50$ the approximation is good to a few percent, and as N is increased to 100 the agreement between the exact and the approximate coefficients becomes excellent.

```
T=1;
N=10;
ts=T/N;
t=Ts*(0:N-1);
x=cos(2*pi*t);
plot(t,x,'k');
hold on;
plot(t,x,'ko');
xlabel('Time (sec)');
ylabel('Amplitude');
X=fft(x)/N
```

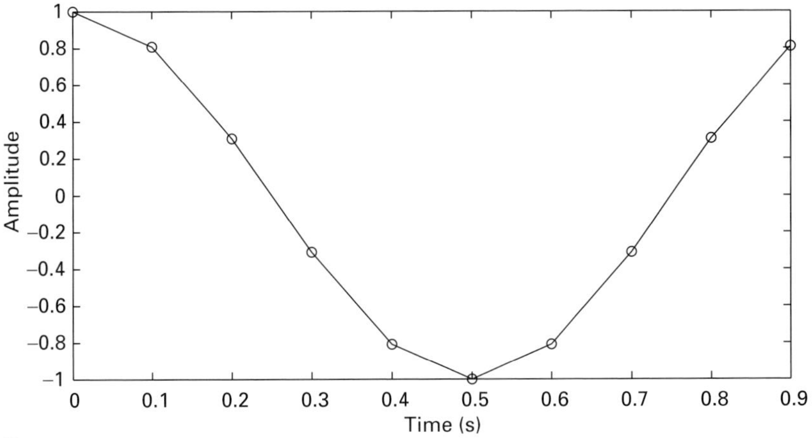

```
X=
Columns 1 through 5
-0.0000    0.5000-0.0000i    0.0000-0.0000i    -0.0000+0.0000i
  -0.0000+0.0000i
Columns 6 through 10
0.0000    0.0000-0.0000i    0.0000-0.0000i    -0.0000+0.0000i
  0.5000+0.0000i
```

Fig. 2.36. Sampled 1 Hz cosine for $N = 10$.

Example 2: Let us try using the FFT for a simple case of a 1 Hz cosine where we already know the answer. The following code creates a cosine with a period of $T = 1$ s and $N = 10$ samples. The sampling interval is $T_s = T/N = 0.1$ s. The signal is therefore sampled at times $t = 0.0, 0.1, 0.2, \ldots, 0.9$ s. Now $x(t) = \cos(2\pi t)$, shown in Figure 2.36. The symbols on the plot with circles 'o' represent the 10 samples.

The $X[k]$s are also listed in Figure 2.36. Remember that we needed to divide by N to get the conventional definition of $X[k]$ compared to what MATLAB produces with FFT. The coefficients c_1 and c_{-1}, which correspond to $X[2]$ and $X[10]$, are both $1/2$ and all the rest are 0 (at least to four decimal places). This matches our expectation that

$$\cos(2\pi t) = \frac{1}{2}[e^{j2\pi t} + e^{-j2\pi t}]. \tag{2.76}$$

Example 3: What if we had a 3 Hz cosine $x(t) = \cos(2\pi(3)t)$? Figure 2.37 shows the samples of this new signal. Note that it is a pretty poor representation because we only have 10 samples over three periods. The coefficients computed by FFT are $c_3 = c_{-3} = 1/2$. They correspond to $X[4]$ and $X[8]$, and match exactly what we expect. Thus, despite the fact of having what appeared to be a poorly represented cosine, we were still able to determine the coefficients correctly. In the next chapter we will discuss how sampling affects the results of the DFT.

Example 4: What if we consider a complex signal $x(t) = e^{j2\pi t}$? The coefficients are $c_1 = 1$ (in $X[2]$) and 0 everywhere else, as we would expect.

```
>> x=exp(j*2*pi*t);
>> X=fft(x)/N
X =
Columns 1 through 5
-0.0000 1.0000 - 0.0000i 0.0000 - 0.0000i 0.0000
+ 0.0000i -0.0000 + 0.0000i
Columns 6 through 10
0.0000 -0.0000 + 0.0000i -0.0000 - 0.0000i 0.0000
+ 0.0000i -0.0000 - 0.0000i
```

Example 5: If $x(t) = 42e^{j2\pi(4)t}$, the coefficient c_4 (in $X[5]$) is 42 and all the other coefficients are 0.

```
>> x=42*exp(j*2*pi*4*t);
>> X=fft(x)/N
X =
Columns 1 through 5
-0.0000 1.0000 - 0.0000i 0.0000 - 0.0000i 0.0000
+ 0.0000i 42.0000 - 0.0000i
Columns 6 through 10
0.0000 -0.0000 + 0.0000i -0.0000 - 0.0000i 0.0000
+ 0.0000i -0.0000 - 0.0000i
```

```
% Example 3.
T=1;
N=10;
ts=T/N;
t=Ts*(0:N-1);
x=cos(2*pi*3*t);
plot(t,x,'k');
hold on;
plot(t,x,'ko');
xlabel('Time (sec)');
ylabel('Amplitude');
X=fft(x)/N
```

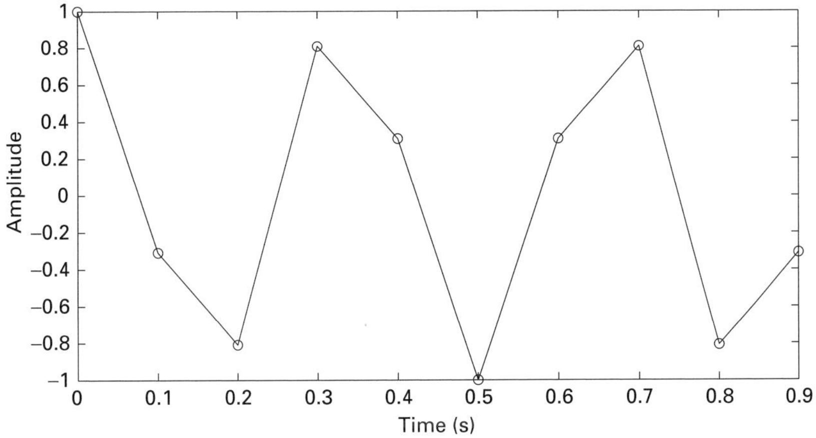

```
X=
Columns 1 through 5
0.0000    -0.0000-0.0000i    0.0000-0.0000i    0.5000+0.0000
 -0.0000-0.0000i
Columns 6 through 10
0.0000    -0.0000+0.0000i    0.5000-0.0000i    -0.0000+0.0000i
 -0.0000+0.0000i
```

Fig. 2.37. Sample 3 Hz cosine for $N = 10$.

Example 6: Consider now a cosine wave at a frequency of 11 Hz, $x(t) = \cos(2\pi(11)t)$. The FFT results are listed below. Observe that these are exactly the same as the results for 1 Hz (see Figure 2.36). This seems peculiar and further experimentation shows that a 13 Hz cosine has the same coefficients as a 3 Hz and so forth. This phenomenon is known as *aliasing*, which constitutes one of the most important topics in the study of signals. It will be explored in detail later.

```
>> x=cos(2*pi*11*t);
>> X=fft(x)/N
X =
Columns 1 through 5
-0.0000 0.5000 - 0.0000i 0.0000 - 0.0000i -0.0000
   + 0.0000i -0.0000 + 0.0000i
Columns 6 through 10
0.0000 -0.0000 - 0.0000i -0.0000 - 0.0000i 0.0000
   + 0.0000i 0.5000 + 0.0000i
```

Fig. 2.38. (a) Pulse train and (b) Fourier coefficients for $N = 1000$.

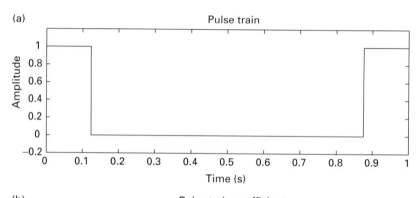

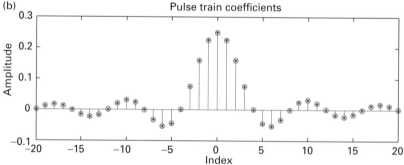

Figure 2.38 plots the coefficients from $k = -20$ to 20 of a pulse train with duty cycle $d = 1/4$. These include the coefficients listed in Table 2.8. Recall that these coefficients are all real because the pulse train is symmetric around the y-axis. The lollipop diagram shows the results of the FFT using $N = 1000$ samples. The asterisks show the theoretical values obtained from integrating Eq. (2.36) directly. Figure 2.39 shows the same comparison with $N = 50$ samples. With these few samples, the approximation is not so good.

Figure 2.40 shows a more complicated periodic exponential signal. The Fourier coefficients are complex so the magnitude and phase are plotted separately for $N = 500$ samples. Notice that the magnitude approximation

Fig. 2.39. (a) Pulse train and (b) Fourier coefficients for $N = 50$.

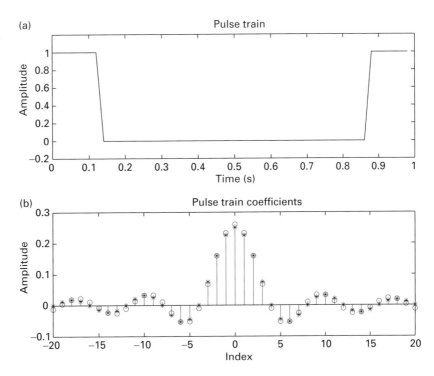

is very good but the phase is not quite as good. This is a common situation: more points are often required to accurately match the phase.

Recall that the $X[k]$ vector from FFT is in a scrambled order. In Figure 2.38, the $X[k]$ vector was reordered into the c_k for plotting purposes using the MATLAB command

```
cint=[c(N-K+1:N),c(1:K+1)]
```

The fdomain function shown below rearranges the indices on the $X[k]$ to match the indices on the c_k, and produces a second vector of the associated frequencies. It is handy because it works for both even- and odd-length vectors and because it computes the frequencies associated with each coefficient. This routine simplifies the task of mapping the coefficients with the appropriate frequencies.

```
function [X,f]=fdomain(x,Fs)
% FDOMAIN computes the Fourier coefficients from
% vector x
% and the corresponding frequencies (two-sided)
N=length(x);
```

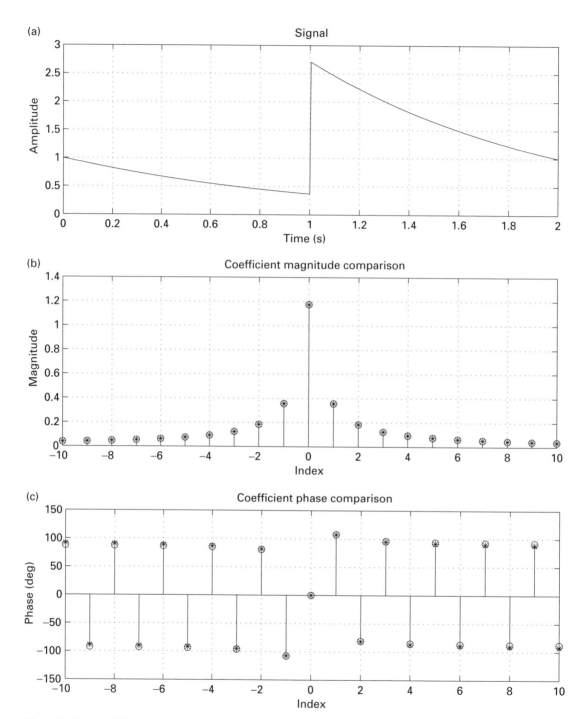

Fig. 2.40. Exponential and
Fourier coefficient
magnitude and phase.

```
if mod(N,2)==0
k=-N/2:N/2-1; % N even
else
k=-(N-1)/2:(N-1)/2; % N odd
end
T=N/Fs;
f=k/T;
X=fft(x)/N;
X=fftshift(X);
```

For example, let us revisit the coefficients for the 3 Hz cosine viewed earlier. Figure 2.41 plots both the X vector obtained with the ordinary fft and the reordered results using fdomain. Note that the MATLAB commands abs and stem are used to plot the magnitude of the coefficients in the stem diagrams. As before, the $X[k]$s are $1/2$ for $k = 4$ and 8. The spectrum is more intuitive because it is $1/2$ at ± 3 Hz.

2.12 Proof of the continuous-time orthogonality principle

We wish to construct an approximation to a continuous-time signal $x(t)$ over a time interval T using a linear combination of the building blocks or basis functions $\phi_k(t)$, that is,

$$x(t) \approx \hat{x}(t) = \sum_k c_k \phi_k(t). \tag{2.77}$$

Problem: Determine the c_k such that the following integrated squared error (ISE) is minimized:

$$\text{ISE} = \int_T |x(t) - \hat{x}(t)|^2 \, dt. \tag{2.78}$$

Solution: The coefficients c_k must satisfy the following orthogonality principle (OP):

$$\int_T [x(t) - \hat{x}(t)] \, \phi_k^*(t) dt = 0, \ \text{ for all } \ k. \tag{2.79}$$

Proof: Let the set of coefficients c_k satisfy the OP, that is,

$$\int_T \left[x(t) - \sum_n c_n \phi_n(t) \right] \phi_k^*(t) dt = 0, \ \text{ for all } \ k. \tag{2.80}$$

Fig. 2.41. $X[k]$ and the spectra of a 3 Hz cosine.

```
T=1;
N=10;
ts=T/N;
t=Ts*(0:N-1);
x=cos(2*pi*3*t);
X1=fft(x)/N;
[X2,f]=fdomain(x,/Ts);1
subplot(2,1,1);
stem(abs(X1), 'k');
title('X[k]');
xlabel('Index');
ylabel('Magnitude');
axis([1,10,0,0.6]);
subplot(2,1,2);
h=stem(f,abs(X2),'k');
% code below between begin and end is for
% display purposes only
% begin
    set(h, 'LineWidth',0.5, 'Marker', 'none');
% end
title('Spectrum');
xlabel('Frequency) (Hz)');
ylabel('Magnitude');
axis([-5,4,0,0.6]);
```

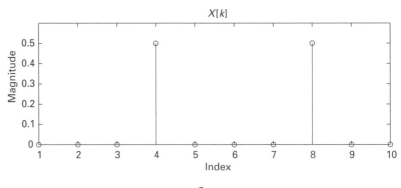

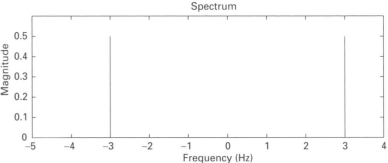

We now compute the ISE for an *arbitrary* set of coefficients b_k as follows:

$$\text{ISE} = \int_T \left| x(t) - \sum_k b_k \phi_k(t) \right|^2 dt. \tag{2.81}$$

Subtracting and adding $\sum_n c_n \phi_n(t)$ (notice the use of dummy indices) yields

$$\text{ISE} = \int_T \left| x(t) - \sum_n c_n \phi_n(t) + \sum_k c_k \phi_k(t) - \sum_k b_k \phi_k(t) \right|^2 dt$$

$$= \int_T \left| \left[x(t) - \sum_n c_n \phi_n(t) \right] + \sum_k (c_k - b_k) \phi_k(t) \right|^2 dt. \tag{2.82}$$

Now we use the following identity for complex numbers or expressions z_1 and z_2

$$|z_1 + z_2|^2 = (z_1 + z_2)(z_1 + z_2)^* = (z_1 + z_2)(z_1^* + z_2^*)$$

$$= |z_1|^2 + z_1 z_2^* + z_1^* z_2 + |z_2|^2 \tag{2.83}$$

to expand the expression for ISE, yielding

$$\text{ISE} = \int_T \left| x(t) - \sum_n c_n \phi_n(t) \right|^2 dt$$

$$+ \int_T \left[x(t) - \sum_n c_n \phi_n(t) \right] \left[\sum_k (c_k - b_k) \phi_k(t) \right]^* dt \tag{2.84}$$

$$+ \int_T \left[x(t) - \sum_n c_n \phi_n(t) \right]^* \left[\sum_k (c_k - b_k) \phi_k(t) \right] dt$$

$$+ \int_T \left| \sum_k (c_k - b_k) \phi_k(t) \right|^2.$$

Consider the second integral in Eq. (2.84). Upon rearranging, we obtain

$$\int_T \left[x(t) - \sum_n c_n \phi_n(t) \right] \left[\sum_k (c_k - b_k) \phi_k(t) \right]^* dt$$

$$= \sum_k (c_k - b_k)^* \int_T \left[x(t) - \sum_n c_n \phi_n(t) \right] \phi_k^*(t) dt = 0 \tag{2.85}$$

because the c_k satisfy the OP (see Eq. (2.80)). Similarly, the third integral in Eq. (2.84) can also be shown to be zero. Thus,

$$\text{ISE} = \int_T \left| x(t) - \sum_n c_n \phi_n(t) \right|^2 dt + \int_T \left| \sum_k (c_k - b_k) \phi_k(t) \right|^2 dt. \tag{2.86}$$

This means that the ISE using the arbitrary set of coefficients b_k is equal to the ISE using the set of coefficients c_k that satisfy the OP plus a non-negative term. The best we can possibly do is make the non-negative term zero by setting $b_k = c_k$ for all k. Thus, the set of coefficients c_k that satisfies the OP minimizes the ISE. Note that no differentiation was required in the above derivation and it is clear that we have achieved a minimum.

2.13 A note on vector spaces

Orthogonality of the basis functions can be best understood by examining vectors in three-dimensional space. Any three-dimensional vector $\mathbf{w}$ can be expressed as

$$\mathbf{w} = w_x\mathbf{i} + w_y\mathbf{j} + w_z\mathbf{k}, \tag{2.87}$$

where w_x, w_y and w_z are the components of $\mathbf{w}$, and $\mathbf{i}$, $\mathbf{j}$ and $\mathbf{k}$ are the unit vectors along the x-, y-, and z-axis, respectively. These unit vectors, as the name implies, have unit magnitude and are defined to be orthogonal (perpendicular) to each other. Mathematically, these unit vectors satisfy the following:

$$\mathbf{i} \cdot \mathbf{i} = \mathbf{j} \cdot \mathbf{j} = \mathbf{k} \cdot \mathbf{k} = |\mathbf{i}|^2 = |\mathbf{j}|^2 = |\mathbf{k}|^2 = 1 \quad \text{and}$$
$$\mathbf{i} \cdot \mathbf{j} = \mathbf{i} \cdot \mathbf{k} = \mathbf{j} \cdot \mathbf{k} = 0, \tag{2.88}$$

where $|\mathbf{a}|$ represents the Euclidean norm of the vector $\mathbf{a}$. Given any vector $\mathbf{w}$, we can easily find its components w_x, w_y and w_z since

$$\mathbf{w} \cdot \mathbf{i} = (w_x\mathbf{i} + w_y\mathbf{j} + w_z\mathbf{k}) \cdot \mathbf{i} = w_x\mathbf{i} \cdot \mathbf{i} + w_y\mathbf{j} \cdot \mathbf{i} + w_z\mathbf{k} \cdot \mathbf{i} = w_x. \tag{2.89}$$

Similarly, $\mathbf{w} \cdot \mathbf{j} = w_y$ and $\mathbf{w} \cdot \mathbf{k} = w_z$.

Of course, any set of three linearly independent vectors can serve as basis vectors. They do not have to be either orthogonal or of unit length. Consider an arbitrary set of basis vectors $\boldsymbol{\phi}_1$, $\boldsymbol{\phi}_2$ and $\boldsymbol{\phi}_3$. Then any vector $\mathbf{w}$ can be expressed as

$$\mathbf{w} = w_1\boldsymbol{\phi}_1 + w_2\boldsymbol{\phi}_2 + w_3\boldsymbol{\phi}_3. \tag{2.90}$$

To find w_1, w_2 and w_3 using this arbitrary set of basis vectors, we note that

$$\mathbf{w} \cdot \boldsymbol{\phi}_1 = w_1|\boldsymbol{\phi}_1|^2 + w_2\boldsymbol{\phi}_2 \cdot \boldsymbol{\phi}_1 + w_3\boldsymbol{\phi}_3 \cdot \boldsymbol{\phi}_1$$
$$\mathbf{w} \cdot \boldsymbol{\phi}_2 = w_1\boldsymbol{\phi}_1 \cdot \boldsymbol{\phi}_2 + w_2|\boldsymbol{\phi}_2|^2 + w_3\boldsymbol{\phi}_3 \cdot \boldsymbol{\phi}_2 \tag{2.91}$$
$$\mathbf{w} \cdot \boldsymbol{\phi}_3 = w_1\boldsymbol{\phi}_1 \cdot \boldsymbol{\phi}_3 + w_2\boldsymbol{\phi}_2 \cdot \boldsymbol{\phi}_3 + w_3|\boldsymbol{\phi}_3|^2$$

which yields three equations with three unknowns (remember $\mathbf{w}$, $\boldsymbol{\phi}_1$, $\boldsymbol{\phi}_2$ and $\boldsymbol{\phi}_3$ are given and specified) that can be readily solved for the components w_1, w_2 and w_3. Clearly, it becomes much easier to solve for these components if $\boldsymbol{\phi}_1$, $\boldsymbol{\phi}_2$ and $\boldsymbol{\phi}_3$ are mutually orthogonal, in which case

$$
\begin{aligned}
w_1 &= \frac{1}{|\boldsymbol{\phi}_1|^2} \mathbf{w} \cdot \boldsymbol{\phi}_1 \\
w_2 &= \frac{1}{|\boldsymbol{\phi}_2|^2} \mathbf{w} \cdot \boldsymbol{\phi}_2 \\
w_3 &= \frac{1}{|\boldsymbol{\phi}_3|^2} \mathbf{w} \cdot \boldsymbol{\phi}_3 .
\end{aligned}
\tag{2.92}
$$

The previous derivation can be easily extended to n-dimensional vectors. Consider a set of n basis vectors, $\boldsymbol{\phi}_i$, for $i = 1, \ldots, n$, whose linear combination yields $\mathbf{w}$ as shown below:

$$
\mathbf{w} = \sum_{i=1}^{n} w_i \boldsymbol{\phi}_i .
\tag{2.93}
$$

To determine the components w_i, the following n simultaneous equations (the readers are encouraged to write them out if there is any uncertainty about the notation) are formulated and solved:

$$
\mathbf{w} \cdot \boldsymbol{\phi}_i = \sum_{k=1}^{n} w_k \boldsymbol{\phi}_k \cdot \boldsymbol{\phi}_i , \quad \text{for } i = 1, \ldots, n .
\tag{2.94}
$$

On the other hand, if the basis vectors $\boldsymbol{\phi}_i$ are orthogonal, then the components w_i are simply given by

$$
w_i = \frac{1}{|\boldsymbol{\phi}_i|^2} \mathbf{w} \cdot \boldsymbol{\phi}_i , \quad \text{for } i = 1, \ldots, n .
\tag{2.95}
$$

Note that the components can be readily obtained individually without having to solve a set of simultaneous equations. Thus, it is easy to see why orthogonal basis vectors are useful and generally preferred.

Now let us extend the previous discussion from basis vectors to basis functions. Suppose we have a signal $w(t)$ that can be expressed as a weighted sum of a set of signals $\phi_i(t)$, for $i = 1, \ldots, n$, over the time interval T:

$$
w(t) = \sum_{i=1}^{n} w_i \phi_i(t) .
\tag{2.96}
$$

Given $w(t)$ and the basis functions $\phi_i(t)$, the coefficients w_i can be calculated as follows. Instead of performing the dot product with the basis vectors like we did previously, here we multiply the signal $w(t)$ by the basis functions and integrate over the time interval of interest as shown,

$$\int_T w(t)\phi_k(t)\mathrm{d}t = \sum_{i=1}^{n} w_i \int_T \phi_i(t)\phi_k(t)\mathrm{d}t, \text{ for } k = 1, \ldots, n \qquad (2.97)$$

which yields n simultaneous equations with n unknowns (again, the readers are encouraged to expand the equations, and remember $w(t), \phi_i(t)$ are known functions). Furthermore, if the following correspondences or analogies are made:

$$w(t) \leftrightarrow \mathbf{w}$$
$$\int_T w(t)\phi_i(t)\mathrm{d}t \leftrightarrow \mathbf{w} \cdot \phi_i \qquad (2.98)$$
$$\int_T \phi_i(t)\phi_k(t)\mathrm{d}t \leftrightarrow \phi_i \cdot \phi_k,$$

we note that Eq. (2.97) has the same form as Eq. (2.91) for calculating the n-dimensional vector coefficients. In addition, if the basis functions $\phi_i(t)$ are orthogonal, that is if they satisfy

$$\int_T \phi_i(t)\phi_k(t)\mathrm{d}t = \begin{cases} K_i, & k = i \\ 0, & k \neq i, \end{cases} \qquad (2.99)$$

then

$$w_i = \frac{1}{K_i} \int_T w(t)\phi_i(t)\mathrm{d}t. \qquad (2.100)$$

Again it becomes clear why orthogonal basis functions are usually preferred.

Moreover, for any set of orthogonal basis functions $\phi_i(t)$, a related set of *orthonormal* basis functions can be constructed using the following expressions:

$$\psi_i(t) = \frac{\phi_i(t)}{\sqrt{K_i}}, \qquad (2.101)$$

in which case the new basis functions $\phi_i(t)$ satisfy

$$\int_T \psi_i(t)\psi_k(t)\mathrm{d}t = \begin{cases} 1, & k = i \\ 0, & k \neq i \end{cases}. \qquad (2.102)$$

In this case, suppose we have two functions $w(t)$ and $v(t)$ that can be expressed as a weighted sum of the $\phi_i(t)$, that is

$$v(t) = \sum_{i=1}^{n} v_i \psi_i(t)$$
$$w(t) = \sum_{i=1}^{n} w_i \psi_i(t),$$

(2.103)

then we can now think of these functions as vectors $\mathbf{v} = [v_1, v_2, \ldots, v_n]$ and $\mathbf{w} = [w_1, w_2, \ldots, w_n]$ in an n-dimensional vector space with the following relationships:

$$\int_T v^2(t)\mathrm{d}t = |\mathbf{v}|^2$$
$$\int_T w^2(t)\mathrm{d}t = |\mathbf{w}|^2$$
$$\int_T v(t)w(t)\mathrm{d}t = \mathbf{v} \cdot \mathbf{w}.$$

(2.104)

So what is the big deal? First, it may be easier to construct a linear combination of the $\phi_i(t)$ than to generate the signal $w(t)$ directly. Second, if the $\phi_i(t)$ (whether orthogonal or not) are known, then any continuous-time signal $w(t)$ can be represented as a vector $\mathbf{w} = [w_1, w_2, \ldots, w_n]$, which is a discrete-time signal that can be stored in a computer and that, as will be shown later in the text, is very advantageous. Third, if it is known how certain systems respond to the $\phi_i(t)$, then how these systems respond to any signal can be easily determined, provided the signal can be expressed as a linear combination of the $\phi_i(t)$. And that is a *really big deal*.

2.13.1 Complex vectors and functions

The only modification to the above derivation for complex vectors and complex functions is a change in the definition of the dot product of any two vectors $\mathbf{v}$ and $\mathbf{w}$ or any two functions $v(t)$ and $w(t)$ to

$$\mathbf{v} \cdot \mathbf{w} = \sum_{i=1}^{n} v_i w_i^* = (\mathbf{w} \cdot \mathbf{v})^* \iff \int_T v(t)w^*(t)\mathrm{d}t. \qquad (2.105)$$

2.13.2 Approximations and the orthogonality principle

Usually the basis functions, or the building blocks for signals, are chosen to have some special properties and thus it may require a very large or even an infinite number of them to construct a given signal "exactly." Because sometimes it may be difficult and impractical to use a large number of terms in the approximation, we usually must approximate the signal with a finite number of basis functions or building blocks. That is

$$w(t) \approx \hat{w}(t) = \sum_i w_i \phi_i(t) \tag{2.106}$$

with i taking on a finite number of integer values. In this case, we can calculate the coefficients w_i that minimize the integrated squared error (ISE) by using the orthogonality principle (OP) as shown earlier in this chapter. Interestingly, we get the same set of n equations with n unknowns as in Eq. (2.97). In fact, some texts derive them that way but do not indicate that the results were obtained by minimizing the ISE. Of course, if the basis functions are complex, we must use the definition of the dot product for complex vectors and complex functions given in Eq. (2.105).

2.13.3 The discrete Fourier transform (DFT)

In Sections 2.8 through 2.10 we discovered that the discrete Fourier transform (DFT) could be used to approximate the Fourier coefficients of a periodic continuous-time signal. It is so useful in this and other applications that a number of efficient algorithms (generically called fast Fourier transforms or FFTs) have been developed. Fundamentally, however, the DFT transforms a finite sequence of numbers into another finite sequence of the same length. In this section we show that this transform is equivalent to representing a vector with a particular set of orthogonal basis vectors.

Consider a periodic sequence $x[n] = x[n + N]$ and the elements of a basis vector

$$\phi_k[n] = \phi_{k+N}[n] = \phi_k[n + N] = e^{jk(2\pi/N)n}. \tag{2.107}$$

Because $\phi_k[n] = \phi_{k+N}[n]$, there are only N distinct sequences $\phi_k[n]$. Moreover, because both the discrete-time signal and the elements of the basis vector are periodic with period N, we need only concern ourselves with a single period, that is a finite sequence of length N. For convenience, we choose the period $0 \leq n \leq N - 1$ and introduce the following vectors

$$\mathbf{x} = (x[0], x[1], x[2], \ldots, x[N-1]) \tag{2.108}$$

and

$$\boldsymbol{\phi}_k = (\phi_k[0], \phi_k[1], \phi_k[2], \ldots, \phi_k[N-1]) \,, \ k = 0, \ldots, N-1. \tag{2.109}$$

Thus, we have an N-dimensional vector $\mathbf{x}$ that we want to express in terms of N basis vectors $\boldsymbol{\phi}_k$. That is, we wish to write

$$\mathbf{x} = \sum_{k=0}^{N-1} a_k \boldsymbol{\phi}_k \tag{2.110}$$

or equivalently

$$x[n] = \sum_{k=0}^{N-1} a_k e^{jk(2\pi/N)n}, \ n = 0, \ldots, N-1, \tag{2.111}$$

where equality requires that the basis vectors ϕ_k be linearly independent.

To show that these basis vectors are orthogonal and linearly independent, we must prove that

$$\boldsymbol{\phi}_m \cdot \boldsymbol{\phi}_k^* = \sum_{n=0}^{N-1} \phi_m[n]\phi_k^*[n] = \begin{cases} N, & m = k \\ 0, & m \neq k. \end{cases} \tag{2.112}$$

Substituting Eq. (2.107) into Eq. (2.112), we have

$$\boldsymbol{\phi}_m \cdot \boldsymbol{\phi}_k^* = \sum_{n=0}^{N-1} e^{jm(2\pi/N)n} e^{-jk(2\pi/N)n} = \sum_{n=0}^{N-1} e^{j(m-k)(2\pi/N)n}. \tag{2.113}$$

For $m = k$, Eq. (2.113) reduces to

$$\boldsymbol{\phi}_m \cdot \boldsymbol{\phi}_k^* = \sum_{n=0}^{N-1} 1 = N \tag{2.114}$$

and for $m \neq k$, Eq. (2.113) becomes

$$\boldsymbol{\phi}_m \cdot \boldsymbol{\phi}_k^* = \sum_{n=0}^{N-1} \left(e^{j(m-k)(2\pi/N)} \right)^n$$
$$= \frac{1 - e^{j(m-k)(2\pi/N)N}}{1 - e^{j(m-k)(2\pi/N)}} = \frac{1 - e^{j(m-k)(2\pi)}}{1 - e^{j(m-k)(2\pi/N)}} = 0. \tag{2.115}$$

The results of Eqs. (2.114) and (2.115) clearly indicate that the basis vectors are orthogonal and linearly independent.

The dot product of vectors $\mathbf{x}$ and $\boldsymbol{\phi}_k^*$ is given by

$$\mathbf{x} \cdot \boldsymbol{\phi}_k^* = a_0 \boldsymbol{\phi}_0 \cdot \boldsymbol{\phi}_k^* + a_1 \boldsymbol{\phi}_1 \cdot \boldsymbol{\phi}_k^* + \cdots + a_k \boldsymbol{\phi}_k \cdot \boldsymbol{\phi}_k^*$$
$$+ \cdots + a_{N-1} \boldsymbol{\phi}_{N-1} \cdot \boldsymbol{\phi}_k^* = a_k N. \tag{2.116}$$

Thus, we find

$$a_k = \frac{1}{N} \mathbf{x} \cdot \boldsymbol{\phi}_k^* \tag{2.117}$$

or equivalently

$$a_k = a_{k+N} = \frac{1}{N} \sum_{k=0}^{N-1} x[n] e^{-jk(2\pi/N)n}. \tag{2.118}$$

Finally, defining $a_k = X[k]$ gives the usual form of the discrete Fourier transform (DFT) and the inverse discrete Fourier transform (IDFT) as follows:

$$X[k] = X[k + N] = \frac{1}{N} \sum_{n=0}^{N-1} x[n] e^{-jk(2\pi/N)n}$$

$$x[n] = x[n + N] = \sum_{k=0}^{N-1} X[k] e^{jk(2\pi/N)n}, \tag{2.119}$$

where $X[k]$ and $x[n]$ denote the DFT and IDFT, respectively.

2.13.4 Finding orthogonal basis functions for known signals

Finally, there are applications (such as digital communication systems) where we want to represent each of a finite number of *known* signals (say n of them) as a weighted sum of orthogonal basis functions. Given the signals, can we find a set of orthogonal basis functions and, if so, how many do we need? Surprisingly, it turns out that we never need more than n (and sometimes, depending on the signals, far fewer) to represent the signals *exactly*! Furthermore, we can construct a set of basis functions (they are not unique) by using the Gram–Schmidt orthogonalization process.

2.14 Problems

2-1 Determine an expression for the complex Fourier series coefficients of the periodic pulse train shown in Figure 2.42.

Fig. 2.42. Figure for
Problem 2-1.

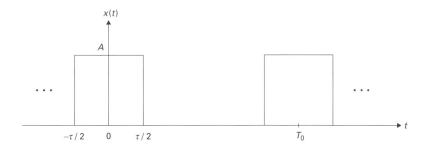

2-2 A Hadamard matrix consists of elements ± 1 and has the property that its rows are orthogonal to each other (recall that if two vectors are orthogonal their dot product is zero). Starting with a 2×2 matrix $[H_2]$, a matrix $[H_{2n}]$ of size $2n \times 2n$ can be constructed recursively using the equation given below:

$$[H_2] = \begin{bmatrix} 1 & 1 \\ 1 & -1 \end{bmatrix} \quad [H_{2n}] = \begin{bmatrix} [H_n] & [H_n] \\ [H_n] & -[H_n] \end{bmatrix}$$

(a) Construct $[H_4]$ and $[H_8]$.

(b) If c_{ik} denotes the element in the ith row and kth column of $[H_4]$, sketch the following waveforms:

$$\phi_i(t) = \sum_{k=1}^{4} c_{ik}\, p(t - (k-1)T), \quad i = 1, 2, 3, 4,$$

where $p(t)$ is a pulse of unit amplitude on the interval $0 \le t < T$, and T is an arbitrary time interval.

(c) Are the $\phi_i(t)$ of part (b) orthogonal to each other? Why or why not? The waveforms $\phi_i(t)$ are known as *Walsh functions* and among other things are used to separate channels in Code Division Multiple Access (CDMA) cell phones.

2-3 The signal $x(t)$ is to be approximated by $x(t) \approx \hat{x}(t) = a_1\phi_1(t) + a_2\phi_2(t)$ as shown in Figure 2.43.

(a) Determine a_1 and a_2 such that the integrated square error (ISE) is minimized. Sketch the approximated signal $\hat{x}(t)$. Label the relevant amplitudes and times.

(b) Sketch another non-trivial basis function $\phi_3(t)$, defined in the region $1/4 \le t \le 1/2$, that is orthogonal to both $\phi_1(t)$ and $\phi_2(t)$. Label the relevant amplitudes and times.

Fig. 2.43. Figure for
Problem 2-3.

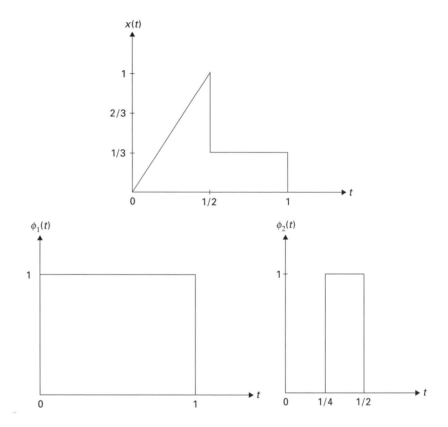

(c) Consider a system that is linear and time-invariant. If the response to $\phi_1(t)$ is $y_1(t)$ and the response to $\phi_2(t)$ is $y_2(t)$, what is the response of the system to $\hat{x}(t)$? Please explain your results.

2-4 The signal $x(t) = e^{2t} + 5$, defined in the interval $0 \le t \le 1$, is to be approximated by $\hat{x}(t) = a_1\phi_1(t) + a_2\phi_2(t)$, where the basis functions $\phi_1(t)$ and $\phi_2(t)$ are shown in Figure 2.44.

(a) Determine the coefficients a_1 and a_2 such that the integrated squared error (ISE) is minimized. Sketch $x(t)$ and $\hat{x}(t)$. Label the appropriate amplitudes.

(b) Construct a third basis function $\phi_3(t)$ that is orthogonal to both $\phi_1(t)$ and $\phi_2(t)$ and is defined to be non-zero in the interval $1/2 \le t \le 1$.

(c) In words, describe the advantage of using a set of orthogonal functions as basis functions.

Fig. 2.44. Figure for
Problem 2-4.

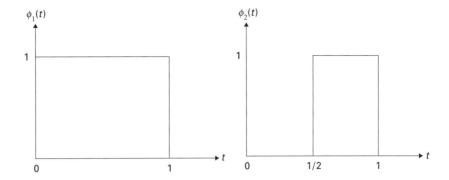

2-5 In Section 2.5.1, we determined an expression for the Fourier coefficients, c_k, for a periodic pulse train $x(t)$, where

$$x(t) = x(t + T_0) \approx \hat{x}(t) = \sum_{k=-K}^{K} c_k e^{jk\omega_0 t}, \qquad \omega_0 = \frac{2\pi}{T_0}.$$

    (a) Write an M-file to generate graphs of $\hat{x}(t)$ for A $= 1, \tau = 1, T_0 = 4$
and $K = 10, 20$ and 50. Put all your plots in one window using
MATLAB's `subplot` command.

    (b) Why do we use $-K \leq k \leq K$ in part **(a)**? Does the orthogonal-
ity principle still work if $0 \leq k \leq K$? Explain *briefly* why or why
not?

2-6 A Fourier series can be used to represent a non-periodic signal $x(t)$
over a given time interval by constructing the periodic extension of the
signal $x_p(t)$, where

$$x_p(t) = \sum_{n=-\infty}^{\infty} x(t - nT_0).$$

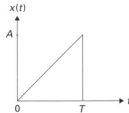

Fig. 2.45. Figure for
Problem 2-6.

    (a) For the signal $x(t)$ shown in Figure 2.45, determine an expres-
sion for the Fourier coefficients of $x_p(t)$ for $T_0 = T$. Repeat for
$T_0 = 2T$.

    (b) Modify the M-file you created in Problem 2-5(a) to generate graphs
of $\hat{x}_p(t)$ for $A = 2, T = 1, T_0 = 1$ and $K = 10, 20,$ and 50. Repeat
for $T_0 = 2$.

2-7 Consider the signal

$$x(t) = 10 + 20\cos(2\pi(100)t + \pi/4) + 10\cos(2\pi(250)t).$$

Using Euler's identity, the signal $x(t)$ can be expressed as a sum of complex exponential signals in the form

$$x(t) = X_0 + \text{Re}\left\{\sum_{k=1}^{N} X_k e^{jk2\pi f_0 t}\right\}.$$

(a) Determine f_0 and N? What are the values of X_k? *It is not necessary to evaluate any integrals to obtain X_k.*
(b) Is the signal $x(t)$ periodic? If so, what is its period?
(c) Plot the two-sided spectrum of this signal.

2-8 A signal composed of sinusoids is given by the equation

$$x(t) = 10\cos(800\pi t + \pi/4) + 7\sin(1200\pi t) - 3\cos(1600\pi t).$$

(a) Sketch the spectrum of this signal, indicating the complex amplitude of each frequency component.
(b) Is $x(t)$ periodic? If so, what is its period?
(c) Suppose $y(t) = x(t) + 5\cos(1000\pi t + \pi/2)$. How is the spectrum changed? Is $y(t)$ periodic? If so, what is its period?

2-9 An amplitude-modulated (AM) signal (like those used in AM radio) is given by

$$x(t) = [10 + 8\cos(2\pi t - \pi/3)]\cos(2\pi(10)t).$$

(a) Show that

$$x(t) = A_1\cos(2\pi f_1 t + \phi_1) + A_2\cos(2\pi f_2 t + \phi_2)$$
$$+ A_3\cos(2\pi f_3 t + \phi_3)$$

and determine the values of A_i, f_i and ϕ_i, for $i = 1, 2, 3$.
(b) Sketch the two-sided spectrum of this signal.
(c) Use MATLAB to plot $x(t)$ for $0 \leq t \leq 4$.

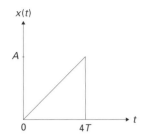

$x(t)$

A

0 $4T$ t

Fig. 2.46. Figure for Problem 2-10.

2-10 Determine the coefficients c_k that minimize the ISE in the series approximation of the signal shown in Figure 2.46:

$$x(t) \approx \hat{x}(t) = \sum_{k=1}^{4} c_k \phi_k(t),$$

where $\phi_k(t)$, for $k = 1, 2, 3, 4$, are the *Walsh functions* of Problem 2-2.

2-11 A signal $x(t)$ has the two-sided spectrum representation shown in Figure 2.47.
   (a) Write an equation for $x(t)$ as the sum of cosines.
   (b) Is $x(t)$ a periodic signal? If so, what is its period?
   (c) Explain why "negative" frequencies are needed in the spectrum.

2-12 The signal $x(t) = e^t$, for $0 \le t < 1$, is to be approximated by $\hat{x}(t) = a_1\phi_1(t) + a_2\phi_2(t)$, where the basis functions $\phi_1(t)$ and $\phi_2(t)$ are shown in Figure 2.48.
   (a) Determine the coefficients a_1 and a_2 such that the integrated squared error (ISE) is minimized.
   (b) The approximation is changed to include a third basis function $\phi_3(t)$, so that $\hat{x}(t) = a_1\phi_1(t) + a_2\phi_2(t) + a_3\phi_3(t)$. Determine a_3 such that the ISE is minimized. Is it necessary to recompute a_1 and a_2? Why or why not?

Fig. 2.47. Figure for Problem 2-11.

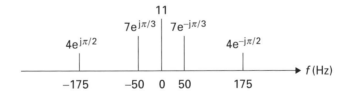

Fig. 2.48. Figure for Problem 2-12.

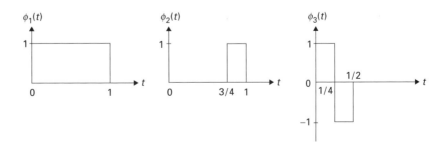

2-13 Consider the following continuous-time signal:

$$x(t) = 12 \cos \left(200\sqrt{2}\, t \right) - 8 \sin \left(140\sqrt{2}\, t \right)$$
$$+ 19 \cos \left(90\sqrt{2}\, t + \frac{\pi}{4} \right) + 16.$$

(a) Determine the fundamental frequency ω_0 and the period T_0 of the signal.

(b) Express $x(t)$ in the phasor form

$$x(t) = X_0 + \text{Re} \left\{ \sum_{k=1}^{N} X_k e^{jk\omega_0 t} \right\}.$$

(c) Sketch and label the spectrum $x(t)$.

(d) If

$$y(t) = x(t) - 5 \sin \left(5\sqrt{2}\, t - \frac{\pi}{3} \right)$$

determine the fundamental frequency ω_0 of $y(t)$.

2-14 A system has the property that sinusoidal components of the input $x(t)$ with frequencies between 4500 and 5500 Hz are allowed to pass through to the output $y(t)$ unchanged while components at all other frequencies are completely blocked. Write an expression for the output $y(t)$ if the input $x(t)$ is a square wave with period $T_0 = 1 \times 10^{-3}$ s, and the Fourier coefficients of the square wave are given by

$$X_k = \begin{cases} \dfrac{4}{j\pi k} & k = \pm 1, \pm 3, \pm 5, \ldots \\ 0 & k = 0, \pm 2, \pm 4, \pm 6, \ldots . \end{cases}$$

3

Sampling and data acquisition

In virtually every field of science and engineering there is a need to acquire, record, process, and interpret data. In the real world, data comes from continuous-time signals. Not so long ago, the data were often recorded on a strip chart. For example, a seismometer records earthquake data on a rotating drum. Now it is often more convenient to sample the data periodically and record the information as a series of numbers in a computer.

The recording may be modeled with a *continuous-to-discrete* (or C/D) *converter* shown in Figure 3.1. The converter accepts a continuous-time signal $x(t)$ and produces a discrete-time signal $x[n]$ with samples spaced by T_s, that is with a sampling frequency of $f_s = 1/T_s$. Of great interest is how frequently we should sample the signal. If we sample the signal too slowly, we may miss some interesting behaviors and get erroneous results. If we sample the signal too fast, we may get lost in a sea of unnecessary data, need huge disk space, and waste unnecessary processing time. The C/D converter is also sometimes represented with a switch as shown in Figure 3.2. Figures 3.1 and 3.2 are different representations of the same converter.

Fig. 3.1. Continuous to discrete (C/D) converter.

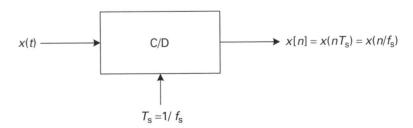

Fig. 3.2. Switch symbol for
C/D converter.

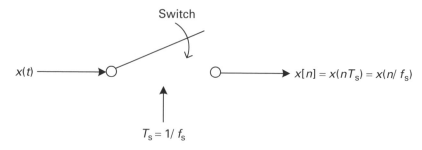

An *analog-to-digital* (or A/D) *converter* is a practical implementation
of a C/D converter. In the C/D converter, $x[n]$ is a real number that
preserves the value of $x(t)$ to infinite precision. This is obviously unre-
alistic. An A/D converter records $x[n]$ using a finite number of bits that
represents set values. Every data point sampled is rounded to one of the
predetermined values. As a result, for every data point we lose some accuracy. How much accuracy is lost depends on how many bits are used.
For example, in a music compact disc, audio signals are stored using 16-bit
samples at 44.1 kHz. Thus, the music for a compact disc is recorded using
65 536 or 2^{16} set values, which has sufficient resolution to produce excellent
sound quality. If enough bits are used, the error becomes negligible, so an
A/D converter with sufficient resolution may be viewed as an ideal C/D
converter.

The rate at which a signal is sampled is also very important. A simple
example consists of a spinning bicycle wheel illuminated by a strobe light.
Depending on the rate at which the strobe light is fired, the spokes of the
wheel may appear to be rotating clockwise, standing still, or even rotating
counterclockwise.

This simple experiment illustrates how the sampling rate affects the sampled or the perceived signal when compared to the actual signal. The flashes
of the strobe may be viewed as samples of the continuous-time signal, that
is the position of the spinning wheel. This suggests that a sampled signal
may behave in peculiar ways if the sampling rate is too low compared to
the rate at which the actual signal changes.

A similar effect occurs in Western movies as a stagecoach starts to roll.
At first the spokes appear to turn forward. As the wheel picks up speed,
the spokes may appear to spin backward. This is related to the fact that the
movie is shot at 24 frames per second and each frame may be viewed as a
sample of an image that is changing continuously in time.

3.1 Sampling theorem

To understand the effects of sampling, consider a complex exponential $p(t)$ given by

$$p(t) = Ae^{j(2\pi f_r t + \theta)} = Ae^{j\theta}e^{j2\pi f_r t}, \tag{3.1}$$

which may be viewed as a rotating vector with a length of A, phase θ, and a counterclockwise rotation rate of f_r cycles/second (or hertz). Consider now a spoke on a bicycle wheel rotating at 13 Hz in the clockwise direction, or $f_r = -13$ Hz. For simplicity, assume the spoke starts in the vertical position, where $\theta = \pi/2$. Suppose we sample the position of the spinning wheel with a strobe flashing at $f_s = 15$ Hz. Mathematically, we may represent the spoke with a phasor of the following form:

$$p(t) = Ae^{j\frac{\pi}{2}}e^{j2\pi(-13)t}. \tag{3.2}$$

Sampling the spinning wheel with a strobe at a sampling period of $T_s = 1/f_s = 1/15$ s, we obtain the following samples:

$$p[n] = p(nT_s) = Ae^{j\frac{\pi}{2}}e^{j2\pi(-13)nT_s} = Ae^{j\frac{\pi}{2}}e^{j2\pi(-\frac{13}{15})n} = Ae^{j\frac{\pi}{2}}e^{j2\pi \hat{f}_r n}, \tag{3.3}$$

where $\hat{f}_r = f_r/f_s = -13/15$ is known as the *discrete frequency*, which is unitless. If we change this frequency to $\hat{f}_r + k$, where k is some arbitrary integer, then the set of new samples becomes

$$p_k[n] = Ae^{j\frac{\pi}{2}}e^{j2\pi(-\frac{13}{15}+k)n} = Ae^{j\frac{\pi}{2}}e^{j2\pi(-\frac{13}{15})n}e^{j2\pi kn} = Ae^{j\frac{\pi}{2}}e^{j2\pi(-\frac{13}{15})n}. \tag{3.4}$$

These are exactly the samples of Eq. (3.3) that we obtained previously with a different sampling rate. Thus, we see that changing the discrete frequency by any integer has no effect whatsoever on the sampled values. In other words, continuous-time complex exponentials with frequencies $f_r = (\hat{f}_r + k)f_s$ will all produce the same set of samples $p[n]$.

The eye and the brain interpret the rotation as the lowest possible frequency, that is the frequency closest to 0, associated with the observed samples. For example, in the case of the rotating wheel, this would correspond to $k = 1$, which leads to

$$\hat{f}_r = -\frac{13}{15} + 1 = \frac{2}{15}. \tag{3.5}$$

The apparent frequency of rotation is then

$$f_r = \hat{f}_r f_s = 2 \text{ Hz}. \tag{3.6}$$

Fig. 3.3. Continuous-time spectrum of a complex exponential.

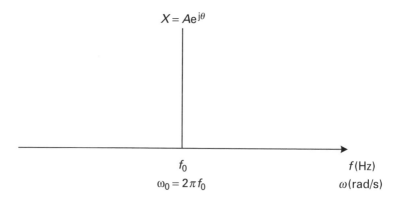

In other words, the wheel appears to be spinning counterclockwise at 2 Hz rather than the actual clockwise rotation of 13 Hz.

This phenomenon constitutes one of the most important concepts in the study of signals, so let us examine the general case where f_r and f_s are arbitrary. Consider a continuous-time signal given by

$$p_k(t) = A e^{j\theta} e^{j2\pi(f_r + kf_s)t}, \tag{3.7}$$

where k is some integer. When sampled at f_s, the samples are given by

$$p_k[n] = A e^{j\theta} e^{j2\pi(\frac{f_r}{f_s})n} e^{j2\pi kn} = A e^{j\theta} e^{j2\pi(\frac{f_r}{f_s})n} = p[n], \tag{3.8}$$

where $p[n]$ denote the samples of $p(t)$ of Eq. (3.1). Thus, we see that the samples remain the same for a phasor rotating at f_r, $f_r + f_s$, $f_r + 2f_s$, $f_r + 42f_s$, $f_r - 23f_s$, etc. This means that utmost care must be taken when analyzing data because there are theoretically an infinite number of possible rotation rates that would yield the same set of sampled results.

3.2 Discrete-time spectra

The aforementioned effect can be better understood by looking at the spectrum of signals. Suppose we have a signal

$$x(t) = A e^{j\theta} e^{j2\pi f_0 t}. \tag{3.9}$$

The continuous-time spectrum is given in Figure 3.3. It may be labeled with frequency f in hertz or angular frequency ω in rad/s, where $\omega = 2\pi f$.

Fig. 3.4. Discrete-time
spectrum of a complex
exponential.

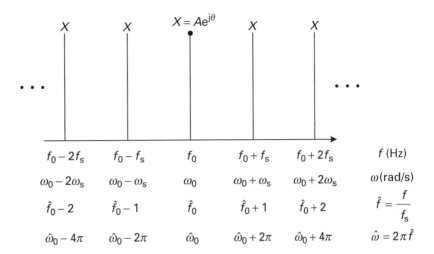

Now let us sample $x(t)$ at a rate f_s to obtain $x[n]$. A *discrete-time spectrum* for $x[n]$ is shown in Figure 3.4. The original frequency, indicated with the dot, is at f_0, but frequencies of $f_0 \pm k f_s$, where k is any integer, would also give the identical set of samples, and they are also shown on the spectrum. In other words, the discrete-time spectrum is a plot of *all* the possible frequencies that would give the particular samples. The graph may be plotted as a function of frequency (in hertz), or as a function of the unitless digital frequency. It also may be plotted as a function of angular frequency (in rad/s) or unitless digital angular frequency.

3.3 Aliasing, folding and reconstruction

Consider now a more familiar signal. For example, suppose we have the following continuous-time signal:

$$x(t) = A\cos(2\pi f_0 t + \theta) = \underbrace{\frac{1}{2}Ae^{j\theta}\,e^{j2\pi f_0 t}}_{X} + \underbrace{\frac{1}{2}Ae^{-j\theta}\,e^{-j2\pi f_0 t}}_{X^*}. \qquad (3.10)$$

The continuous-time spectrum of $x(t)$ is shown in Figure 3.5(a). The complex amplitudes are $X = Ae^{j\theta}/2$ and $X^* = Ae^{-j\theta}/2$. If we sample $x(t)$ at $f_s = 4f_0$, the corresponding discrete-time spectrum of $x[n]$ is shown in Figure 3.5(b). The original frequencies, f_0 and $-f_0$, are again indicated

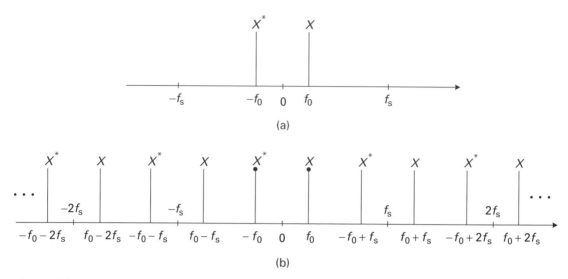

Fig. 3.5. (a) Continuous-
and (b) discrete-time
spectra of a cosine signal
($f_s = 4 f_0$).

with dots and the other frequencies, $f_0 \pm k f_s$, are also shown and are spaced at a frequency of f_s apart.

As a second example, consider the same continuous-time signal, $x(t)$, but sampled more slowly at $f_s = 4 f_0/5$. The continuous-time (Figure 3.6(a)) and the discrete-time (Figure 3.6(b)) spectra are shown in Figure 3.6. Now observe that some of the extra discrete-time frequencies are closer to the origin than the actual frequencies.

Suppose we wished to build a device that converts discrete-time samples back to a continuous-time signal. Such a discrete to continuous (D/C) converter is shown in Figure 3.7. The converter needs to know the sampling rate in order to reconstruct the original continuous-time signal correctly. However, notice that the D/C converter must make some assumptions: it will always reconstruct the signal based on the discrete-time spectrum between $-f_s/2$ and $f_s/2$. The portions of the spectrum in Figure 3.5(b) and Figure 3.6(b) within $-f_s/2$ and $f_s/2$ are redrawn in the dotted boxes as shown in Figure 3.8. In the first case, the D/C converter will reconstruct the signal correctly as

$$x(t) = A \cos(2\pi f_0 t + \theta). \tag{3.11}$$

But in the second case, the D/C converter will incorrectly reconstruct the signal as

$$x(t) = A \cos(2\pi (f_0 - f_s)t + \theta). \tag{3.12}$$

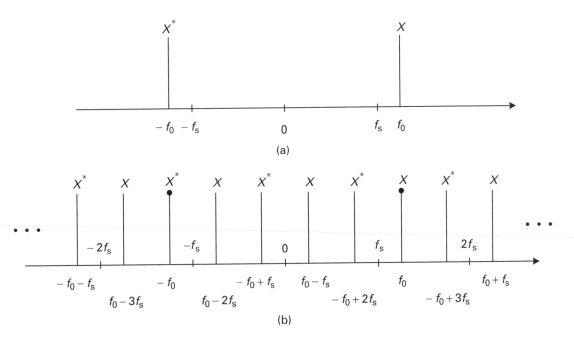

Fig. 3.6. (a) Continuous- and (b) discrete-time spectra of a cosine signal ($f_s = 4f_0/5$).

In other words, if the original continuous-time cosine signal were at 125 Hz and the sampling frequency were 100 Hz, the cosine signal would be reconstructed using the D/C converter to be at 25 Hz!

This effect is called *aliasing*. It may occur when the sampling frequency is less than twice the highest frequency of the continuous-time signal. The sampling frequency is known as the *Nyquist rate* when it is twice the signal frequency. Aliasing has caused no end of grief in sampling data because it causes the original signal to be reconstructed incorrectly, just as the stage-coach wheels seem to spin in the wrong direction. However, if we understand the aliasing phenomenon we can be sure to sample fast enough to avoid it. Moreover, knowing what causes aliasing, we can exploit it to our advantage in certain applications such as detecting known high frequency components

Fig. 3.7. Discrete to continuous (D/C) converter.

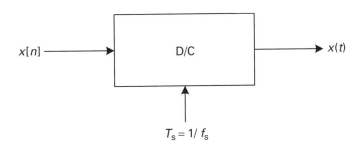

Fig. 3.8. Continuous-time spectra of the reconstructed signals ($f_s = 4f_0$ and $f_s = 4f_0/5$).

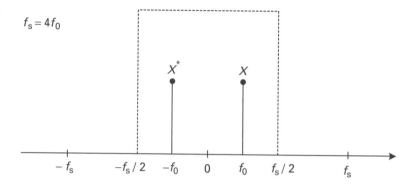

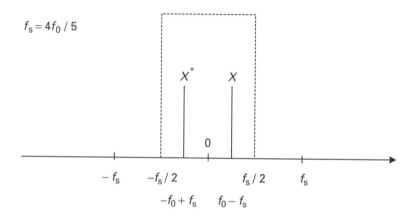

of a signal at sampling rates well below the Nyquist rate, as we will explore later.

Figure 3.9 shows an example in which $f_s = 4f_0/3$. Again, note that the sampling rate is less than twice the frequency of the continuous-time signal, so we expect something interesting to occur. When the signal is reconstructed using the frequencies between $-f_s/2$ and $f_s/2$, the reconstructed signal is found to be

$$x(t) = A\cos(2\pi(f_s - f_0)t - \theta). \tag{3.13}$$

The frequency is clearly incorrect. In addition, the phase is also inverted (becomes negative). This effect is called *folding* and is intimately related to aliasing.

Previously we used different examples to introduce the concepts of aliasing and folding, and to motivate the importance of choosing the proper sampling frequency. We now examine the relationship between the signal

Fig. 3.9. Continuous- and discrete-time spectra of a cosine signal ($f_s = 4 f_0/3$).

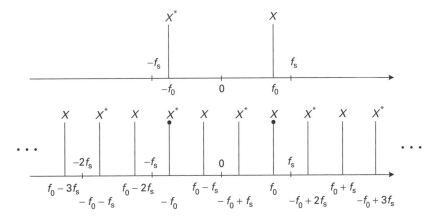

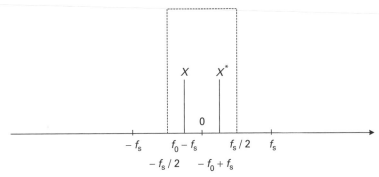

frequency f_0 and the sampling frequency f_s mathematically. Consider the following continuous-time sinusoid with a signal frequency of f_0, amplitude A and phase θ:

$$x(t) = A \cos(2\pi f_0 t + \theta). \tag{3.14}$$

If we sample the signal $x(t)$ with a sampling period of $T_s = 1/f_s$, we have

$$x[n] = A \cos(2\pi f_0 n T_s + \theta). \tag{3.15}$$

Consider now a different sinusoid $y(t)$, with the same amplitude and phase as $x(t)$, but with a frequency of $f_0 + l f_s$, where l is an arbitrary integer. Then we can describe $y(t)$ as

$$y(t) = A \cos(2\pi (f_0 + l f_s)t + \theta). \tag{3.16}$$

If the signal $y(t)$ is sampled with a period T_s, we get

$$\begin{aligned} y[n] &= A \cos(2\pi f_0 n T_s + 2\pi l f_s n T_s + \theta) \\ &= A \cos(2\pi f_0 n T_s + 2\pi n l + \theta). \end{aligned} \tag{3.17}$$

By inspection, note that

$$y[n] = x[n] \tag{3.18}$$

because nl is simply an integer. Although $x(t)$ and $y(t)$ are clearly distinct continuous-time signals (they contain different frequencies), $x[n]$ and $y[n]$ have identical sampled values and they are indistinguishable from one another when they are both sampled at a period of T_s. Since l was specified only to be an integer (either positive or negative), this implies that there are an infinite number of continuous-time sinusoids that will lead to the same sequence of samples $x[n]$. The frequencies $f_0 + lf_s$ are known as *aliases* of f_0 with respect to the sampling frequency f_s, because all of them appear to be the same when sampled at a period of T_s.

A second source of aliased signals originates from signals with frequencies of $-f_0 + lf_s$, where l is an arbitrary positive or negative integer. Consider now a third signal of the form

$$w(t) = A\cos(2\pi(-f_0 + lf_s)t - \theta) \tag{3.19}$$

whose phase is the negative of $x(t)$. Sampling with a sampling period of T_s, we get

$$\begin{aligned} w[n] &= A\cos(-2\pi f_0 n T_s + 2\pi l f_s n T_s - \theta) \\ &= A\cos(-2\pi f_0 n T_s + 2\pi n l - \theta). \end{aligned} \tag{3.20}$$

Again, we note that

$$w[n] = x[n] \tag{3.21}$$

because cosine is an even function $(\cos(x) = \cos(-x))$. When the three distinct continuous-time signals $x(t)$, $y(t)$ and $w(t)$ are all sampled at a rate of $f_s = 1/T_s$, their sampled values are all indistinguishable from one another, that is $x[n] = y[n] = w[n]$.

In summary, aliasing occurs when we sample below the frequency of the original signal. Folding occurs when we sample above the frequency but not above twice the frequency of the original signal. The original signal can only be reconstructed correctly if we sample above twice the highest frequency in the signal. Twice the highest frequency is also known as the *Nyquist frequency*. Table 3.1 summarizes the effects of various sampling frequencies on the reconstructed signal.

Table 3.1. The effects of sampling frequency on the reconstructed signal

Signal frequency / sampling frequency	Reconstructed signal
$0 < f_0/f_s < 1/2$	good reconstruction
$n + 1/2 < f_0/f_s < n + 1, \quad n = 0, 1, 2, \ldots$	folding
$n + 1 < f_0/f_s < n + 3/2, \quad n = 0, 1, 2, \ldots$	aliasing

Example: We can hear the effects of aliasing and folding by listening to an audio sweep from 100 Hz to 2000 Hz over a 2-second duration. If the sweep is reconstructed from samples taken at a rate of 8 kHz, it sounds like a tone rising in pitch. If we reduce the sampling rate to 4 kHz, it still sounds like a rising tone because we are still sampling at twice the highest frequency. If we reduce the sampling rate to 2 kHz, we are sampling at the highest frequency. The tone rises then falls. If we further reduce the sampling rate to 1 kHz, we hear the tone rise and fall and rise and fall again as we alias and fold the signal. Note that the frequency of the reconstructed tone never exceeds $f_s/2$. After the first rise the succeeding falls and rises are due to folding and aliasing, respectively, as shown in Table 3.1 (see Problem 3-11). These sounds can be heard using the following MATLAB code:

```
function [t,x] = playsweep(f1, f2, dur, fs);
% Sweep sinusoidal signal from f1 to f2
% for duration dur sampled at fs.
t = 0:1/fs:dur;
x = cos(2*pi*((f2 - f1)/(2*dur)*t.*t + f1*t));
soundsc(x,fs);
>> playsweep(100, 2000, 2, 8000);
>> playsweep(100, 2000, 2, 4000);
>> playsweep(100, 2000, 2, 2000);
>> playsweep(100, 2000, 2, 1000);
```

We have come to the end of the first segment of the text covering signals. At this point, we have examined the following key points:

- Representing continuous-time and discrete-time signals
  - Transformations
  - Shift, expand, compress, reverse, upsample, downsample
- Constructing signals from building blocks

– Orthogonality principle

– Orthogonal building blocks

– Fourier series

• Spectrum of continuous-time signals

• Using the DFT and FFT to compute Fourier coefficients of complicated signals

• Sampling and the discrete-time spectrum

– C/D and D/C converters

– Aliasing and folding

In the next segment of the text, we will examine mechanical and electrical systems. In the final segment of the text, we will use digital signal processing to analyze signals and systems using the computer.

3.4 Continuous- and discrete-time spectra

To review and tie together everything we have done so far, let us consider an example of data acquisition. Suppose our input signal is a simple continuous-time sinusoid signal. Using Euler's formula, the sinusoid could also be written as the sum of complex exponentials as shown:

$$
\begin{aligned}
x(t) &= 2\sin(2\pi f_0 t) \\
&= 2\cos\left(2\pi f_0 t - \tfrac{\pi}{2}\right) \\
&= \underbrace{e^{-j\frac{\pi}{2}}}_{c_1} e^{j2\pi f_0 t} + \underbrace{e^{j\frac{\pi}{2}}}_{c_{-1}} e^{-j2\pi f_0 t},
\end{aligned} \tag{3.22}
$$

where the coefficients of the exponentials are simply the Fourier coefficients of the signal. We could also represent the signal as the real part of a complex exponential

$$
x(t) = \mathrm{Re}\left\{ \underbrace{2e^{-j\frac{\pi}{2}}}_{X} e^{j2\pi f_0 t} \right\}. \tag{3.23}
$$

The continuous-time spectrum of $x(t)$ is shown in Figure 3.10. If we sample the signal at the rate $f_s = 1/T_s$, we obtain

$$
\begin{aligned}
x[n] &= x(nT_s) \\
&= 2\cos\left(2\pi f_0 n T_s - \tfrac{\pi}{2}\right) \\
&= 2\cos\left(2\pi \underbrace{(f_0 T_s)}_{\hat{f}_0} n - \tfrac{\pi}{2}\right).
\end{aligned} \tag{3.24}
$$

Fig. 3.10.
Continuous-time spectra
of $x(t)$.

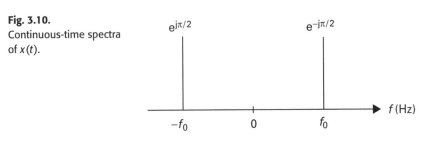

Fig. 3.11. Discrete-time
spectra of $x[n]$.

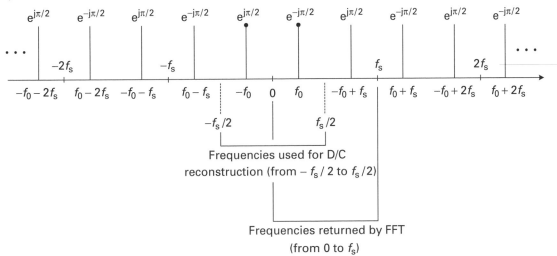

The discrete-time spectrum of $x[n]$ is shown in Figure 3.11. If we sample fast enough, that is for $f_s > 2f_0$, there will be no aliasing or folding. Recall that a D/C converter reconstructs the signal from the coefficients of the frequencies that lie between $-f_s/2$ and $f_s/2$. However, the FFT returns coefficients whose frequencies lie between 0 and f_s. Thus, we must be careful when selecting the correct coefficients, as we will review later.

3.5 Aliasing and folding (time domain perspective)

If the sampling frequency is too slow ($f_s < 2f_0$), we may experience aliasing or folding. These effects may be understood in both the frequency and the time domains. In the frequency domain, we used the discrete-time spectra to study the underlying causes of aliasing or folding. We can also illustrate the aliasing or folding phenomenon in the time domain through the following examples that are coded in MATLAB.

Fig. 3.12. A 10 Hz sinusoid sampled at 25 Hz.

```
>> aliasfold
frequency of sinusoid in Hz:10
sampling frequency in Hz:25
```

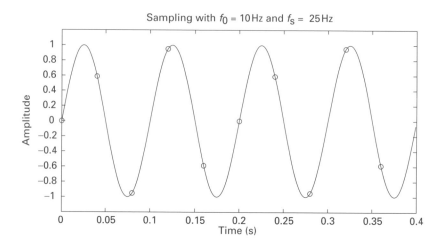

Fig. 3.13. Aliasing: a 52 Hz sinusoid sampled at 50 Hz.

```
>> aliasfold
frequency of sinusoid in Hz:52
sampling frequency in Hz:50
```

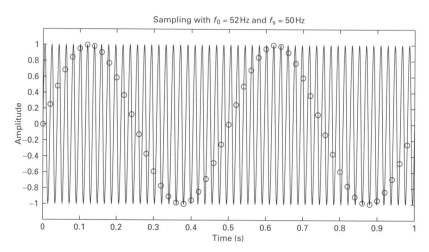

Figure 3.12 shows a 10 Hz sinusoid sampled at 25 Hz. The line represents the sinusoid and the circles represent the samples. We have enough information to reconstruct the original sinusoid. Figure 3.13 shows a 52 Hz sinusoid sampled at 50 Hz. If one connects the circles, one would probably say that they look like a 2 Hz sinusoid, and not the actual 52 Hz sinusoid! This is an example of aliasing.

Fig. 3.14. Folding: a 48 Hz
sinusoid sampled at 50 Hz.

```
>> aliasfold
frequency of sinusoid in Hz:48
sampling frequency in Hz:50
```

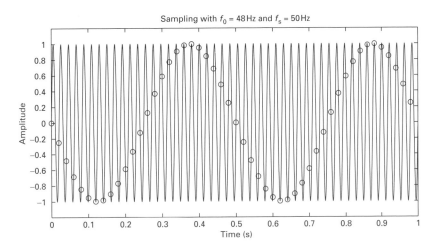

Figure 3.14 shows a 48 Hz sinusoid sampled at 50 Hz. This clearly illustrates folding and we observe a -2 Hz sinusoid. This can also be viewed as a 2 Hz sinusoid with the phase reversed.

For a simple signal given by Eq. (3.22), the coefficients are easy to find by inspection. In a more complicated signal, we could use the FFT or DFT to compute the coefficients (both perform the same function, but the FFT is faster and is therefore more popular). In the special case of signals that are sums of sinusoids, we could still find the coefficients by inspection because each sinusoid can be written as the sum of complex exponentials.

Let us analyze the sine wave more thoroughly. Suppose we wished to calculate the Fourier coefficients of Eq. (3.22) with the DFT using the following MATLAB code. Let the sinusoid be at $f_0 = 10$ Hz ($T_0 = 0.1$ s). To correctly reconstruct the signal, we must have sampled at least $f_s > 20$ Hz. Suppose we take $N = 3$ samples over one period, that is sample at $f_s = N f_0 = 30$ Hz, for a duration of $T = \text{periods}/f_0$.

```
% fftdemo.m
% demonstrate use of FFT to compute Fourier
% coefficients of a sinusoid
N=3;
periods=1;
fo=10;
```

```
T=periods/fo;
dt=T/N;
fs=1/dt;
t=dt*(0:N-1);
x=2*cos(2*pi*fo*t-pi/2);
X=fft(x)/N
```

The MATLAB results are

```
>> fftdemo
X =
0.0000 -0.0000 - 1.0000i -0.0000 + 1.0000i
```

These coefficients correspond to $c_0 = 0$, $c_1 = -j$ and $c_{-1} = j$, respectively, where j denotes the complex unity. Note that the MATLAB results exactly match the theoretical solutions. For bookkeeping purposes, however, it would be more convenient to rearrange the MATLAB coefficients so that they appear in the order of increasing index, that is c_{-1}, c_0 and c_1. The MATLAB fftshift command does this for us automatically, as shown below:

```
>> fftshift(X)
ans =
-0.0000 + 1.0000i 0.0000 -0.0000 - 1.0000i
```

Alternatively, we could write a function that automates this shift and also returns the correct frequency associated with each coefficient. Such a function, fdomain, was developed in Section 2.11. It computes the FFT and shifts the coefficients into the correct order. It also returns a vector of the corresponding frequencies.

Consider now the signal of Eq. (3.22), with $f_0 = 10$ Hz or $T_0 = 0.1$ s. In the following examples we will study how T and N affect the spectrum of the signal. We could apply fdomain to the signal of Eq. (3.22) using three points per period for a sampling rate of $f_s = N f_0 = N/T_0 = 3/0.1 = 30$ Hz. The spectrum is shown in Figure 3.15. The two peaks are at ± 10 Hz with a magnitude of 1, as expected. Note that the frequency resolution (the spacing between the computed frequencies that are indicated with the stems) is $\Delta f = 1/T = 10$ Hz.

Suppose we consider the same function over five periods by changing $T = 5/f_0$. Now we need at least two points per period, so let us use $N = 11$.

Fig. 3.15. Spectrum for
$T = 0.1$ s and $N = 3$.

```
>> fftdemo
>> [X,f] = fdomain (x, fs)
>> stem (f, abs (X), 'k');
>> axis ([-11, 11, 0, 1.2]);
>> title ('Spectrum');
>> xlabel ('Frequency (Hz)');
>> ylabel ('Magnitude');

X =
  -0.0000+1.0000i   0.0000   -0.0000-1.0000i
f =
  -10   0   10
```

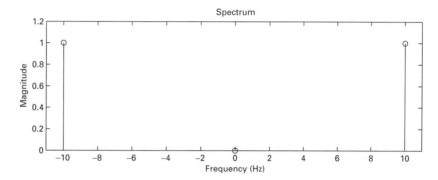

Executing the MATLAB code fftdemo, we obtain:

```
>> fftdemo
X =
Columns 1 through 6
-0.0000 -0.0000 - 0.0000i -0.0000 - 0.0000i
  -0.0000 - 0.0000i 0.0000 - 0.0000i 0.0000
  -1.0000i
Columns 7 through 11
0.0000 + 1.0000i 0.0000 + 0.0000i -0.0000
  + 0.0000i -0.0000 + 0.0000i -0.0000 + 0.0000i
```

If we use fdomain and plot the magnitude of the coefficients, as seen in
Figure 3.16, we find the peaks are still at ± 10 Hz. However, the frequency
resolution has improved to 2 Hz.

```
>> fftdemo
>> [X,f] = fdomain (x, fs)
>> stem (f, abs (X), 'k');
>> axis ([-11, 11, 0, 1.2]);
>> title ('Spectrum');
>> xlabel ('Frequency (Hz)');
>> ylabel ('Magnitude');


X =
Columns 1 through 6
0.0000+1.0000i    0.0000+0.0000i   -0.0000+0.0000i   -0.0000
 +0.0000i   -0.0000+0.0000i -0.0000
Columns 7 through 11
-0.0000-0.0000i    -0.0000-0.0000i   -0.0000-0.0000i   0.0000
 -0.0000i   0.0000-1.0000i
f =
   -10   -8   -6   -4   -2   0   2   4   6   8   10
```

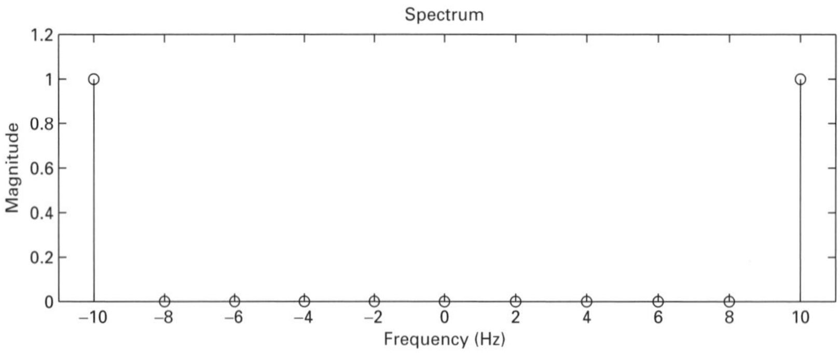

Fig. 3.16. Spectrum for $T = 0.5$ s and $N = 11$.

Suppose we use more points (for example, $N = 32$) and plot the spectrum again. Figure 3.17 shows that the coefficients are still found at ± 10 Hz and have a magnitude of 1. The frequency resolution remains unchanged at 2 Hz because $T = 0.5$ s. However, the maximum frequency in the spectrum has increased because we have sampled more often. On the other hand, suppose we create the function over 5.5 periods. The coefficients are shown in Figure 3.18. Notice that the coefficients are no longer what we would expect. This can be best explained by graphing the time trajectory of the

Fig. 3.17. Spectrum for $T = 0.5\,\text{s}$ and $N = 32$.

```
>> fftdemo
>> [X,f] = fdomain (x, fs)
>> stem (f, abs (X), 'k');
>> axis ([-34, 32, 0, 1.2]);
>> title ('Spectrum');
>> xlabel ('Frequency (Hz)');
>> ylabel ('Magnitude');
```

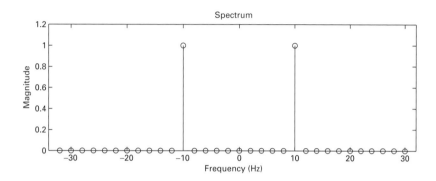

Fig. 3.18. Spectrum for $T = 0.55\,\text{s}$ and $N = 32$.

```
>> fftdemo
>> [X,f] = fdomain (x, fs)
>> stem (f, abs (X), 'k');
>> axis ([-30, 30, 0, 0.7]);
>> title ('Spectrum');
>> xlabel ('Frequency (Hz)');
>> ylabel ('Magnitude');
```

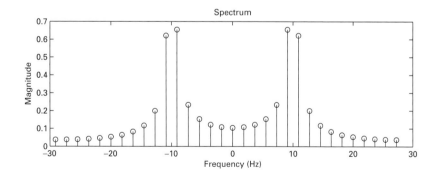

Fig. 3.19. Sinusoid
repeating after 5.5 periods.

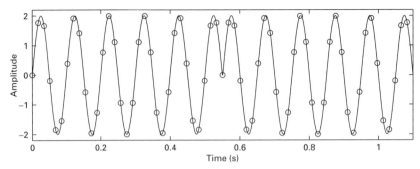

signal. Figure 3.19 plots the sinusoid in the time domain. Recall that in applying FFT we implicitly assume the sampled data repeats after every interval T. Therefore, the sinusoid repeats each 5.5 periods (0.55 s). By inspection (see Figure 3.19), we see that there is now a discontinuity where the signal repeats. Observe that the coefficients still have a peak near 10 Hz, as we would expect (see Figure 3.18). However, the discontinuities in the time domain mean that we are computing the coefficients of a signal that is no longer a pure sinusoid.

3.6 Windowing

Real data rarely, if ever, repeat after some prescribed sampling interval. Thus, when we assume the data to repeat outside the time interval of interest, we will very likely introduce discontinuities. To eliminate these unavoidable discontinuities, we could taper the signal at both ends, as shown in Figure 3.20. For this example, we have multiplied the actual signal by a triangle waveform. This process is called *windowing*. While windowing distorts the signal, it eliminates the discontinuity at each end by forcing the endpoints to be exactly zero, which in turn makes the signal continuous as it repeats from one period to the next. The FFT of the windowed sinusoid is shown in Figure 3.21. We see that the spectrum of a windowed signal is closer to what we would expect compared to the spectrum of signal without windowing (see Figure 3.18). There are many ways to window a signal, and windows are commonly used when analyzing real data to eliminate the undesirable effects of discontinuities.

When we took data over five periods, in principle it would seem that we could get away with only $N = 10$ points. The results are given below:

Fig. 3.20. Tapering
sinusoid with a triangular
window.

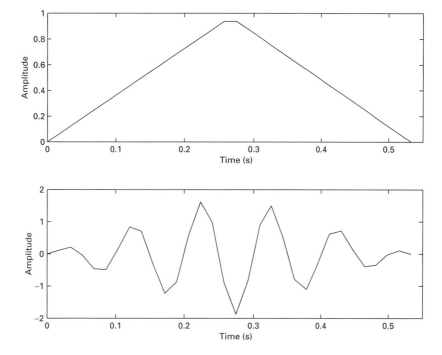

Fig. 3.21. Spectrum of a
windowed sinusoid for
$T = 0.55$ s and $N = 32$.

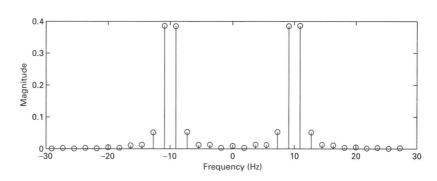

```
>> fftdemo
X =
1.0e-014 *
Columns 1 through 6
-0.0233 -0.0917 + 0.0120i 0.1204 - 0.1471i 0.0274
  + 0.1053i -0.0782 - 0.0121i 0.0797
Columns 7 through 10
-0.0782 + 0.0121i 0.0274 - 0.1053i 0.1204
  + 0.1471i -0.0917 - 0.0120i
```

Fig. 3.22. Sinusoid sampled at $2f_0$.

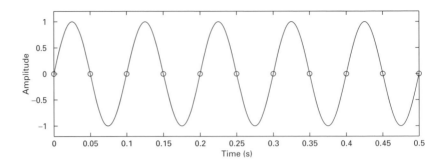

At first glance the results appear to be very poor, because all of the coefficients are nonzero. But closer inspection reveals that all of the coefficients are multiplied by 10^{-14} ($1.0e-014$), so that they are all essentially zero. This can be best explained by analyzing the signal and the samples in the time domain. Figure 3.22 shows the sampled sinusoid. Observe that all the samples $x[n]$ fall at nearly 0 (see x below), so our sampled signal consists of a string of 0s. Thus, we cannot sample at exactly twice the frequency of the original signal, but at a slightly faster rate, that is we need to sample at a rate that is at least twice the frequency of the original signal. This again has to do with aliasing and folding.

```
x =
1.0e-014 *
0.0122 0.0122 -0.0367 -0.2940 -0.0857 0.1102
  0.9311 -0.1961 -0.5390 -0.1471
```

3.7 Aliasing and folding (frequency domain perspective)

Let us return to the aliased signal produced by a 52 Hz sinusoid sampled at 50 Hz, shown in Figure 3.13. Figure 3.23 plots the spectrum. Observe that the frequencies show up at ± 2 Hz, as we would expect from our understanding of aliasing. Figure 3.24 shows the same process for the 48 Hz sinusoid sampled at 50 Hz (from Figure 3.14). Again, we see coefficients at ± 2 Hz. Because we have ignored the phase; we do not notice that the phase was flipped.

In another example, let us look at a dial tone in Figure 3.25. It consists of 0.5 seconds of tone followed by 0.1 seconds of silence. Our objective is to determine which telephone button or key on the telephone pad was

Fig. 3.23. Spectrum of a 52 Hz sinusoid sampled at 50 Hz.

```
>> aliasfold
frequency of sinusoid in Hz:52
sampling frequency in Hz:50
>> [X,f]=fdomain (xs,fs);
>> stem (f,abs(X), 'k');
>> xlabel ('Frequency (Hz)');
>> ylabel ('Magnitude');
```

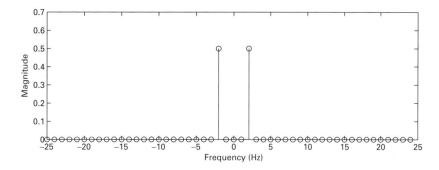

Fig. 3.24. Spectrum of a 48 Hz sinusoid sampled at 50 Hz.

```
>> aliasfold
frequency of sinusoid in Hz:48
sampling frequency in Hz:50
>> [X,f]=fdomain (xs,fs);
>> stem (f,abs(X), 'k');
>> xlabel ('Frequency (Hz)');
>> ylabel ('Magnitude');
```

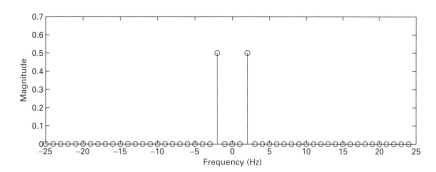

Table 3.2. Touch-tone telephone tones

Frequencies	1209 Hz	1336 Hz	1447 Hz
697 Hz	1	2	3
770 Hz	4	5	6
852 Hz	7	8	9
941 Hz	*	0	#

Fig. 3.25. Dial tone (entire signal and zoomed section).

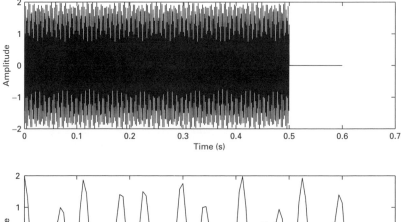

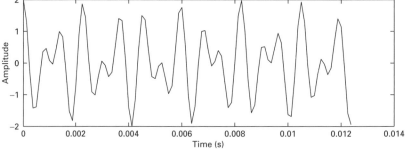

pressed. Figure 3.26(a) plots the spectrum of the signal. Zooming in the spectrum, we notice that there are components at about 1336 Hz and 852 Hz. Figure 3.26(b) shows the spectrum from 830 Hz to 920 Hz. Note a distinct peak near 852 Hz. Utilizing the matrix of dial tones shown in Table 3.2, we recognize that this tone most likely corresponds to key 8 on the telephone dial. Notice that the spectrum contains other frequencies as well. Because of the silence, the signal was 0.6 seconds long. Therefore, all of the frequencies that we can find must be multiples of 1/0.6, which do not fall at exactly 1336 Hz or 852 Hz.

We could avoid this phenomenon by truncating the signal at exactly 0.5 seconds so we only analyze the tone and not the silence. Now the spectrum would be found for multiples of $1/0.5 = 2$ Hz, so they will fall exactly on the frequencies in the signal. The spectrum is shown in Figure 3.27(a).

Fig. 3.26. Spectrum of dial
tone (entire section and
zoomed section).

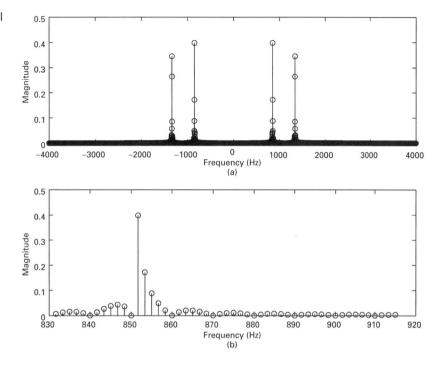

Fig. 3.27. Spectrum of dial
tone truncated before
silence.

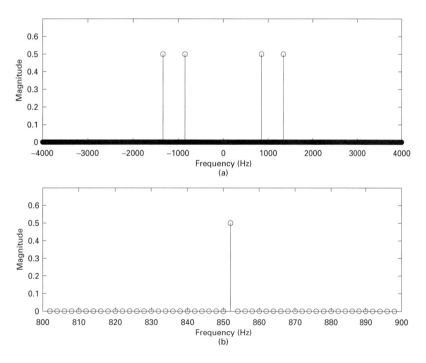

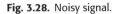

Fig. 3.28. Noisy signal.

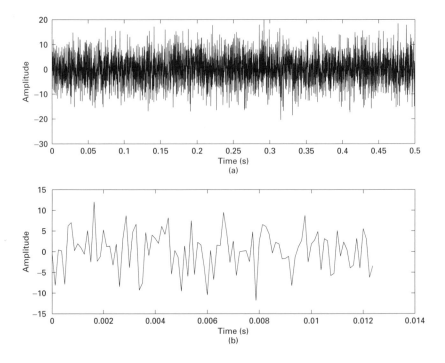

Zooming in shows that the spectrum has exactly the anticipated frequency components at 1336 Hz and 852 Hz. Figure 3.27(b) shows the spectrum from 800 Hz to 920 Hz. Note the distinct peak at 852 Hz. Together with the 1336 Hz component, the tone corresponds to key 8.

Figure 3.28 shows a signal, *ntone*, that looks like noise. The signal still appears as noise as we zoom in. Playing it also sounds like noise. However, the spectrum in Figure 3.29(a) shows that there are two distinct peaks near 1336 Hz and 852 Hz that seem to dominate and rise above the noise. Figure 3.29(b) shows the spectrum from 800 Hz to 920 Hz. Note that there is a clear peak in the vicinity of 852 Hz. Therefore, we expect that the dial tone corresponding to key 8 is still present in the signal, even though it cannot be detected by the ear or by visually inspecting the signal in the time domain.

3.8 Handling data with the FFT

Suppose you collected some data and that the information you want is embedded in the frequencies contained within the data. How can you

111   **3.8 Handling data with the FFT**

Fig. 3.29. Spectrum of noisy signal.

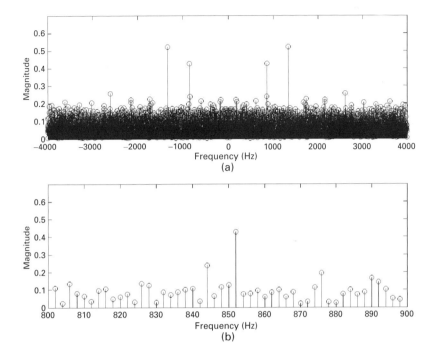

Fig. 3.30. Data acquisition system.

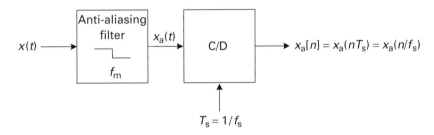

extract this information? Figure 3.30 shows a data acquisition system. In the real world, our signal may not be conveniently confined to the frequency band we want. Thus, before we sample the signal, we need to eliminate any frequencies greater than half the sampling frequency so these frequencies do not alias into the data. This is accomplished with an anti-aliasing filter that eliminates frequencies beyond a specified limit, say f_m. Later in the text we will learn about how to build anti-aliasing filters. The filter is followed by a continuous-to-discrete converter to produce our anti-aliased sampled data $x_a[n] = x_a(nT_s)$. We then take the FFT to determine the spectrum in the signal. This step eliminates the aliasing phenomenon.

In the FFT, we have the following parameters:

- T = time interval over which data is sampled;
- $\Delta f = 1/T = f_s/N$ = frequency resolution;
- f_s = sampling frequency;
- $N = f_s T$ = number of data points.

The procedure is summarized as follows:

1. Determine the maximum frequency of interest, f_m. Set $f_s > 2 f_m$ and choose anti-aliasing filter to cut off frequencies *above* f_m.
2. Choose T (if possible) to get desired frequency resolution Δf.
3. $N = f_s T$.
4. Compute FFT and index the results.

If we want a particular frequency resolution, we need to sample for a sufficiently long time T. If we want to examine high frequencies f_m, we need to sample at a high frequency f_s. The number of sampled data is $N = T f_s$, so obtaining detailed frequency resolution at high frequencies requires a large number of data samples. Finally, note that the frequency resolution is *only* proportional to the sampling interval. Taking more data over the same time interval does *not* make the frequency resolution any better.

3.9 Problems

3-1 A signal $x(t)$ has the two-sided spectrum representation shown in Figure 3.31.

(a) Write an equation for $x(t)$ as the sum of cosines.

(b) A new signal is defined as

$$y_1(t) = x(t) + 6\cos(\alpha t + \pi).$$

Fig. 3.31. Figure for Problem 3-1.

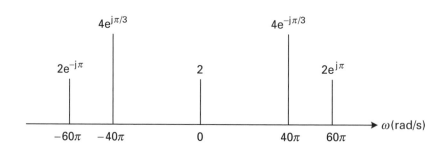

It is known that $y_1(t)$ is periodic with period $T_0 = 0.2$ s. Determine two positive values for the frequency α that will satisfy this condition.

(c) Using either of the frequencies α found in part **(b)**, modify the spectrum plot so that it becomes the spectrum of $y_1(t)$.

(d) Consider another new signal

$$y_2(t) = x(t) + \cos(\sqrt{2}\pi t + \pi/3).$$

Is $y_2(t)$ periodic? If so, what is its period? If not, why not?

3-2 Consider the following continuous-time signal

$$x(t) = \sin^6(400\pi t).$$

(a) Sketch and label the spectrum of $x(t)$. The problem can be solved by the direct application of Euler's identity, that is express $x(t)$ as the sum of two complex exponentials and then raise the result to the appropriate power. This approach, while conceptually simple, is algebraically intensive. Alternatively, one can make use of the following binomial expansion:

$$(a + b)^n = \sum_{k=0}^{n} \binom{n}{k} a^k b^{n-k} = \sum_{k=0}^{n} \binom{n}{k} a^{n-k} b^k,$$

where

$$\binom{n}{k} = \frac{n!}{k!(n-k)!}.$$

(b) Use MATLAB's fft command to determine the Fourier coefficients of $x(t)$ in part **(a)**. How do these coefficients compare with the exact coefficients found in part **(a)**?

3-3 A vibration sensor mounted on a rocket during a test firing produces an output $x(t)$ that is proportional to the acceleration. The sensor output $x(t)$ is sampled at a rate of $f_s = 1000$ Hz (or samples per second) and a total of 10 000 samples are recorded. The sequence $x[n] = x(nT_s)$ is then processed by using MATLAB's fft algorithm and divided by the number of samples to compute the sequence $X[k]$, where $k = 1, 2, \ldots, 10\,000$.

(a) If $X[501] = 2 + 2j$, $X[3251] = 3 + 4j$, $X[6751] = 3 - 4j$, $X[9501]$ $= 2 - 2j$ and $X[k] = 0$ otherwise. Sketch and dimension the spectrum of $x[n]$ and give an expression for $x(t)$.

(b) An engineer familiar with the dynamics of the rocket points out that the highest frequency component determined in part (a) is to be expected but the lowest frequency component does not make sense. Further, a frequency component somewhere in the range of $500 < f < 1000$ Hz should be present. Can you determine the frequency of this component? If so, explain how and give an expression for $x(t)$.

3-4 Consider a signal $x(t)$ that is sampled over a 2 second interval. The resulting sequence $x[n] = x(nT_s)$ is then processed using MATLAB's fft algorithm and divided by the number of samples to give the following vector:

$$X = [24, 0, -2.5j, 0, 0, -6, -6, 0, 0, 2.5j, 0].$$

(a) Determine an expression for the signal $x(t)$. What is the sampling frequency f_s? Sketch and dimension the spectra of the continuous-time signal $x(t)$ and the discrete-time signal $x[n] = x(nT_s)$. What assumption are you making about $x(t)$? Find a different signal $y(t)$ that would produce the same results and sketch its spectrum.

(b) The signal $x(t)$ from part (a) is now sampled by taking ten uniformly spaced samples over a 2 second interval. Sketch and dimension the spectrum of the discrete-time signal $x[n] = x(nT_s)$. Repeat if the phase of the highest frequency component is shifted by $\pi/2$ and explain what happens.

(c) Determine and graph the discrete-time spectrum if the signal $x(t)$ from part (a) is sampled over a 1 second interval with a sampling frequency $f_s = 20$ Hz. Can you determine $x(t)$ exactly from this spectrum? Explain.

3-5 Consider the system shown in Figure 3.32 with ideal C/D and D/C converters.

(a) Suppose that the discrete-time signal $x[n]$ is given by the formula

$$x[n] = 10\cos(0.13\pi n + \pi/13).$$

Fig. 3.32. Figure for
Problem 3-5.

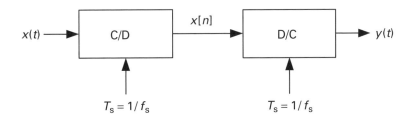

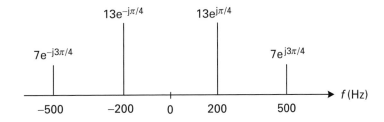

If the sampling rate is $f_s = 1000$ samples/s, determine two *different* continuous-time signals $x(t) = x_1(t)$ and $x(t) = x_2(t)$ that could have been inputs to the above system, that is find $x_1(t)$ and $x_2(t)$ such that $x[n] = x_1(nT_s) = x_2(nT_s)$, if $T_s = 0.001$ s.

(b) If the input $x(t)$ is given by the two-sided spectrum representation in Figure 3.32, determine a simple formula for $y(t)$ when $f_s = 700$ samples/s (for both the C/D and D/C converters).

3-6 A signal

$$x(t) = 4\cos\left(200\pi t - \frac{\pi}{4}\right) + 10\cos(400\pi t)$$

is sampled at $f_s = 500$ Hz.

(a) Sketch and label the spectrum of the continuous-time signal $x(t)$.

(b) Sketch and label the spectrum of the discrete-time signal $x[n] = x(nT_s)$.

(c) Samples of $x(t)$ are collected over a 0.5 second time interval and the DFT $X[k]$ is computed. List the numerical values of each nonzero $X[k]$.

(d) Repeat part (c) if samples of $x(t)$ are collected over a 1 second time interval.

3-7 The discrete Fourier transform (DFT) of a sequence $x[n]$ is given by

$$X[k] = \frac{1}{N}\sum_{n=0}^{N-1} x[n]e^{-jk(2\pi/N)n}.$$

(a) Suppose $x[0] = 0$, $x[1] = 1$, $x[2] = 0$ and $x[3] = 3$. Determine the discrete Fourier transform of the periodic sequence $x[n] = x[n + 4]$.

(b) If the sequence is the result of sampling a continuous-time signal $x(t)$ at a sampling frequency of $f_s = 20\,\text{Hz}$, sketch the discrete-time spectrum. Give values for all frequencies and complex amplitudes.

3-8 Consider the following continuous-time signal:

$$x(t) = 10\cos\left(20\pi t + \frac{\pi}{6}\right).$$

(a) Sketch the spectrum of $x(t)$ and $x[n] = x(nT_s)$ if the sampling frequency is $f_s = 7\,\text{Hz}$. Specify all frequencies and amplitudes. Given an expression for the signal $y(t)$ reconstructed by a D/C converter using the same sampling frequency.

(b) Sketch the discrete-time spectrum of $x[n]$ if $f_s = 15\,\text{Hz}$. Specify all frequencies and complex amplitudes. Give an expression for the signal $y(t)$ reconstructed by a D/C converter using the same sampling frequency.

3-9 Consider a signal $x(t)$ that is sampled over a 2 second interval. The resulting sequence $x[n] = x(nT_s)$ is then processed using MATLAB's fft algorithm and divided by the number of samples to give the following vector:

$$X = [-17, 0, -4.5j, 0, 1.2j, -2, -2, -1.2j, 0, 4.5j, 0].$$

(a) Determine an expression for the signal $x(t)$. What assumptions are you making regarding the sampling frequency?

(b) The signal $x(t)$ from part (a) is now sampled at $f_s = 2\,\text{Hz}$. Sketch and dimension the spectrum of the discrete-time signal $x[n] = x(nT_s)$.

(c) Repeat part (a) if the signal $x(t)$ is sampled over a 0.5 s interval.

3-10 The spectrum of a continuous-time signal $x(t)$ is shown in Figure 3.33.

(a) Determine an expression for the signal $x(t)$ consisting of constants and sinusoids.

(b) Determine an expression for the signal $y(t) = 2x(5t) - 20$.

(c) Sketch and dimension the spectrum $y(t)$.

Fig. 3.33. Figure for
Problem 3-10.

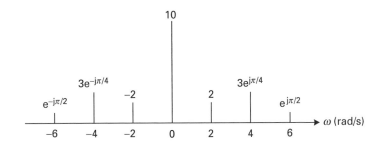

(d) What sampling frequency, f_s, is required to accurately reconstruct
   $x(t)$ from $x[n]$? To accurately reconstruct $y(t)$ from $y[n]$?

(e) What is the period of $x(t)$? Of $y(t)$?

3-11 For a single sinusoid with frequency f_0 and sampling frequency f_s
   plot the reconstructed frequency f_r as a function of the ratio f_0/f_s.
   Compare to Table 3.1.

4 Lumped element modeling of mechanical systems

Until now we have been examining signals. In particular, we have learned how to model continuous-time and discrete-time signals, how to construct these signals from building blocks, and how to convert between continuous- and discrete-time. Now we will begin examining *systems*. Specifically, we would like to understand how to model and characterize the response of physical systems to various input signals or excitations. Figure 4.1 shows the schematic of an arbitrary system with an input signal $x(t)$ and output signal $y(t)$. Thus far we have focused our attention on the analysis of the signals. Now we will find ways to characterize the system so that the output may be predicted given the input. This is called modeling the system. In this text, we may regard modeling as the translation of the physical behavior of components and collections of components into a mathematical representation. This representation must include descriptions of the individual components as well as descriptions of how the components interact. Finally, a good model will provide an accurate description of the physical behavior of the system, while at the same time remain sufficiently simple mathematically to permit easy calculations. Accuracy and simplicity are often at odds. Thus, modeling often involves some trade-offs. There are many types of systems that may be modeled, including mechanical, electrical, hydraulic, and thermal. In this text, we will focus on the first two, starting with mechanical systems.

4.1 Introduction

In modeling any mechanical system, we will be primarily interested in the motion of the system. Often this involves repetitive motions, called

Fig. 4.1. The input–output relationship for a general system.

Input : $x(t)$ ⟶ System ⟶ Output : $y(t)$

vibrations. For example, the motions of a swinging pendulum or a plucked guitar string are vibrations. The vibration characteristics are often the limiting factor in the performance of physical devices.

The analysis of any physical system generally involves the following steps:

1. Construct a mathematical model that adequately describes the system.
2. Derive the governing equations for the system (usually in the form of a differential equation).
3. Solve the governing equation.
4. Interpret the results, redesign if necessary.

4.1.1 Mathematical modeling

The objective of mathematical modeling is to create a model that is complete enough to describe all of the important features of the system without being unduly complicated. A great deal of engineering judgment goes into constructing the simplest mathematical model. This is generally dictated by what results we seek.

There are two types of model that are commonly used to describe physical systems: *lumped element model* and *distributed element model*. In a lumped element model, we assume the physical parameter of interest is concentrated or lumped at a specific point. In a distributed model, the parameter is varied over space. For example, consider modeling the mass of an automobile. Suppose we are only interested in the location of the vehicle as it traverses along the freeway. In this case, we may model the vehicle as a lumped point mass. Suppose we are now interested in the effects of dynamic weight shift of the automobile during hard braking. In this case, we may need to understand the distribution of the mass in space.

Lumped element models are generally described with ordinary differential equations in which time is the independent variable. Distributed element models are often described with partial differential equations in which both time and space are independent variables. Although all physical systems are in reality distributed, a great number of them can be accurately approximated by using the lumped element models. Thus, in this text, we will restrict our attention to the simpler case of lumped element models.

4.1.2 The governing equation

Given the mathematical model, we must derive the differential equation called the *governing equation* that describes the physical behavior of the system. For a mechanical system, this usually involves applying Newton's Second Law ($F = ma$ for a translational mechanical system[1] or $T = I\alpha$ for a rotational mechanical system, where F and T denote the force and torque; m and I are the mass and mass moment of inertia; a and α represent the acceleration and angular acceleration) to obtain the governing equation. The solution of the governing equation describes the motion or vibration of the system. The governing equation for a mechanical system is often called the *equation of motion*.

Before we can apply Newton's Second Law, we need to sketch a *free body diagram* (FBD) to visualize the forces or torques. The free body diagram is a schematic of the mass (or inertia) with all the forces (or torques) directly acting on it properly identified. Sketching the free body diagram of a mechanical system constitutes one of the most important first steps in analyzing any mechanical system.

4.1.3 Solve the governing equation

We now solve the governing equation to find the response of the system to an input. The most common methods include the analytical approach of directly solving the differential equation and the numerical approach of solving by computer (for example, MATLAB). Both approaches are viable in solving for the response of a given system. The analytical approach usually gives more physical insight. The numerical approach usually allows finding the response to more complicated inputs.

---

[1] Newton's Second Law is actually a vector relationship. However, since we will be considering only rectilinear motion in this text, where the displacement, velocity and acceleration have a known and fixed direction, we only need to specify the sense and magnitude of the force and acceleration; this may be conveniently accomplished by using a scalar quantity with a plus or minus sign to indicate the direction. The same remark also will apply to Newton's Second Law in rotation about a fixed axis or a fixed point.

4.1.4 Interpret the results

Once we have solved for the output, we must interpret the result and perform a check whenever possible to see if the solution makes sense physically. If it is reasonable and meets the specifications for the system, then our task is complete. If the results do not satisfy the specifications, we may need to redesign the system by modifying the system parameters.

4.2 Building blocks for lumped mechanical systems

Vibration of mechanical systems involves the transfer of energy between *kinetic* and *potential* forms. If the total energy (sum of the kinetic and potential energies) remains constant for all times, the system is said to be *conservative*. In real (*nonconservative*) systems, energy will be lost or *dissipated* in each cycle of vibration through losses to heat, friction or sound.

There are three basic building blocks or lumped elements that are used to represent mechanical systems. They are *masses*, *springs*, and energy dissipating elements called *dampers*. Using these simple building blocks we can model very complicated systems such as the suspension of automobiles, accelerometers used to measure vibrations, and large civil engineering structures such as buildings and bridges. We can also model very complex phenomena such as the response of dams in an earthquake, the motion of airplane wings during flight, and the behavior of machines on foundations.

The mass element stores kinetic energy because of its motion. One might expect it to also have potential energy because of its position. However, potential energy must be defined with respect to a *datum* (that is, a reference location). Often the datum is the ground, but one could equally well choose the ceiling or any other point. Later we will see that if we choose the datum to coincide with the point of *static equilibrium*, the potential energy of the mass can be treated as zero. Therefore, we will say that the potential energy is stored in the spring instead. The damper dissipates energy, much like a shock absorber in an automobile.

4.2.1 Springs

Any structural member that deforms under the application of a load or force may be modeled as a spring. The corresponding spring constant can be obtained by examining the force–displacement relationship, which can be determined either analytically (using engineering mechanics) or experimentally (by measuring the deflection to a known input force).

A translational spring exerts a restoring force when it is displaced. If the force–displacement relationship is linear, it can be modeled with Hooke's Law as follows:

$$F_s = kx, \tag{4.1}$$

where F_s denotes the force exerted by the spring, x is the deformation of the spring from its undeformed configuration, and k represents the spring constant, which has units of force/length (N/m). Similarly, a torsional spring exerts a restoring torque when angularly displaced. For example, if one twists a rubber rod, the rod will exert a restoring torque. If the torque-angular displacement relationship is linear, it is given by

$$T_s = k_t\theta, \tag{4.2}$$

where T_s denotes the torque exerted by the torsional spring, θ is the angular deformation of the spring from its undeformed configuration, and k_t represents the torsional spring constant, which has units of force $\times$ length (N m). Using theories in engineering mechanics, we can show that a translational or torsional spring constant depends on the geometric and material properties of the structural member, and the way in which the force or torque is applied.

For example, consider a diving board. It may be modeled as a cantilever or fixed-free *beam*, whose left end is attached to a wall and whose right end is completely free to move, as shown in Figure 4.2(a). The hashed lines indicate that the wall is considered immovable or fixed. Physically, if we exert a force at its free end, we expect the diving board to deflect. Therefore, we may model the diving board as a spring suspended from a vertical position as shown in Figure 4.2(b), with one end attached to the wall or ceiling and the other end free to move. Intuitively, we expect the spring constant for the diving board to depend on its geometric and material properties. From engineering mechanics, we can show that when the fixed-free beam is lumped at its free end, the spring constant for the cantilever beam is given by

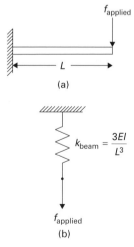

Fig. 4.2. (a) A cantilever or fixed-free beam and (b) its model.

$$k_{beam} = \frac{3EI}{L^3}, \tag{4.3}$$

Fig. 4.3. (a) A fixed-free rod and (b) its model.

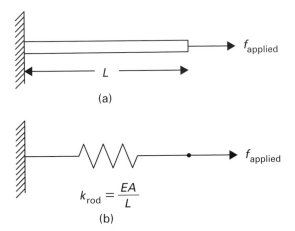

where I denotes the *area moment of inertia* (a geometric property) of the beam, E is its *Young's modulus* (a material property) and L is its length.

A structural member that deforms longitudinally when it is loaded axially is called a *rod*. Figure 4.3(a) shows a fixed-free rod of length L. Because the rod elongates when a load is applied, when the rod is lumped at its free end, it may again be modeled as a spring (see Figure 4.3(b)) with a spring constant of

$$k_{rod} = \frac{EA}{L}, \tag{4.4}$$

where A corresponds to the cross-sectional area of the rod.

A structural member that twists under the action of a torque is called a *shaft*. Figure 4.4(a) shows a fixed-free shaft of length L. Because the shaft twists when a torque is applied, when the shaft is lumped at its free end, it may be modeled as a torsional spring (see Figure 4.4(b)) with a torsional spring constant of

$$k_{shaft} = \frac{GJ}{L}, \tag{4.5}$$

where G is the *shear modulus* (analogous to the Young's modulus) of the shaft, and J is its *polar moment of inertia*, which depends on the geometric properties of the cross section of the shaft. Finally, note that unlike k_{beam} and k_{rod}, which have units of force/length, k_{shaft} has units of force $\times$ length (N m).

The appropriate model for a structural element depends on what we seek. For example, consider a long structural member (say, a long piece of metal). If we apply a force perpendicular to the major axis of the structure, we must

Fig. 4.4. (a) A fixed-free shaft and (b) its model.

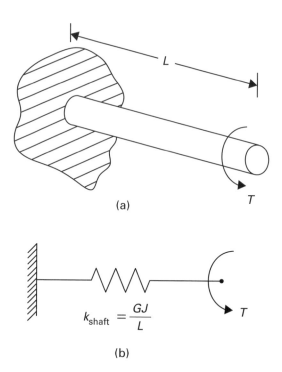

(a)

(b)

$$k_{shaft} = \frac{GJ}{L}$$

model it as a beam. If we apply a force parallel to its major axis, we must model it as a rod. If we apply a torque about its major axis, we must model it as a shaft. Thus, the same structure can be modeled differently depending on what is sought and how the load is applied.

An *ideal* spring element is a type of mechanical linkage that is assumed to be massless and to have no damping. In other words, a spring does not possess the properties of the other two elements. This assumption is made to avoid any confounding effects among the lumped elements, and this is the key part of lumped element modeling. Clearly this assumption is highly idealized, because physical springs may also exhibit some energy dissipating mechanism (damping), and all real springs certainly have mass, which may or may not be negligible. Thus, in order to accurately model real springs, we may need to include dampers and/or masses in addition to the idealized spring elements.

Finally, a lumped element model is sufficient when it captures the behavior of the real system accurately. If a lumped element model misses important details of the system behavior, then a different lumped element model may be required, or a more sophisticated distributed modeling may be necessary.

Fig. 4.5. Schematic of a
spring attached to a fixed
wall.

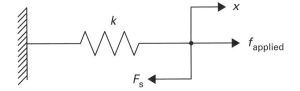

The constitutive equation for a spring relates the force and the displacement. From physics, recall that the force–displacement equation for a linear spring is given by Hooke's law:

$$F_s = kx, \tag{4.6}$$

where F_s denotes the force exerted by the spring, k is the spring constant, and x is the deflection of the spring from its undeformed or equilibrium configuration. The spring is represented schematically as shown in Figure 4.5. Deflection x shows how far the spring has been stretched from its undeformed position. The force F_s is shown acting in the opposite direction to x, as the spring *always* exerts a restoring force that opposes the displacement. The force f_{applied} is the force applied to the right end or the free end of the spring. Note that f_{applied} and F_s must always be equal and opposite. This is a direct consequence of Newton's Third Law, and can be easily validated by sketching the free body diagram for the free end of the spring, and then applying Newton's Second Law. Figure 4.6 shows a free body diagram for the spring at its right end. The dot represents the point mass at the free end of the spring. To define its position, we must choose a coordinate system. We

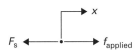

Fig. 4.6. Free body
diagram for the free end of
the spring of Figure 4.5.

measure the displacement of the free end and record it as x with a minus or a plus sign, according to whether the free end moves to the left or to the right. The distance x, with the appropriate sign, completely defines the position of the free end of the spring. If the free end moves to the right, we define x as positive; when the free end moves to the left, we define x as negative. The choice for a coordinate system is completely arbitrary. However, once we have selected a coordinate system (in this case with x to the right being positive), we have to obey it throughout the analysis. Because F_s is a restoring force, it points to the left (the restoring spring force always opposes the displacement). Finally, f_{applied} is a positive applied force. Thus, it points to the right, as indicated by the associated arrow. Applying Newton's Second Law, we know the sum of the forces equals the mass times its acceleration. Using our chosen coordinate system, f_{applied} is positive because it acts in the

$+x$ direction and F_s is negative because it acts in the $-x$ direction. Because the free endpoint of the spring is massless (a point by itself is massless, that is, $m = 0$), the sum of the forces must be zero as follows:

$$\sum F = ma = m\frac{d^2x}{dt^2} = 0 = f_{\text{applied}} - F_s. \tag{4.7}$$

Hence, we find that the restoring force and the applied force have the same magnitude but opposite directions ($F_s = f_{\text{applied}}$), obeying Newton's Third Law. Finally, it should be noted that f_{applied} and x are not independent variables. A force applied at the spring's free end causes a displacement x. Similarly, in order to have a displacement of x at its free end we must apply a force.

Spring force arises whenever the spring is stretched or compressed. It is caused by the displacement of one end of the spring relative to the other. If both ends of the spring move in the same direction with the same amount, no restoring force will be exerted. While compact, Hooke's Law, as given by Eq. (4.6), does not show explicitly that a spring force is developed only when there is relative displacement between the two ends of the spring, nor does it give us a means to easily evaluate the spring force when both ends of the spring move. Motivated by the above shortcomings, we aim to develop a more general expression for Hooke's Law that considers the displacement of both ends of the spring.

Consider a spring whose two ends move freely as shown in Figure 4.7. Let us define the positions of the left and right ends of the spring as x_L and x_R, respectively. Furthermore, let us define a coordinate system such that to the right is positive. The restoring force acting on the right end of the spring points to the leftward direction because it opposes the displacement x_R. If the left end is held fixed ($x_L = 0$) and the right end is stretched by the amount x_R, the restoring spring force is

$$F_s = kx_R. \tag{4.8}$$

If both ends move uniformly, $x_R = x_L$, then the restoring spring force must be

$$F_s = 0 \tag{4.9}$$

because the spring has simply moved as a rigid body. Combining the results of Eqs. (4.8) and (4.9), we obtain the following general expression:

$$(F_s)_R = k(x_R - x_L). \tag{4.10}$$

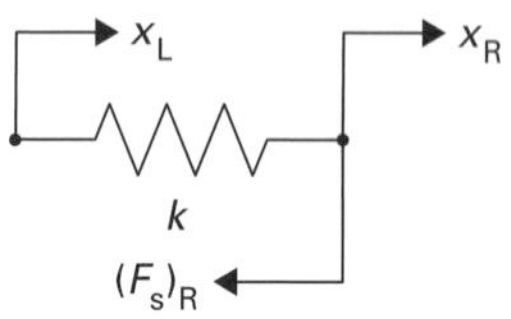

Fig. 4.7. Restoring force on the right end of a spring.

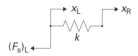

Fig. 4.8. Restoring force on the left end of a spring.

Using a similar argument, we find that the restoring force at the left end of the spring points to the left (see Figure 4.8; again, the restoring force opposes displacement) with a magnitude given by

$$(F_s)_L = k(x_L - x_R). \tag{4.11}$$

Based on the above discussion, we propose an expression for the generalized form of Hooke's Law as follows:

$$F_s = k(x_{near} - x_{far}), \tag{4.12}$$

where x_{near} represents the displacement of the endpoint at which the restoring force F_s is being considered (hence "near"), and x_{far} represents the displacement of the other endpoint of the spring (hence "far"). Equation (4.12) is also known as the constitutive equation for a spring element. While complicated at first glance, the generalized Hooke's Law is easy to apply in practice, as long as we remember the following:

- A spring *always* exerts a restoring force that *opposes* displacement.
- The magnitude of the restoring force is given by Eq. (4.12), where x_{near} and x_{far} are defined according to which endpoint or node is being analyzed.

A force is a vector; it possesses both a magnitude and a direction. The spring force opposes displacement (this captures the direction of the force), and its magnitude (which can be positive or negative) is given by Eq. (4.12). In a complicated spring assemblage system, it can often be difficult to visualize the direction of each of the restoring forces. However, as long as the generalized Hooke's law is used and the proper book-keeping procedure is followed, we will always be able to solve the problem correctly.

One could also define a *torsional* or *rotational spring* which exerts a restoring torque that opposes angular displacement. For a torsional spring, its constitutive equation is given by

$$T_s = k_t(\theta_{near} - \theta_{far}), \tag{4.13}$$

where T_s represents the restoring torque exerted by the torsional spring, θ_{near} is the angular displacement of the endpoint at which the restoring torque is being considered, θ_{far} is the angular displacement of the other endpoint of the spring, and k_t is the torsional spring constant, measured in units of force $\times$ length (N m).

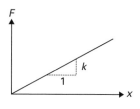

Fig. 4.9.
Force–displacement
relationship for a linear
spring.

In summary, any structural member that deforms under the application of a load can be modeled as a spring. If the force–displacement plot is shown in Figure 4.9, the slope of the line corresponds to the spring constant k. In this text, we will only consider linear springs in which the line has a constant slope. This is a good approximation for most physical systems subjected to small displacements. Of course, when large displacements are applied, the springs can become nonlinear, and the resulting problem may become difficult if not impossible to treat analytically.

The spring force or restoring force always opposes displacement with a magnitude given by Eq. (4.12). For example, consider stretching a spring by pulling on both ends. The restoring or the spring forces at both ends are shown in Figure 4.10, and f_{applied} is applied to each end. The spring force on the right end is $(F_s)_R = k(x_R - x_L)$, and the spring force on the left end is $(F_s)_L = k(x_L - x_R)$. Both restoring forces point to the left because a spring always exerts a force that opposes displacement. Note again that what is considered the near end and what is considered the far end depends on which end is under consideration.

The generalized Hooke's Law of Eq. (4.12) shows that the force applied to one end of an ideal spring is transmitted through to the other end without any change. Consider the right end of the spring of Figure 4.10, where we have already shown that $f_{\text{applied}} = (F_s)_R$. The applied and restoring forces are equal in magnitude but opposite in direction; consistent with Newton's Third Law. At the left end of the spring element, we see that $f_{\text{applied}} = -(F_s)_L$. The forces are still equal in magnitude, but this time opposite in sign because the forces point in the same direction. Also, note that the spring forces on the two ends of the spring are exactly equal and opposite, as one would physically expect. In other words, we find that the force applied at one end of an ideal spring element will be transmitted to the other end without any change.

Fig. 4.10. A spring with a
force applied at both ends.

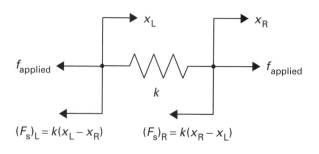

Fig. 4.11. (a) A viscous damper and (b) its symbolic representation.

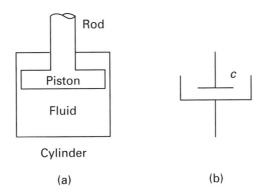

(a)   (b)

4.2.2 Dampers

In many physical systems, the vibrational energy is gradually converted to heat or sound, or lost to friction during every cycle of vibration. Due to the reduction in energy, the response, such as the displacement of the mass, gradually decreases. The *damper* models this energy loss mechanism. A damper is assumed to have neither mass nor elasticity (that is, springiness characteristics). Such a damper is said to be ideal. The *viscous damper* is the most commonly used damping mechanism. A physical damper is shown in Figure 4.11(a) and schematically in Figure 4.11(b). It consists of a piston that is allowed to move inside a cylinder which is filled with some viscous fluid, hence its name. A commonly used damper is the shock absorber in an automobile.

Like a spring, a viscous damper also exerts a restoring force. However, the restoring force opposes motion or velocity, not displacement. Empirically, the damping force is found to develop whenever there is a relative velocity between the two ends of the damper. Thus, the constitutive equation for a damper element can be expressed as

$$F_d = c \frac{d}{dt} (x_{\text{near}} - x_{\text{far}}) = c(\dot{x}_{\text{near}} - \dot{x}_{\text{far}}), \tag{4.14}$$

where F_d represents the restoring force exerted by the damper, $\dot{x}_{\text{near}}$ is the velocity of the endpoint at which the restoring force is being considered, $\dot{x}_{\text{far}}$ is the velocity of the other endpoint of the damper, and c is the viscous damping coefficient, measured in units of force $\times$ time/length (N s/m). Notice that Eq. (4.14) is nearly identical to the generalized Hooke's Law for a spring, except that the proportionality constant is c rather than k, and the restoring force is proportional to the time derivative of the relative displacement rather than the relative displacement itself. Using the same argument as for a spring element, we can easily show that for an ideal damper element, it pulls on whatever it is attached to when it is pulled on the other end. Alternatively, we say that a force applied at one end of an ideal damper element is transmitted through to the other end without any change.

One could also define a *torsional* or *rotational* damper which exerts a restoring torque that opposes angular velocity. For a torsional damper, its constitutive equation is given by

$$T_d = c_t \frac{d}{dt}(\theta_{near} - \theta_{far}) = c_t(\dot{\theta}_{near} - \dot{\theta}_{far}), \tag{4.15}$$

where T_d represents the restoring torque exerted by the rotational damper, $\dot{\theta}_{near}$ is the angular velocity of the endpoint at which the restoring torque is being considered, $\dot{\theta}_{far}$ is the angular velocity of the other endpoint of the damper, and c_t is the torsional viscous damping coefficient, measured in units of force $\times$ length $\times$ time (N m s).

4.2.3 Mass/inertia

The mass element m is considered to be a rigid body with no elasticity or damping characteristics. Recall that for spring and damper elements, their constitutive equations are given by Eqs. (4.12) and (4.14), which show that the restoring forces these elements exert depend on the displacement and the velocity of one endpoint of the element relative to the other endpoint of the element. In the case of the mass element, its constitutive relationship yields the inertia force, which is simply the product of the mass and its acceleration measured relative to a fixed reference frame,

$$F_m = m \frac{d^2 x}{dt^2} = m\ddot{x} = ma, \tag{4.16}$$

where the acceleration a is simply the second derivative of position x. Because the mass element is rigid, its position is uniquely defined with a single point x rather than a pair of endpoints x_{near} and x_{far}. However, because acceleration is measured relative to a Newtonian reference frame (that is, a fixed reference point), if we denote the position of the mass x_{near} and our reference point x_{fixed}, we can write Newton's Second Law alternatively as

$$F_m = m \frac{d^2}{dt^2}(x_{near} - x_{far}) = m(\ddot{x}_{near} - \ddot{x}_{far}) = m(\ddot{x}_{near} - \ddot{x}_{fixed}), \tag{4.17}$$

Table 4.1. Constitutive equations for ideal
translational mechanical elements

Element	Constitutive equation
Spring	$F_s = k(x_{near} - x_{far})$
Damper	$F_d = c(\dot{x}_{near} - \dot{x}_{far})$
Mass	$F_m = m(\ddot{x}_{near} - \ddot{x}_{fixed})$

where $x_{far} = x_{fixed}$ denotes some arbitrary fixed reference point. This form puts F_m on equal footing with F_s and F_d. Of course, x_{fixed} is constant and it becomes zero when the second derivative is taken.

For an *inertia element*, its constitutive equation is given by

$$T_I = I \frac{d^2}{dt^2}(\theta_{near} - \theta_{far}) = I(\ddot{\theta}_{near} - \ddot{\theta}_{far}), \tag{4.18}$$

where T_I represents the torque exerted by the mass moment of inertia I, θ_{near} denotes its angular displacement at the near endpoint, and θ_{far} is the angular displacement of the far endpoint. For an inertia element, $\theta_{far} = \theta_{fixed}$ (an arbitrary fixed reference point), and it reduces to zero when a time derivative is taken. The mass moment of inertia has units of force $\times$ length $\times$ time^2 (N m s^2).

The constitutive equations for an ideal translational spring element, damper element and mass element are summarized in Table 4.1. Observe that the spring, damping and inertia forces are related to the zeroth, first, and second derivative of the differences between two endpoints, respectively. Later we will use this fact, specifically the order of these constitutive equations, to find the analogous elements in electrical systems. Physically, the damper dissipates energy, so we expect its electrical analog to be the resistor. If we arrange the constitutive equations for the mechanical and electrical elements in the order of increasing derivatives, we will see that the resistor will indeed map onto the damping element, making them analogous to one another. This will be discussed in detail in the next chapter.

4.3 Inputs to mechanical systems

Having defined the three building blocks for mechanical systems, we can proceed to construct lumped element models for such systems under external inputs or excitations. There are two types of excitations that may be applied to translational mechanical systems, they include input forces and input displacements. These inputs are generally known functions of time, and they will cause the mechanical system to vibrate. Force inputs are typically applied to the mass, and displacement inputs are usually applied to the endpoint of a spring or a damper. Forces can also be applied to the endpoints of springs or dampers. In these cases, however, because we assume the springs and dampers to be massless, the forces applied at one endpoint will be entirely transmitted through to the other endpoint without any change. Thus, when forces are applied at the endpoints of springs or dampers, we can immediately move the applied forces to the other endpoints and ignore the intermediate springs or dampers.

For rotational or torsional mechanical systems, the excitations include input torques (or moments) and input angular displacements. Torque inputs are generally applied to the mass moment of inertia, and angular displacement inputs are typically applied to the endpoint of a torsional spring or a torsional damper. Torques applied at one endpoint of a torsional spring or torsional damper will be transmitted through to the other endpoint without any change because torsional springs and dampers are assumed to have no inertias.

4.4 Governing equations

Now that we have the basic building blocks and the possible inputs, we can derive the governing equation for a mechanical system that describes the motion of the system. The governing equation is also known as the equation of motion.

4.4.1 Newton's Second Law

For a translational mechanical system, the equation of motion is obtained by the direct application of Newton's Second Law:

$$\sum F = ma = m\ddot{x}. \tag{4.19}$$

For a rotational mechanical system, the torque T is the analog of the force F, the mass moment of inertia I is the analog of mass m, and the angular

Fig. 4.12. A spring–mass system.

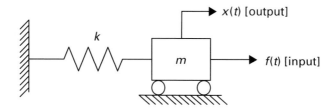

$x(t)$ [output]

$f(t)$ [input]

Fig. 4.13. Free body diagram for the spring–mass system of Figure 4.12.

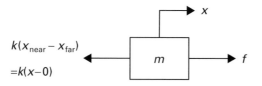

$k(x_{near} - x_{far})$

$= k(x - 0)$

acceleration α is the analog of the linear acceleration a. Therefore, Newton's Second Law for a rotational system is simply

$$\sum T = I\alpha = I\ddot{\theta}. \tag{4.20}$$

The governing equations for mechanical systems can be readily obtained by first sketching the free body diagram associated with the mass or inertia of the system, and then applying the appropriate form of Newton's Second Law, depending on if the system undergoes translational or rotational motion. The following examples will illustrate the required steps to analyze mechanical systems.

4.4.2 Examples

We now derive the governing equations for various translational and rotational mechanical systems.

Example 1: Figure 4.12 shows a mass connected to a rigid wall through a spring. A force, $f(t)$, is applied to the mass. The hashed symbol on the left represents a fixed wall. The wheels above the fixed floor indicate a frictionless surface. Determine the governing equation for the horizontal displacement of the mass.

Solution: We begin by drawing a free body diagram, shown in Figure 4.13. Let us arbitrarily choose a coordinate system in which x is defined to be positive to the right, as indicated with the right arrow showing x. Incidentally, the choice of the coordinate system is completely arbitrary. However, once chosen, we have to obey its sign convention throughout the problem. The spring exerts a restoring force of magnitude $k(x_{\text{near}} - x_{\text{far}})$ that opposes displacement, thus it points to the left. Because we are determining the restoring force on the mass, the endpoint of interest of the spring is simply the displacement of the mass itself. Thus, $x_{\text{near}} = x$ and x_{far} is the displacement of the other endpoint, which is zero because it is simply a point attached to the fixed wall. Thus, the restoring force simplifies to kx and it points to the left. Applying Newton's Second Law, we find

$$\sum F = m\frac{\mathrm{d}^2 x}{\mathrm{d}t^2} = f(t) - kx. \tag{4.21}$$

It is very important to observe the correct sign for each of the forces in the free body diagram. The applied force $f(t)$ is positive because it acts to the right, and the spring force kx is negative because it acts to the left (restoring force). We can rearrange the terms to obtain the governing equation with the position (the dependent variable) on the left-hand side and the input on the right-hand side as follows:

$$m\frac{\mathrm{d}^2 x}{\mathrm{d}t^2} + kx = f(t). \tag{4.22}$$

Equation (4.22) represents the equation of motion or the governing equation for the system of Figure 4.12.

    Example 2: Consider the same system of Figure 4.12, except now that the spring–mass system is rotated by 90 degrees. Thus, we have a mass hanging from a ceiling on a spring. A force is applied to the mass in the downward direction, as shown in Figure 4.14. Determine the governing equation for the vertical displacement of the mass.

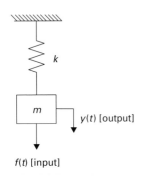

Fig. 4.14. A spring–mass system suspended from a ceiling.

Solution: This system differs from the system of Figure 4.12 because it is also subjected to a gravitational force. Because there is gravity, the spring will be stretched by some amount y_s called the static deflection. The static deflection is simply the deflection of the spring due to the weight of the attached mass. This is shown in Figure 4.15. Because the force of gravity is mg, the displacement or the amount of stretch of the spring due to gravity must be

$$y_s = \frac{mg}{k}. \tag{4.23}$$

The position of the mass in the absence of an input force is called the *static equilibrium*. When an input force is applied, the mass will be displaced by

Fig. 4.15. Static equilibrium and the displacement from static equilibrium.

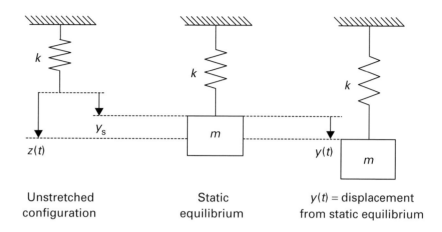

Unstretched configuration   Static equilibrium $y(t)$ = displacement from static equilibrium

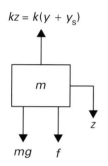

Fig. 4.16. Free body diagram for the system of Figure 4.14.

y from its static equilibrium. We now sketch a free body diagram for the system, shown in Figure 4.16, noting that a coordinate system is defined such that z is positive in the downward direction. Thus, the spring stretches by $z = y + y_s$ from its undeformed configuration, and it exerts a restoring force pointing upwards. Applying Newton's Second Law, we obtain

$$\sum F = m\frac{\mathrm{d}^2 z}{\mathrm{d}t^2} = f(t) + mg - kz \qquad (4.24)$$

or alternatively, in terms of the coordinate y

$$m\frac{\mathrm{d}^2 y}{\mathrm{d}t^2} + ky + (ky_s - mg) = f(t), \qquad (4.25)$$

which reduces to

$$m\frac{\mathrm{d}^2 y}{\mathrm{d}t^2} + ky = f(t) \qquad (4.26)$$

because the force exerted by the stretched spring in the static equilibrium position exactly cancels the force caused by gravity, consistent with what we learned in physics (see Eq. (4.23)). Thus, by conveniently changing the coordinate from z (whose origin coincides with the undeformed position of the spring) to y (whose origin coincides with the static equilibrium position), the weight of the mass does not appear in the equation of motion. Note that Eqs. (4.22) and (4.26) are identical, except for the variables used to denote the displacement of the masses. This observation leads us to the following rule of thumb: *if the direction of motion is parallel to gravity at all time, we may ignore the effect of gravity by performing an appropriate coordinate transformation.*

Fig. 4.17. A base
excitation system.

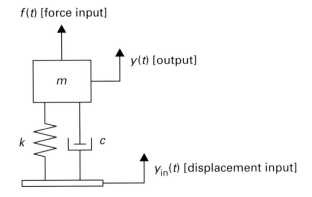

Example 3: Consider a mass excited by an applied force and a base that is allowed to move. The mass is attached to the moving base by a spring and a damper as shown in Figure 4.17. One end of the spring and one end of the damper are attached to the mass, while their other ends are connected to the movable base. Determine the governing equation for the vertical displacement of the mass.

Solution: This system is known as *base excitation*, and it models the dynamics of many interesting physical objects. For example, an automobile suspension can be modeled as a spring and damper. The rigid vehicle body is viewed as the mass and the bumps in the road are the displacement of the base. Other examples include large machines on foundations and vibration-measuring devices. For this system, $y_{in}(t)$ is the input displacement, and $f(t)$ is the input force. Notice that both types of input are present in this system.

Let us measure $y(t)$ from the static equilibrium so we can ignore gravity. The free body diagram is shown in Figure 4.18. For this problem, we choose a coordinate system that is defined to be positive upwards. The spring and damper forces act downward to oppose the displacement and the velocity, respectively. The near node is the mass (at position y), and the far node is the base (at position y_{in}). Applying Newton's Second Law, we get

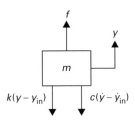

Fig. 4.18. Free body
diagram for the system of
Figure 4.17.

$$\sum F = m\frac{d^2 y}{dt^2} = f(t) - k(y - y_{in}) - c\frac{d}{dt}(y - y_{in}). \tag{4.27}$$

We will always simplify the governing equations so that the output variable (in this case the position y of the mass) and its derivatives all appear on the

Fig. 4.19. Propeller of a ship.

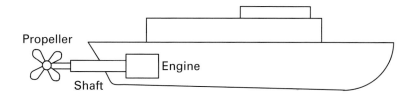

Fig. 4.20. A simple mathematical model for the propeller of Figure 4.19.

$\theta(t)$ [angular displacement output]

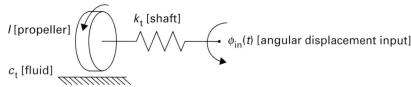

left side of the equation, and all the inputs appear on the right side. Upon rearranging, Eq. (4.27) becomes

$$m\frac{d^2 y}{dt^2} + c\frac{dy}{dt} + ky = c\frac{dy_{\text{in}}}{dt} + ky_{\text{in}} + f(t). \tag{4.28}$$

Equation (4.28) is the governing equation for the system of Figure 4.17.

Example 4: Consider the propeller of a ship and the connecting shaft as shown in Figure 4.19, which is an example of a rotational system. Figure 4.20 shows a mathematical model of the propeller and the connecting shaft. We treat the propeller as having a rotational inertia I (note that I in a rotational system is analogous to m in a translational system). The shaft has some torsional flexibility so we model it as a torsional spring of stiffness k_{t}, with one end of the torsional spring attached to the propeller and the other end attached to the engine. The engine drives the far end of the shaft (see Figure 4.20) with an input angular displacement of $\phi_{\text{in}}(t)$. The surrounding water also exerts a retarding torque on the propeller, and we model this as a torsional damper with a torsional damping coefficient of c_{t}. Determine the governing equation for the angular displacement, $\theta(t)$, of the propeller.

Solution: We have modeled a very complicated physical system using simple lumped elements. In some circumstances, the resulting model will give a sufficiently accurate description of the motion of the propeller. In other circumstances, we might need a more elaborate model. There is clearly a need for engineering judgment when selecting the appropriate model for a problem. Here, we assume the simple lumped element model is sufficient to describe the behavior of the propeller.

As usual, we begin with drawing the free body diagram. When sketching the free body diagram, we must keep track of the signs for the various torques. Any error in sign will render the governing equation incorrect, and any subsequent analysis useless. By inspection (see Figure 4.20), we note that the input angular displacement and the response of the mass moment of inertia are out of the page, while the restoring torques exerted by the shaft and the fluid are into the page. For a complicated system with many different angular displacements and restoring/applied torques, it may be difficult to visualize whether they are into the page or out of the page. For convenience and clarity, they are replaced by doubled-headed arrows along the axis of rotation, with the direction of the arrows governed by the right-hand rule. For the system of Figure 4.20, note that both $\theta(t)$ and $\phi_{in}(t)$ are out of the page. Thus, using the right-hand rule, we curl our fingers about the x-axis in the direction shown (out of the page), and our thumb will point to the right. Thus, we may replace all the components that rotate out of the page with translational components that point to the right. Such a free body diagram is shown in Figure 4.21. Because the rotational spring and rotational damper exert restoring torques that are into the page, we can replace them with translational components that point to the left, as shown in Figure 4.21. Summing the torques, we obtain the following equation of motion:

$$\sum T = I\frac{d^2\theta}{dt^2} = -c_t\frac{d\theta}{dt} - k_t(\theta - \phi_{in}).\tag{4.29}$$

Upon rearranging, Eq. (4.29) becomes

$$I\frac{d^2\theta}{dt^2} + c_t\frac{d\theta}{dt} + k_t\theta = k_t\phi_{in}.\tag{4.30}$$

Equation (4.30) represents the governing equation for the system of Figure 4.20.

Example 5: Figure 4.22 shows a pendulum suspended from a ceiling. Determine the governing equation for the angular displacement of the pendulum.

Solution: While both Figure 4.20 and Figure 4.22 are rotational systems, they are different in that the rotation for the system of Figure 4.20 is into or out of the page, while the rotation for the system of Figure 4.22 is always confined to the plane of the page. Let us model the pendulum as a point

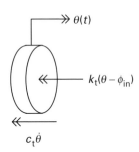

Fig. 4.21. Free body diagram for the system of Figure 4.20.

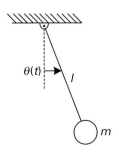

Fig. 4.22. A pendulum system.

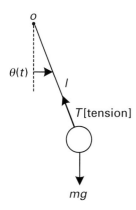

Fig. 4.23. Free body diagram for the system of Figure 4.22.

mass m lumped at the free end of a massless rod. The rod is of length l and the angular displacement of the pendulum is denoted by the angle θ. Assume the pivot is frictionless and the air resistance is negligible. Figure 4.23 shows the free body diagram of the simple pendulum. We will take the clockwise direction to be positive, as indicated by the direction of the arrow on θ.

For this pendulum system, the direction of motion is not exclusively parallel to gravity. Therefore, we cannot ignore the effects of gravity. In fact, for this pendulum system, gravity serves as the restoring torque, without which the pendulum will not oscillate when displaced. The forces acting on the system include gravity acting downward and tension acting in the direction of the rod. Because the motion of this system involves purely rotations, we need to apply Newton's Second Law in rotation to obtain the system's governing equation. Summing the torques about the pivot point o, we get

$$\sum T_o = I\frac{\mathrm{d}^2\theta}{\mathrm{d}t^2} = -mgl\sin\theta. \tag{4.31}$$

The torques are also called moments. Let us define all the moments about the pivot point. The free body diagram of Figure 4.23 shows that gravity exerts a force downwards with a moment arm of $l\sin\theta$, where the moment arm is defined as the perpendicular distance from the point about which the moment is taken, that is point o, to the line of action of the force. In the configuration shown, because the torque exerted by gravity will tend to rotate the pendulum clockwise, it will appear as negative in Eq. (4.31). The tension that acts in the direction of the rod does not appear in the governing equation, because any force whose line of action passes through the pivot point will have a zero moment about that point. Finally, recall from physics that the mass moment of inertia of a lumped mass on a massless rod is given by ml^2. Rearranging Eq. (4.31), we obtain the following governing equation for a simple pendulum:

$$ml^2\frac{\mathrm{d}^2\theta}{\mathrm{d}t^2} + mgl\sin\theta = 0 \quad \text{or} \quad l\frac{\mathrm{d}^2\theta}{\mathrm{d}t^2} + g\sin\theta = 0. \tag{4.32}$$

This differential equation is highly nonlinear and has no simple solution. However, if the angular displacement $\theta(t)$ is assumed to be small, we can make the following approximation:

$$\sin\theta \approx \theta \tag{4.33}$$

and reduce the nonlinear differential equation of Eq. (4.32) to a linear differential equation of the form

$$l\frac{\mathrm{d}^2\theta}{\mathrm{d}t^2} + g\theta = 0 \tag{4.34}$$

whose solution we can easily obtain (see Chapter 6). Equation (4.34) corresponds to the governing equation for the system of Figure 4.22. It is valid only for small angular displacements or small motions, because we obtained the equation by using the small angle approximation of Eq. (4.33).

We have considered five different mechanical systems, and have outlined the steps required to obtain the corresponding governing equations. We have shown how to properly sketch the free body diagram, and how to correctly apply Newton's Second Law. In Chapter 6 we will solve these differential equations for the purpose of obtaining the response of the system.

4.4.3 A note on stability

Most physical systems that we consider will be *stable*. A *stable* system is one in which the system's output variables return to their equilibrium positions if slightly disturbed. Conversely, an *unstable* system is one in which the system's output variables continue to recede from their equilibrium positions if slightly perturbed. Finally, there is an intermediate case, known as *neutrally stable*, in which the system's output variables do not return to their equilibrium positions, but neither do they continue to recede. A ball in a pit or well is an example of a stable system. If the ball is displaced slightly from the bottom of the well, it will roll back and forth. On account of friction between the ball and the surface of the well, the ball will eventually stop at the bottom of the pit, that is we say that the ball has returned to its equilibrium position. The system is said to be stable. A ball on top of an infinite hill (a hill without a bottom) is an example of an *unstable* system. If the ball is displaced slightly from the top, it will roll off the hill and never return. The ball in the pit becomes an example of a *neutrally stable* system if there is no friction between the ball and the surface of the well; when slightly perturbed, the ball will roll back and forth forever in the pit without coming to a stop.

For a system to be stable, a necessary condition is that all of the coefficients in the governing equation must have the same sign. This is called the

Routh-Hurwitz criterion.[2] In general, this criterion is necessary but not sufficient to guarantee stability. However, for first- and second-order systems,[3] having the same sign is not only necessary but is also sufficient to ensure stability. This gives us a simple check on the correctness of our governing equation. We know intuitively that the automobile suspension system is stable; when we hit a bump the suspension system will not cause the vehicle's amplitude to increase without bound. Therefore, when we model an automobile suspension system, if we find that the coefficients are not all of the same sign in the equation of motion, we know immediately that either the model is incorrect or we have made an error in determining the governing equation.

4.5 Parallel combination

Springs are usually manufactured only in certain increments of stiffness values depending on such things as the numbers of turns, material, and so on. Because mass production and large sales can bring down the price of a product, the designer is often faced with a limited choice of spring constants when designing a system. It may thus be cheaper to use several off-the-shelf springs to create the stiffness value necessary rather than order a special spring with specific stiffness. Combination rules for elements in parallel and series can then be used to obtain the desired or acceptable stiffness. The same also holds for dampers.

In many practical applications, several like elements, that is elements of the same type, may be used in combination to achieve a desired stiffness or damping value. Often, the effects of like elements can be combined into a single equivalent element, thus greatly simplifying the analysis. There are two ways in which springs and dampers may be combined: either in *parallel* or in *series*. Springs and/or dampers are in parallel if they share the same endpoints, that is node variables, or if they have relative displacements of equal magnitude. They are in series if they share a common force, a direct consequence of Newton's Third Law.

Consider the pair of springs in parallel shown in Figure 4.24. The springs are in parallel because they share a relative displacement of equal magnitude, $|x_2 - x_1|$. The stiffnesses are k_1 and k_2 and the node variables on the left

---

[2] See *Modern Control Engineering* by K. Ogata, Prentice Hall, New Jersey, 2002.
[3] By definition, the order of a system corresponds to the highest derivative in the system's governing equation.

Fig. 4.24. Parallel combination of two springs.

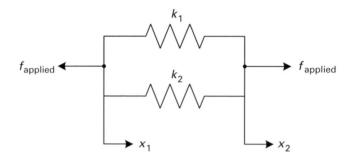

Fig. 4.24. Parallel combination of two springs.

Fig. 4.25. Free body diagram for the system of Figure 4.24.

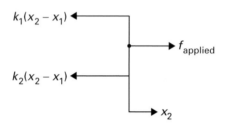

and the right are x_1 and x_2, respectively. Assume the springs are in tension, thus a force, f_{applied}, is applied to each end as shown. Figure 4.25 shows the free body diagram at node 2 (defined by x_2). The forces acting on the node include the applied force and the two restoring forces of the springs. The restoring forces act to the left to oppose displacement, and they have magnitudes of $k_1(x_2 - x_1)$ and $k_2(x_2 - x_1)$ as shown. Applying Newton's Second Law at node 2, we find

$$m_2 \frac{d^2 x_2}{dt^2} = f_{\text{applied}} - k_1(x_2 - x_1) - k_2(x_2 - x_1). \tag{4.35}$$

Because the node is massless, $m_2 = 0$ and Eq. (4.35) simplifies to

$$f_{\text{applied}} = k_1(x_2 - x_1) + k_2(x_2 - x_1) = (k_1 + k_2)(x_2 - x_1). \tag{4.36}$$

It would be helpful if we could combine the two springs in parallel into one equivalent spring, that is a spring that has the same displacements at the two endpoints when subjected to the same applied force, with a spring stiffness of k_p (subscript p denotes parallel), as shown in Figure 4.26. Figure 4.27 shows the free body diagram at x_2 again. Applying Newton's Second Law, we get

$$f_{\text{applied}} = k_p(x_2 - x_1), \tag{4.37}$$

where k_p represents the stiffness of the equivalent spring.

Fig. 4.26. Equivalent spring for the spring system of Figure 4.24.

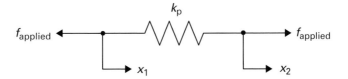

Fig. 4.27. Free body diagram for the system of Figure 4.26.

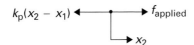

Fig. 4.28. Parallel combination of N springs.

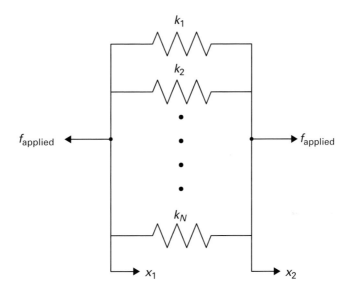

Comparing Eqs. (4.36) and (4.37), we find that the two springs in parallel can be replaced with a single spring whose stiffness is

$$k_p = k_1 + k_2. \tag{4.38}$$

Using a similar approach, we can easily extend the result to N springs in parallel. Specifically, we can replace the parallel combination of N springs (as shown in Figure 4.28) with an equivalent spring of stiffness

$$k_p = \sum_{i=1}^{N} k_i, \tag{4.39}$$

where k_i represents the spring stiffness of the ith spring in the parallel combination. Similarly, we can show that for N dampers in parallel, we can replace the dampers with an equivalent damper of damping coefficient

$$c_\mathrm{p} = \sum_{i=1}^{N} c_i, \tag{4.40}$$

where c_i represents the damping coefficient of the ith damper in the parallel combination.

4.6 Series combination

Figure 4.29 shows two springs mounted end to end. Because every spring exerts the same restoring force by virtue of Newton's Third Law (equal and opposite reactions), the two springs are in series. Again, let us assume the springs are under tension as a force is applied to both ends. Our goal is to collapse the series combination into a single spring and find its equivalent spring constant.

To determine the equivalent spring stiffness, we begin by considering the right node of each spring. At node 3 (labeled x_3), spring k_2 exerts a restoring force pointing to the left with a magnitude of

$$F_{\mathrm{s}2} = k_2(x_3 - x_2). \tag{4.41}$$

This restoring force must be equal to f_applied because the node is massless. Thus, $F_{\mathrm{s}2} = f_\mathrm{applied}$. Since force applied at one end is transmitted through to the other end for an ideal spring, performing the same analysis on the right endpoint of spring k_1, we find

$$F_{\mathrm{s}1} = k_1(x_2 - x_1) = f_\mathrm{applied}. \tag{4.42}$$

Consider a single spring with spring constant k_s (subscript s denotes series) as shown in Figure 4.30. The spring systems of Figure 4.29 and Figure 4.30

Fig. 4.29. Series combination of two springs.

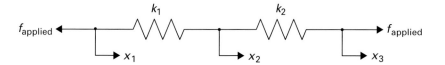

Fig. 4.30. Equivalent
spring for the spring
system of Figure 4.29.

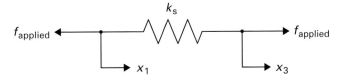

are said to be equivalent because they share the same relative displacements
when subjected to the same applied force. At the right node of spring k_s,
the restoring force is given by

$$F_s = k_s(x_3 - x_1) = f_{applied}. \tag{4.43}$$

If the two systems are equivalent, the relative deformation of the equivalent spring equals the sum of the relative deformations of the individual
springs. The previous statement is often referred to as the *consistency rule*.
Thus,

$$x_3 - x_1 = (x_3 - x_2) + (x_2 - x_1). \tag{4.44}$$

Solving for the relative deformation of each spring using Eqs. (4.41) to
(4.43) and substituting the result into Eq. (4.44), we obtain an expression
for the equivalent spring constant when the springs are in series:

$$\frac{f_{applied}}{k_s} = \frac{f_{applied}}{k_2} + \frac{f_{applied}}{k_1} \quad \text{or} \quad \frac{1}{k_s} = \frac{1}{k_2} + \frac{1}{k_1}. \tag{4.45}$$

This analysis can be extended to N springs in series. Thus, we can replace
the series combination of N springs with a single spring of equivalent spring
constant

$$\frac{1}{k_s} = \sum_{i=1}^{N} \frac{1}{k_i}, \tag{4.46}$$

where k_i represents the spring stiffness of the ith spring in the series
combination.

Similarly, N dampers in series can be replaced by a single equivalent
damper whose damping coefficient is given by

$$\frac{1}{c_s} = \sum_{i=1}^{N} \frac{1}{c_i}, \tag{4.47}$$

Fig. 4.31. Springs k_1 and k_2 in series.

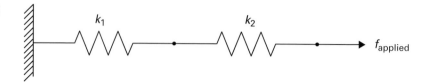

Fig. 4.32. Springs k_1 and k_2 in parallel.

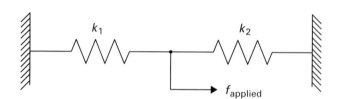

where c_i represents the damping coefficient of the ith damper in the series combination.

4.7 Combination of masses

Recall that masses are modeled as a rigid body and referenced from a fixed point. When they are combined, they can only appear in parallel because they are displaced by the same amount. Hence the equivalent mass of a combination of masses is the sum of the individual masses.

4.8 Examples of parallel and series combinations

Let us consider some examples of parallel and series combinations. Figure 4.31 and Figure 4.32 show two springs connected end to end. To correctly identify how the springs are configured, we must go back to the original definition of series and parallel. Recall that springs are in parallel when they share the same endpoints (that is, both springs share common magnitude in relative displacements). Springs are in series when they share a common force. In Figure 4.31, the springs are in series because they share the same force, while in Figure 4.32, because both springs will deform by the same relative displacement (one spring will stretch and the other spring will compress; both by the same amount), they are in parallel.

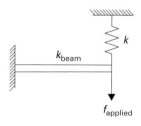

Fig. 4.33. k and k_{beam} in parallel.

Figure 4.33 shows a beam connected to a spring. Recall that a beam is a structural element that bends under the application of a transverse load.

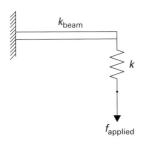

Fig. 4.34. k and k_{beam} in series.

The stiffness of the beam, k_{beam}, can be modeled as a spring acting in parallel with the other spring k, because both k_{beam} and k displace equally. In Figure 4.34, on the other hand, the springs k_{beam} and k act in series because the same force is transmitted to each. Alternatively, both springs will exert the same restoring force. Consider a more complicated system as shown in Figure 4.35. Using the definition of parallel and series combinations, we see that k_1 and k_{beam} act in parallel and their combination appears in series with k_2.

4.9 Division of force in parallel combination

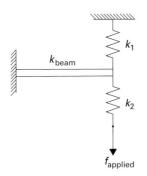

Fig. 4.35. k_1 and k_{beam} in parallel, and the result in series with k_2.

When multiple springs act in parallel, each spring has the same relative displacement but may carry a different fraction of the restoring force. For example, if a steel spring and a rubber band act in parallel, we expect the steel spring to bear most of the restoring force, although both the steel spring and the rubber band experience the same deformation. Our objective is to determine how much restoring force each spring exerts. In general, we expect the fraction of the restoring force borne by a particular spring to be proportional to its spring constant.

Figure 4.36 shows a parallel combination of N springs. The N springs may be replaced by a single spring with an equivalent spring constant of k_{p}, whose stiffness is given by Eq. (4.39). The total restoring force exerted by the equivalent spring at its right node is equal to the applied force

$$f_{\text{applied}} = k_{\text{p}}(x_2 - x_1). \tag{4.48}$$

The restoring force due to the jth spring at its right node is given by

$$F_{sj} = k_j(x_2 - x_1). \tag{4.49}$$

The sum of the restoring forces must equal the applied force. Thus, taking the ratio of the restoring force exerted by the jth spring to the total applied force, we obtain

$$\frac{F_{sj}}{f_{\text{applied}}} = \frac{k_j(x_2 - x_1)}{k_{\text{p}}(x_2 - x_1)} = \frac{k_j}{\sum_{i=1}^{N} k_i}, \tag{4.50}$$

Fig. 4.36. Parallel combination of N springs.

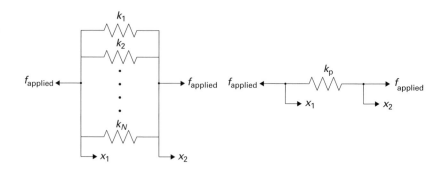

which implies that the restoring force of the jth spring is proportional to its stiffness, consistent with physical intuition. For two springs in parallel, the restoring force exerted by the jth spring is

$$F_{sj} = \frac{k_j}{k_1 + k_2} f_{\text{applied}}. \tag{4.51}$$

Thus, for two springs in parallel, the stiffer spring carries the greater load or exerts the greater restoring force. Finally, the same result can also be extended to N dampers in parallel. Namely,

$$\frac{F_{dj}}{f_{\text{applied}}} = \frac{c_j}{c_p} = \frac{c_j}{\sum\limits_{i=1}^{N} c_i}. \tag{4.52}$$

4.10 Division of displacement in series combination

When multiple springs act in series, each spring bears the same force but may carry a different portion of the overall displacement. Figure 4.37 shows the series combination of N springs and its equivalent single spring. From our previous derivations, we know that for the equivalent spring, its overall deflection or its relative displacement can be expressed as

$$(x_R - x_L)_{\text{overall}} = \frac{f_{\text{applied}}}{k_s}, \tag{4.53}$$

where k_s is given by Eq. (4.46). For the jth spring, its relative displacement is given by

$$(x_R - x_L)_j = \frac{f_{\text{applied}}}{k_j}. \tag{4.54}$$

Fig. 4.37. Series combination of N springs.

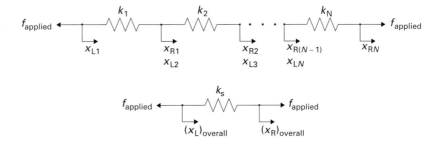

We can find the ratio of the displacement of the jth spring to the total displacement from Eqs. (4.53) and (4.54), yielding

$$(x_R - x_L)_j = \frac{\dfrac{1}{k_j}}{\dfrac{1}{k_s}}(x_R - x_L)_{\text{overall}}. \tag{4.55}$$

Thus, for springs in series, each spring's share of the relative displacement is inversely proportional to its stiffness, that is the stiffer spring will stretch less. Applying this to the special case of two springs, we find

$$(x_R - x_L)_j = \frac{\dfrac{1}{k_j}}{\dfrac{1}{k_1} + \dfrac{1}{k_2}}(x_R - x_L)_{\text{overall}}. \tag{4.56}$$

The result of Eq. (4.56) is consistent with physical intuition. If a steel spring and a rubber band are in series, we would expect the rubber band with the smaller spring constant to bear most of the overall displacement.

Finally, similar results can be derived for dampers in series. Specifically, we can show that the ratio of the relative velocity of the jth damper to the total relative velocity of its equivalent damper is given by

$$(\dot{x}_R - \dot{x}_L)_j = \frac{\dfrac{1}{c_j}}{\dfrac{1}{c_s}}(\dot{x}_R - \dot{x}_L)_{\text{overall}}, \tag{4.57}$$

where c_s denotes the equivalent damping coefficient (see Eq. (4.47)). Thus, for dampers in series, each damper's share of the relative velocity is inversely proportional to its damping coefficient.

Fig. 4.38. Figure for Problem 4-1.

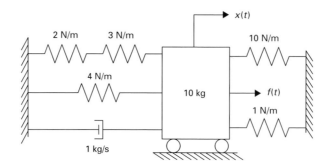

Fig. 4.39. Figure for Problem 4-2.

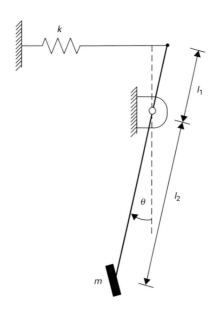

4.11 Problems

4-1 Consider the spring–mass–damper system shown in Figure 4.38. Determine the equation of motion that governs the displacement of the mass.

4-2 A control pedal of an aircraft can be modeled by the system shown in Figure 4.39. Consider the lever as a massless rod and the pedal as a lumped mass at the end of the rod. For small motions, determine the equation of motion for the angular displacement, θ, of the system. Assume the spring to be unstretched at $\theta = 0$.

4-3 Consider the system shown in Figure 4.40.

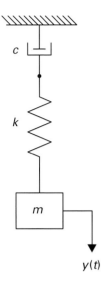

Fig. 4.40. Figure for Problem 4-3.

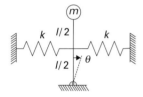

Fig. 4.41. Figure for Problem 4-4.

Fig. 4.42. Figure for Problem 4-5.

(a) Determine the equation of motion for the vertical displacement of the mass.

(b) Determine the equation of motion for the displacement of the node between the damper and the spring.

4-4 Consider an inverted pendulum of length l connected to two springs of equal stiffness k as shown in Figure 4.41. The system can be regarded as a simplified model of a rocket on the launch pad. The supporting trusses for the rocket are modeled as linear translational springs and the rocket itself can be modeled as a lumped mass. For small motions, determine the equation of motion for the angular displacement, θ, of the pendulum.

4-5 For the spring assembly shown in Figure 4.42, find expressions for each of the following in terms of relevant spring constants and the input $x(t)$:

(a) $x_1(t)$;

(b) $x_3(t) - x_1(t)$;

(c) $x_2(t) - x_1(t)$.

4-6 A two-story building during an earthquake is represented by a lumped spring–mass system (Figure 4.43) in which the floors are modeled as lumped masses, and the supporting columns are modeled as massless beams with stiffness k_1 and k_2. Because the masses of the beams are substantially less than the masses of the floors, for all practical purposes, we can model the beams as being massless. The movement of the earth is given by $x_{\text{in}}(t)$, which constitutes a displacement input to the system. Derive the governing equations for $x_1(t)$ and $x_2(t)$.

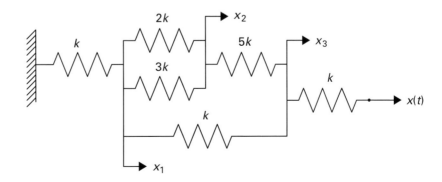

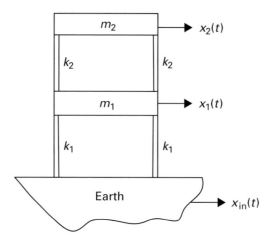

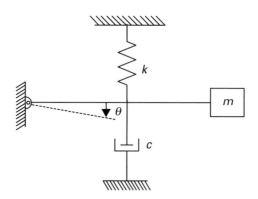

4-7 Consider the system shown in Figure 4.44, where a punctual mass is
attached to a massless rod that is pivoted about its end. The rod is
supported by a spring and damper as shown.

(a) Determine the governing equation for the angular displacement
of the rod. Assume small motion and that the rod is completely
horizontal at static equilibrium.

(b) Now the system is rotated 90° clockwise. Determine the governing
equation for the angular displacement of the rod.

(c) Now rotate the system by 90° counterclockwise from the position in part (a). Determine the governing equation for the angular
displacement of the rod.

4-8 Consider the system shown in Figure 4.45. The spring is linear and
the pulley has a mass moment of inertia I about its center o. The wire

Fig. 4.45. Figure for
Problem 4-8.

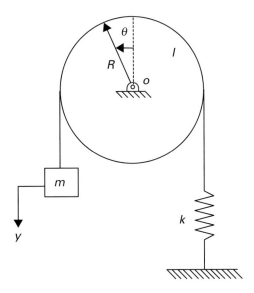

is wrapped around the disk in such a way that there is no slipping in
the system.

(**a**) Find the governing equation for the vertical displacement, $y(t)$, of
the mass m.

(**b**) Find the governing equation for the angular displacement, $\theta(t)$, of
the mass moment of inertia I.

4-9 Consider an inverted pendulum, of length l, connected to two springs of
equal stiffness k as shown in Figure 4.46. The system can be regarded
as a simplified model of a rocket at the launch pad. The supporting
trusses for the rocket are modeled as linear translational springs and
the rocket itself is modeled as a lumped mass. For the system to remain
stable (a stable system is one in which the system's output returns to its
equilibrium position if the system is slightly disturbed), a necessary
condition is that the coefficients of the governing equation must all
have the same sign. Determine the required stiffness k in order for the
system to remain stable.

4-10 Consider the system shown in Figure 4.47, known as the Maxwell's
model of a visco-elastic material, which can be used to describe the
dynamics of rubber. Determine the governing equation that describes
the horizontal motion of the lumped mass.

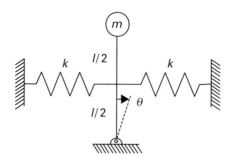

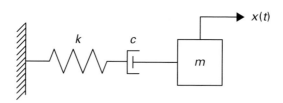

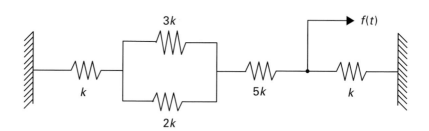

4-11 Consider the spring assemblage shown in Figure 4.48. Find the restoring force in each spring in terms of input force $f(t)$.

4-12 Consider the spring assembly shown in Figure 4.49, with a force input $f(t)$ and a displacement input of $x_{in}(t)$. The springs remain horizontal at all times. The spring stiffnesses are $k_1 = 1$ N/m, $k_2 = 1$ N/m, $k_3 = 1/3$ N/m, $k_4 = 2/3$ N/m, $k_5 = 1$ N/m, $k_6 = \pi$ N/m, $x_{in}(t) = 0.03$ m and $f(t) = 0.02$ N.

(a) Find the displacement x_1.

(b) Find the restoring force exerted by k_4 at its left node. Indicate the direction of the restoring force.

(c) Find the relative displacement of k_2, $x_1 - x_2$.

(d) Find the restoring force exerted by k_1 at its right node. Indicate the direction of the restoring force.

Fig. 4.49. Figure for
Problem 4-12.

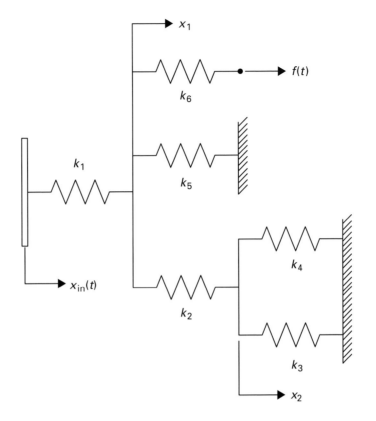

(e) If $x_{in}(t) = 0.04$ m, find $f(t)$ such that the restoring force of k_1 is zero.

4-13 Consider a beam of length L, fixed at one end and free at its other end (such a beam is known as a fixed-free or cantilever beam). From engineering mechanics, when a constant load F is applied at its free end, that is at its tip (see Figure 4.50(a)), the beam's tip deflects by an amount given by

$$x = \frac{FL^3}{3EI},$$

where I represents the area moment of inertia of the cross section of the beam (a geometric property), and E is the Young's modulus of the beam (a material property). When the beam of length L is simply supported (see Figure 4.50(b)), its midspan deflects by

$$x = \frac{FL^3}{48EI}$$

Fig. 4.50. Figure for Problem 4-13.

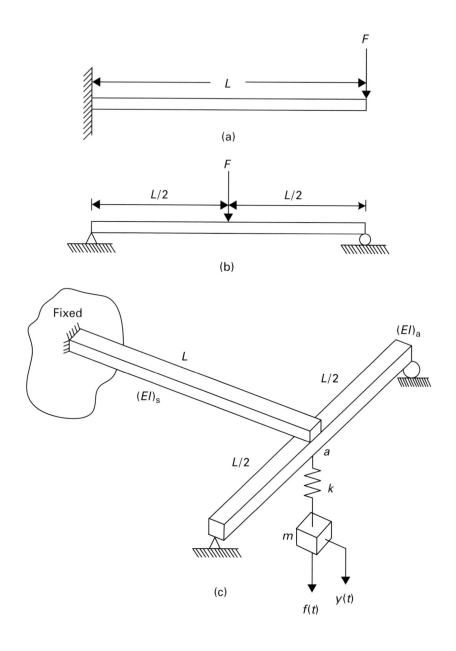

(a)

(b)

(c)

when a constant F is applied at its midpoint. A steel cantilever beam and an aluminum simply supported beam having values of $(EI)_s$ and $(EI)_a$, respectively, are welded at point a as shown in Figure 4.50(c). Attached to the midspan of the simply supported beam is a spring–mass system, of stiffness k and mass m. A force is applied to the lumped mass as shown. Assume the masses of the beams are negligible compared to m, and $(EI)_s = 4(EI)_a$.

(a) Derive the governing equation for the vertical displacement, $y(t)$, of the lumped mass m.

(b) Determine the reactive force exerted by the simply supported beam at its midspan.

5 Lumped element modeling of electrical systems

So far, we have considered developing the mathematical models and deriving the governing equations for various mechanical systems. The basic elements in mechanical systems are springs, dampers, and masses. Each of these elements is described by a force–displacement equation. For a given mechanical system, we sketch a free body diagram and apply Newton's Second Law, $F = ma$ or $T = I\alpha$, to obtain the governing equation. In Chapter 6 we will solve the governing equations and analyze the results. But before we proceed to solving the differential equations, we will look at the mathematical models for electrical systems.

5.1 Building blocks for lumped electrical systems

The concept of lumped and distributed element models can also be extended to electrical systems. For the precise analysis of electrical systems, it may be necessary to consider the resistance, capacitance and inductance to be distributed throughout the system components and wiring. Nevertheless, a great number of electrical systems can be accurately approximated by a simple lumped element model.

Surprisingly, electrical systems display many similarities to mechanical systems. The three basic elements in electrical systems are *inductors*, *resistors*, and *capacitors*, and the physical variables of interest are *voltage* and *current*. Kirchhoff's Voltage and Current Laws relate these quantities. Inputs to electrical systems include input voltage sources and input current sources.

The fundamental quantity in electrical circuits is charge, q (measured in coulombs, C). The charge on an electron is $q_e = -1.6 \times 10^{-19}$ C. Atoms

Fig. 5.1. Current flow
through an element.

have equal amounts of negative charge (in the electrons) and positive charge
(in the nucleus). Moving electrons are called current. The current i is the
rate at which charge moves. The current and charge are related as follows:

$$i = \frac{dq}{dt}.$$
(5.1)

Current is measured in coulombs per second or amperes (A). Figure 5.1
shows current i_a flowing into an electrical element. Unless there is charge
accumulation in the element, the same current i_b must flow out of the element. In mechanical systems, recall that for ideal spring and damper elements, the force applied to one end is also transmitted through to the other
end without any change in magnitude. Therefore, we say that current in
an electrical system plays the role of a force in a translational mechanical
system. Alternatively, we say that current and force are analogous.

Node voltages are always defined relative to other nodes. The voltage
difference $(v_b - v_a)$ is defined as the work required to move a unit positive
charge from point a to b. Recalling from physics, this work is given by

$$v_b - v_a = -\frac{\int_a^b \mathbf{F} \cdot d\mathbf{l}}{q},$$
(5.2)

where $\mathbf{F}$ denotes the force vector[1] and $d\mathbf{l}$ is the displacement vector. From
Eq. (5.2), we deduce that voltage is measured in units of N m/C. This unit
is more commonly called the volt. The force acting on an electron is

$$\mathbf{F} = q\mathbf{E},$$
(5.3)

where $\mathbf{E}$ is the electric field. Hence, Eq. (5.2) simplifies to

$$v_b - v_a = -\int_a^b \mathbf{E} \cdot d\mathbf{l}.$$
(5.4)

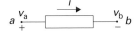

Fig. 5.2. Voltage drop and
node voltages across an
element.

Figure 5.2 shows an element between nodes a and b. Recall that current is
considered positive when it flows from the positive terminal to the negative
terminal. Assuming i is positive in the direction shown (the current flows
from left to right), then the voltage at terminal a is at a higher potential
relative to the voltage at terminal b. Thus, we assign the voltage at node
a positive and the voltage at node b negative. This set of positive and
negative signs allows us to represent the voltage drop across each element

---

[1] We will follow the common notation of using a bold-faced variable to denote a vector. This
convention will be used throughout the text.

unambiguously. We say that there is a *voltage drop* from a to b, and that the voltage *across* the element is

$$v_{\text{element}} = v_a - v_b. \tag{5.5}$$

This is reminiscent of $(x_{\text{near}} - x_{\text{far}})$ that appears in the constitutive equation for all of the mechanical elements.

The direction of the current through any electrical element is completely arbitrary. However, once a direction is assigned, we must obey it throughout the analysis. If we assume the current flows from point a to point b, then terminal a is at a higher electrical potential relative to terminal b. Conversely, if we assume the current flows from point b to point a, then terminal b is at a higher electric potential relative to terminal a. It does not matter which direction we assume the current to flow, because the mathematics will eventually dictate whether our initial assumption was correct or not: any current direction that is incorrectly labeled will just come out as negative in the final analysis.

In a mechanical system, we often measure the potential energy relative to a datum defined as 0. Similarly, in analyzing any electrical circuits, we always have to measure the voltage relative to a particular node that we define to have zero voltage. This particular node is called the *ground node* (or simply *ground*) and is indicated on a schematic with a triangle, as shown in Figure 5.3. The ground node is analogous to a reference point when calculating the gravitational potential. The potential energy of an object at a given location has no physical meaning unless a datum point is specified. Similarly, a voltage at a node is physically meaningless unless there is a specified reference voltage. For convenience, the ground voltage is chosen as the reference voltage because many actual circuits include a connection of one of the nodes to the earth (or ground). Our electrical systems consist of a collection of elements that are connected in some fashion. The connection points are known as nodes. Like mechanical systems, there are also three basic elements which are commonly used to model lumped electrical systems.

Fig. 5.3. Ground symbols.

5.1.1 Inductors

Our first circuit element is the *inductor*. An inductor can be built from a coil of wire, also commonly known as a *solenoid*. As the current i_L flows through the wire, as shown in Figure 5.4, it creates a magnetic field B. The

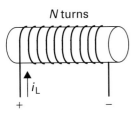

Fig. 5.4. Schematic of an inductor.

field depends on the *magnetic permeability*, μ, of the material inside the coils. If the coil of wire has length l, with the coil having N turns and a cross-sectional area S, the magnetic field can be derived from Ampere's Law to be[2]

$$B = \frac{\mu N i_L}{l}.$$
(5.6)

The *magnetic flux* is then given by

$$\lambda = N \int_S \mathbf{B} \cdot d\mathbf{S},$$
(5.7)

where $\mathbf{B}$ denotes the magnetic field and $d\mathbf{S}$ denotes an infinitesimal element of the surface area, whose direction is that of the outward normal to this infinitesimal element. For the solenoid, this becomes

$$\lambda = \frac{\mu S N^2 i_L}{l}.$$
(5.8)

The inductance is defined to be the ratio of the magnetic flux to the current. Thus, the inductance is given by

$$L = \frac{\lambda}{i_L} = \frac{\mu S N^2}{l}.$$
(5.9)

Inductance has units of V s/A, or *henries* for short. In practice, the inductance of a circuit element is measured experimentally rather than calculated analytically using the expression of Eq. (5.9). Regardless, note that the inductance depends only on the material and geometric properties, and not on the voltages or currents.

Magnetic flux has units of webers. Inductance has units of webers per ampere, called henries. Faraday's Law states that a changing magnetic field creates an electric field. When the magnetic flux enclosed by a loop of wire changes with time, a voltage is induced in the loop of wire. Mathematically, we can express this as

$$v_L = \frac{d\lambda}{dt} = L\frac{di_L}{dt}.$$
(5.10)

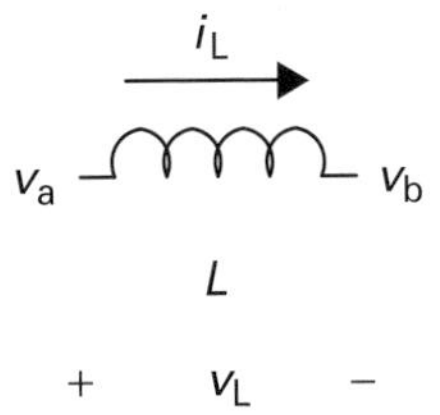

This is the constitutive relationship for the inductor. The circuit symbol is shown in Figure 5.5. To put the relationship in the same form as the other electrical elements, we express the current through the inductor as a

Fig. 5.5. Circuit symbol for an inductor.

[2] See any text on Electricity and Magnetism. For example, *Physics* by D. Halliday, R. Resnick and K. S. Krane, John Wiley & Sons, Inc., New York, 1992.

function of the voltage drop across the inductor. Rearranging Eq. (5.10) and integrating, we obtain

$$i_L(t) = \frac{1}{L}\int_{-\infty}^{t} v_L(\tau)d\tau = \frac{1}{L}\int_{-\infty}^{t} (v_a - v_b)d\tau, \qquad (5.11)$$

where v_L represents the voltage drop across the inductor, and v_a and v_b are the node voltages at the positive and negative terminals, respectively.

The power delivered to an inductor is

$$p_L = i_L v_L = L\, i_L \frac{di_L}{dt}, \qquad (5.12)$$

which may be positive or negative. The energy stored in the inductor is

$$W = \int_{t_0}^{t_1} p_L(\tau)d\tau = \frac{1}{2}L\left[i_L^2(t_1) - i_L^2(t_0)\right]. \qquad (5.13)$$

Finally, we will see that the inductor in an electrical system is analogous to the spring in a mechanical system. The restoring force for a spring is proportional to the relative displacement. In terms of the relative velocity, $u = \dot{x}$, we obtain the following relationship:

$$F_s = k(x_a - x_b) = k\int_{-\infty}^{t} (u_a - u_b)d\tau. \qquad (5.14)$$

Comparing Eqs. (5.11) and (5.14), we see immediately that the reciprocal of the inductance is analogous to the spring constant.

5.1.2 Resistors

Our next circuit element is the *resistor*. A uniform resistor is built from a material with *resistivity* ρ, length l, and cross-sectional area S. We often speak of the conductivity $\sigma = 1/\rho$. Materials may be classified as *conductors* (high conductivity), *semiconductors* (moderate conductivity), and *insulators* (low conductivity). Wires are usually built from materials with high conductivity such as copper.

Figure 5.6 shows a voltage applied across a wire. The electrons flow toward the positive terminal of the voltage source. Current is defined as the movement of charge; since electrons have a negative charge, current flows in the opposite direction to the electrons.

The voltage across the copper wire is

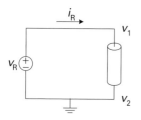

Fig. 5.6. Voltage source and a wire.

$$v_R = -\int_{1}^{2} \mathbf{E} \cdot d\mathbf{l} = El. \qquad (5.15)$$

The current density (current per unit area) in a conductor is

$$\mathbf{J} = \sigma \mathbf{E}. \tag{5.16}$$

Hence, the total current flowing through the wire is

$$i_R = \int_S (\sigma \mathbf{E}) \cdot d\mathbf{S} = \sigma E S. \tag{5.17}$$

The resistance is defined as the ratio of the voltage across the resistor to the current through it, and is given by

$$R = \frac{v_R}{i_R}. \tag{5.18}$$

This is Ohm's law.

In terms of the physical parameters, a resistor of length l, conductivity σ, and cross-sectional area S has resistance

$$R = \frac{\rho l}{S} = \frac{l}{\sigma S}. \tag{5.19}$$

Resistance is measured in units of volts/ampere, more commonly known as *ohms* (written as Ω). The reciprocal of the resistance is known as the *conductance* G, and it is given by

$$G = \frac{1}{R} \tag{5.20}$$

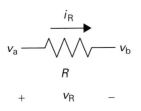

Fig. 5.7. Circuit symbol for a resistor.

which has units of Ω^{-1}, also called *mhos* (written as $\mho$). The circuit symbol for a resistor is shown in Figure 5.7. The current through a resistor is indicated as i_R, and the voltage drop or the voltage across the resistor is given by

$$v_R = v_a - v_b. \tag{5.21}$$

For a resistor, the current–voltage relationship is given by the following constitutive equation:

$$i_R = \frac{v_R}{R} = \frac{v_a - v_b}{R}, \tag{5.22}$$

where v_R represents the voltage drop across the resistor, and v_a and v_b are the node voltages at the positive and negative terminals, respectively. The *power* into a device is the product of the current and the voltage, so for a resistor, this is

$$p_R = i_R v_R = \frac{v_R^2}{R} \geq 0 \tag{5.23}$$

Fig. 5.8. Schematic of a capacitor.

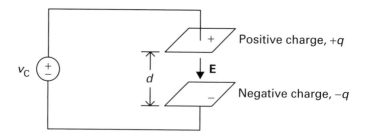

Positive charge, $+q$

Negative charge, $-q$

which implies that a resistor always dissipates power. It is reminiscent of the damper, which also dissipates power. Recall that the damper has the following constitutive relationship:

$$F_d = c\frac{d}{dt}(x_a - x_b) = c(u_a - u_b), \tag{5.24}$$

where $x_a = x_{near}$, $x_b = x_{far}$ and $u = \dot{x}$ represents the velocity. Because both the resistor and damper dissipate energy, we say that the resistor and the damper are analogous elements. Comparing Eqs. (5.22) and (5.24), we note that current plays the role of force (both flow through an element unimpeded) and that voltage plays the role of velocity (both are defined at a node).

5.1.3 Capacitors

Our last circuit element is the *capacitor*. A capacitor can be built from two parallel plates. If a positive voltage v_C is applied across the plates, positive charge $+q$ will accumulate on the top plate and negative charge $-q$ will accumulate on the bottom, as shown in Figure 5.8. Assume the plates have area S and are separated by a distance d. Let the *permittivity* or dielectric constant of the material between the plates be denoted by ϵ. Then Gauss' Law states that the electric field between the plates is

$$E = \frac{q}{\epsilon S}. \tag{5.25}$$

The voltage drop across the plates can be expressed as

$$v_C = \int_{top}^{bottom} \mathbf{E} \cdot \mathbf{dl} = Ed = \frac{qd}{\epsilon S}. \tag{5.26}$$

In other words, the voltage drop across the capacitor is proportional to the charge q. We can define this constant of proportionality as the capacitance, C. Capacitance has units of coulombs/volt or *farads*, and it is given by

$$C = \frac{q}{v_C} = \frac{\epsilon S}{d}. \tag{5.27}$$

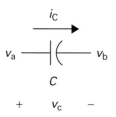

Fig. 5.9. Circuit symbol for a capacitor.

Figure 5.9 shows the circuit symbol for a capacitor. Recalling that current is the derivative of charge with respect to time, we can write the constitutive relationship for the capacitor as follows:

$$i_C = \frac{dq}{dt} = C\frac{dv_C}{dt} = C\frac{d}{dt}(v_a - v_b), \tag{5.28}$$

where v_C represents the voltage drop across the capacitor, and v_a and v_b are the node voltages at the positive and negative terminals, respectively.

Equation (5.28) is reminiscent of the constitutive relationship for a mass element, which has the form

$$F_m = m\frac{d^2}{dt^2}(x_a - x_b) = m\frac{d}{dt}(u_a - u_b), \tag{5.29}$$

where $x_a = x_{near}$ and $x_b = x_{fixed}$. By inspection, we see that the capacitance in the electrical system plays the role of mass in the mechanical system.

The power dissipated by the capacitor is

$$p_C = i_C v_C = C v_C \frac{dv_C}{dt}. \tag{5.30}$$

This quantity may be positive or negative, depending on whether energy is being added to the capacitor or drawn from the capacitor. The total energy stored in the capacitor is the integral of the power delivered since the start of time, t_0:

$$W = \int_{t_0}^{t_1} p_C(\tau)d\tau = \frac{1}{2}C\left[v_C^2(t_1) - v_C^2(t_0)\right]. \tag{5.31}$$

5.2 Summary

To summarize the analogy[3] between electrical and mechanical systems, we note that current i plays the role of force F, and that the voltage at a node v plays the role of the velocity at an endpoint $u = \dot{x}$. The constitutive relationships for the elements are summarized in Table 5.1. Finally, we note that in order to put the mass on equal footing with a spring and a damper, the mass element assumes the motion is measured relative to a stationary node, that is $x_b = x_{fixed}$ (with the velocity $u_b = u_{fixed} = 0$). Thus, while there

---

[3] See *Fundamentals of Modeling and Analyzing Engineering Systems* by P. D. Cha, J. J. Rosenberg and C. L. Dym, Cambridge University Press, Cambridge, 2000 for a detailed discussion on a unifying framework for modeling mechanical, electrical, hydraulic and thermal systems.

Table 5.1. Analogy between electrical and mechanical elements

Electrical	Mechanical	Relationship
$i_L = \dfrac{1}{L}\displaystyle\int_{-\infty}^{t}(v_a - v_b)\mathrm{d}\tau$	$F_s = k\displaystyle\int_{-\infty}^{t}(u_a - u_b)\mathrm{d}\tau$	$\dfrac{1}{L} \Longleftrightarrow k$
$i_R = \dfrac{1}{R}(v_a - v_b)$	$F_d = c(u_a - u_b)$	$\dfrac{1}{R} \Longleftrightarrow c$
$i_C = C\dfrac{\mathrm{d}}{\mathrm{d}t}(v_a - v_b)$	$F_m = m\dfrac{\mathrm{d}}{\mathrm{d}t}(u_a - u_b)$	$C \Longleftrightarrow m$

Fig. 5.10. Voltage and current sources.

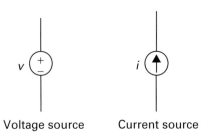

Voltage source   Current source

are many similarities between mechanical and electrical systems, there also exist some major differences. Specifically, the mass is a one-node element because only one variable is required to describe its position (assuming the mass is rigid, and that the inertia force of the mass depends only on the second derivative of its position). The capacitor, on the other hand, is a two-node element, because the current that flows through it depends on the time derivative of the voltage difference between the two nodes. Thus, while mass and capacitor are "analogous," masses can only be combined in parallel, whereas capacitors can combine in either parallel or series. Note also that C is capacitance in electrical systems, and c is the damping coefficient in mechanical systems. They are not analogous.

5.3 Inputs to electrical systems

In a mechanical system, we had inputs of displacement and force. In an electrical system, our inputs will be voltage and current sources, shown in Figure 5.10. A voltage source maintains a voltage $v(t)$ across its terminals independent of the external circuit. It delivers however much current is necessary to hold this voltage. A current source maintains a current $i(t)$ through the source, independent of the external circuit. It establishes whatever voltage across itself is necessary to hold this current.

5.4 Governing equations

To find the governing equation for an electrical circuit, we need to solve for the current or the voltage in the circuit, depending on what is being sought. The governing equations for electrical systems can be obtained by applying Kirchhoff's Current Law or Kirchhoff's Voltage Law.

5.4.1 Kirchoff's current and voltage laws

Because of conservation of charge, charge cannot accumulate at a node in a circuit. This can be written as Kirchhoff's Current Law (KCL) as follows:

$$\sum_{\text{out of a node}} i = \sum_{\text{into a node}} i = \frac{dq}{dt} = 0. \tag{5.32}$$

Kirchhoff's Current Law states that the sum of all the currents flowing *out of a node* or *into a node* is zero. When we use the former, that is "out of a node," the sign convention on the currents is such that currents flowing out of a node are positive, and currents flowing into a node are negative. Theoretically, either "out of" or "into" a node can be used when we sum the currents, as long as we are careful with the signs associated with each current. However, if we use "out of a node," then when we replace the various currents by their constitutive equations that are functions of voltage differences, we find that $(v_a - v_b) = (v_{\text{near}} - v_{\text{far}})$, where v_{near} is the voltage at the node at which the currents are being summed, and v_{far} represents the voltage of the node at the far end of the element. Note the analogy to the relative displacement $(x_{\text{near}} - x_{\text{far}})$ term that appears in all the constitutive equations for mechanical elements.

Kirchhoff's Voltage Law (KVL) states that the sum of voltage drops or differences across the elements in a closed loop is zero. We can write this mathematically as

$$\sum_{\text{around a loop}} (v_a - v_b) = 0. \tag{5.33}$$

Fig. 5.11. Voltage drop
across elements in series.

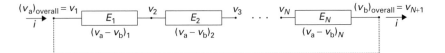

Fig. 5.11. Voltage drop across elements in series.

If we have several elements in series, say $E_1, E_2, \ldots, E_N$, as shown in Figure 5.11, the voltage across the series combination is the sum of the voltages across each element. We can express this using KVL:

$$(v_a - v_b)_{\text{overall}} = \sum_{i=1}^{N} (v_a - v_b)_i, \tag{5.34}$$

where $(v_a - v_b)_{\text{overall}}$ represents the overall voltage drop across the series combination, and $(v_a - v_b)_i$ denotes the voltage drop across the ith element E_i. In expanded form, Eq. (5.34) becomes

$$v_1 - v_{N+1} = (v_1 - v_2) + (v_2 - v_3) + \cdots + (v_N - v_{N+1}), \tag{5.35}$$

where v_i corresponds to the voltage at node i. If the elements are connected into a loop (shown with the dashed line), $v_{N+1} = v_1$, then the sum of the voltages becomes zero, proving KVL.

Finally, as a general rule of thumb, we apply KVL when solving for the current (because we can readily express the voltage drops across the elements in terms the currents), and we use KCL when solving for the voltage (because we can write the currents through the elements in terms of the node voltages). This is merely a guide, and it is not always true. In more complicated systems, we may need to utilize both KCL and KVL to solve for either the current or the voltage.

5.4.2 Examples

We now derive the governing equations for various electrical circuits.

Example 1: Figure 5.12 shows an inductor and capacitor in parallel driven by a current source $i_{\text{in}}(t)$. Determine the voltage drop $v_{\text{out}}(t)$ across the inductor.

Solution: We begin by identifying our nodes. Let us call the ground node 0 and the top node 1. The second step is to define the direction of current flow (like defining a coordinate system in a mechanical system). The choice is completely arbitrary, but once it is made it should be obeyed throughout

Fig. 5.12. Parallel LC circuit with a current input.

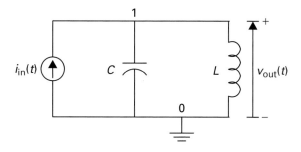

Fig. 5.13. Direction of currents for the circuit of Figure 5.12.

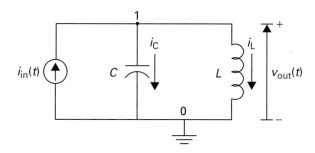

the problem. Let us define positive currents as flowing "out of" node 1. Thus, the current through the inductor and the capacitor are considered positive, and the current from the current source will be negative because it flows "into" node 1 (see Figure 5.13). The final step is to apply Kirchhoff's Current Law (KCL) at node 1, which then yields the governing equation for the voltage drop across the circuit. Applying KCL at node 1 and summing all the currents into and out of the node, we obtain

$$\sum_{\text{out of node 1}} i = -i_{\text{in}}(t) + i_C + i_L = 0, \tag{5.36}$$

where we consider the currents out of node 1 positive, and the current into node 1 negative. Using the constitutive relationships for the inductor and the capacitor, we can rewrite Eq. (5.36) in terms of the node voltages as follows:

$$-i_{\text{in}}(t) + C\frac{\mathrm{d}}{\mathrm{d}t}(v_1 - v_0) + \frac{1}{L}\int_{-\infty}^{t}(v_1 - v_0)\mathrm{d}\tau = 0. \tag{5.37}$$

Node 0 is attached to the ground, so $v_0 = 0$. By inspection, we note that v_{out} is the voltage drop across the capacitor (it is also the voltage drop across the inductor). Thus, $v_{\text{out}} = v_1 - v_0 = v_1$. Substituting $v_1 = v_{\text{out}}$ and differentiating the resulting equation to eliminate the integral, we obtain a

Fig. 5.14. A spring–mass system with a force input.

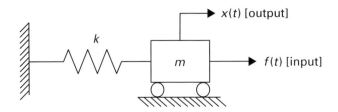

differential equation that relates the output voltage to the input current as follows:

$$-\frac{d}{dt}i_{in}(t) + C\frac{d^2}{dt^2}v_{out} + \frac{1}{L}v_{out} = 0. \qquad (5.38)$$

Upon rearranging, we find the governing equation for the desired voltage drop across the capacitor or inductor:

$$C\frac{d^2}{dt^2}v_{out} + \frac{1}{L}v_{out} = \frac{d}{dt}i_{in}(t). \qquad (5.39)$$

Note that this electrical circuit is just like the spring–mass system with an applied force shown in Figure 5.14, whose equation of motion is given by

$$m\frac{d^2x}{dt^2} + kx = f(t). \qquad (5.40)$$

We see the analogy between the electrical and mechanical systems. Capacitance plays the role of mass, and the reciprocal of the inductance is analogous to the spring constant. Comparing Eqs. (5.39) and (5.40), we notice a minor discrepancy. Earlier, we had related the voltage to the velocity, and the current to the force. The analogy we made earlier appears to conflict with the results of Eqs. (5.39) and (5.40). However, if we differentiate both sides of Eq. (5.40) and introduce $u = \dot{x}$ we obtain

$$m\frac{d^2}{dt^2}u + ku = \frac{d}{dt}f(t). \qquad (5.41)$$

Comparing Eqs. (5.39) and (5.41), we note that our earlier analogy still remains valid.

Example 2: Consider the system in Figure 5.15, which consists of a voltage and a current source along three circuit elements. Determine the governing equation for the output voltage $v_{out}(t)$ (or the voltage drop across the capacitor).

Solution: We begin by numbering the nodes, with the ground node being node 0. The next step is to assign the direction of current. Again, let us

Fig. 5.15. Electrical circuit with current and voltage inputs.

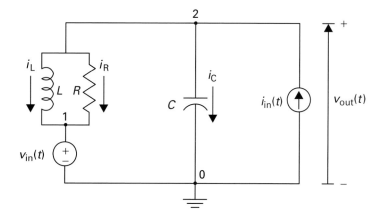

define positive current to be out of a node. Applying KCL at node 2, we find

$$\sum_{\text{out of node 2}} i = -i_{\text{in}}(t) + i_C + i_L + i_R = 0. \tag{5.42}$$

We can rewrite Eq. (5.42) in terms of the node voltages using the constitutive relations for each element as follows:

$$-i_{\text{in}}(t) + C\frac{d}{dt}(v_2 - v_0) + \frac{1}{L}\int_{-\infty}^{t}(v_2 - v_1)d\tau + \frac{1}{R}(v_2 - v_1) = 0. \tag{5.43}$$

Like before, $v_0 = 0$, and $v_{\text{out}} = v_2 - v_0 = v_2$. Because node 1 is connected to the voltage source as shown, $v_1 = v_{\text{in}}$. Substituting

$$v_2 = v_{\text{out}} \quad \text{and} \quad v_1 = v_{\text{in}} \tag{5.44}$$

and taking the derivative with respect to time to eliminate the integral, we obtain

$$C\frac{d^2}{dt^2}v_{\text{out}} + \frac{1}{R}v_{\text{out}} + \frac{1}{L}v_{\text{out}} = \frac{1}{R}\frac{d}{dt}v_{\text{in}} + \frac{1}{L}v_{\text{in}} + \frac{d}{dt}i_{\text{in}}. \tag{5.45}$$

This is analogous to the base excitation system shown in Figure 5.16, where an input displacement is applied at the base and an input force is applied to the mass. The governing equation for the displacement of the mass is given by (see Chapter 4 for detailed derivation)

$$m\frac{d^2}{dt^2}y + c\frac{d}{dt}y + ky = c\frac{d}{dt}y_{\text{in}} + ky_{\text{in}} + f. \tag{5.46}$$

Differentiating Eq. (5.46) and introducing

$$u_{\text{in}} = \frac{dy_{\text{in}}}{dt} \quad \text{and} \quad u = \frac{dy}{dt} \tag{5.47}$$

Fig. 5.16. A base excitation system.

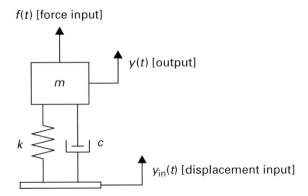

we get

$$m\frac{d^2}{dt^2}u + c\frac{d}{dt}u + ku = c\frac{d}{dt}u_{in} + ku_{in} + \frac{d}{dt}f. \tag{5.48}$$

Comparing Eqs. (5.45) and (5.48), we note again the similarities between the two systems. The spring and damper in parallel driven from below with an input displacement play the role of the inductor and resistor in parallel driven with the voltage source. The mass plays the role of the capacitor, and the external force acts as the current source.

The ability to construct an analogous system is important because it allows us to analyze a system that could otherwise be complicated to study. The base excitation system of Figure 5.16 can be used to model an automobile, where the spring–damper system models the suspension and the mass models the automobile. Suppose we wish to study the effects of different suspensions on the automobile, and how different road profiles and external loads affect the displacement of the vehicle. Instead of constructing a mechanical prototype of the automobile, we could use its electrical analog to simulate how the car would respond to different input forces and displacements (that is, road profiles) by simply changing the current and voltage sources of its analogous circuit. By changing the inductor and resistor values, we could also easily study the effects of the spring and damper, that is the suspension, on the response of the system.

5.5 Parallel combination

Electrical elements are in parallel if they share the same voltage difference. For example, Figure 5.17 shows N resistors in parallel between nodes 1

Fig. 5.17. N resistors in
parallel.

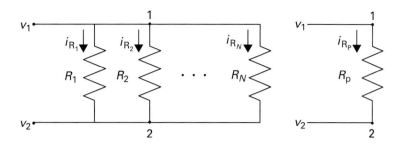

and 2. Because all the elements share the same nodes, the voltage drop across each resistor is $(v_1 - v_2)$. The equivalent circuit is a single resistor of resistance R_p if the currents flowing through the two circuits are equal for the same voltage drop. Applying KCL at node 1, we get

$$i_{\text{overall}} = i_{R_1} + i_{R_2} + \cdots + i_{R_N}$$
$$i_{\text{overall}} = i_{R_p}.$$
(5.49)

Using the constitutive relationship for a resistor, we rewrite Eq. (5.49) as

$$i_{\text{overall}} = \frac{1}{R_1}(v_1 - v_2) + \frac{1}{R_2}(v_1 - v_2) + \cdots + \frac{1}{R_N}(v_1 - v_2)$$
$$i_{\text{overall}} = \frac{1}{R_p}(v_1 - v_2).$$
(5.50)

Equating the right-hand sides of Eq. (5.50), we obtain

$$\frac{1}{R_p} = \sum_{i=1}^{N} \frac{1}{R_i}.$$
(5.51)

Thus, for N resistors in parallel, we can replace the resistors with an equivalent resistor of resistance R_p.

We had mentioned earlier that a resistor and a damper are analogous elements: they both dissipate energy. However, we note that resistors in parallel appear to add differently from dampers in parallel (see Eq. (4.40)). For a damper, the constitutive equation relating the force and the relative velocity is given by

$$F_d = c(u_{\text{near}} - u_{\text{far}}).$$
(5.52)

When N dampers are in parallel, we can replace them with an equivalent damper of damping coefficient

$$c_p = \sum_{i=1}^{N} c_i,$$
(5.53)

where c_i represents the damping coefficient of the ith damper in the parallel combination. For a resistor, the constitutive equation relating the current and the voltage difference is given by

$$i_R = \frac{1}{R}(v_a - v_b). \qquad (5.54)$$

In terms of conductance G, where $G = 1/R$, the constitutive equation for a resistor becomes

$$i_R = G(v_a - v_b) \qquad (5.55)$$

and Eq. (5.51) becomes

$$G_p = \sum_{i=1}^{N} G_i. \qquad (5.56)$$

Thus, comparing Eqs. (5.53) and (5.56), we see that in fact dampers and resistors do indeed combine exactly the same, as long as we equate the through variable (force or current) to a proportionality constant times the difference between the endpoint variables (velocity or voltage).

The same argument can be used to show that for N capacitors and inductors in parallel, we may replace the capacitors and inductors with an equivalent capacitor and inductor of capacitance and inductance

$$C_p = \sum_{i=1}^{N} C_i \qquad (5.57)$$

$$\frac{1}{L_p} = \sum_{i=1}^{N} \frac{1}{L_i}. \qquad (5.58)$$

Again, note that some of the terms appear to add directly while others appear to add in reciprocals, depending on how the term appears in the original constitutive relationship. For the inductor, because the current–voltage is related by the inverse of the inductance, $1/L$, to obtain the equivalent inductance for N inductors in parallel we have to add them as reciprocals. These formulae are analogous to our equations for parallel combinations of masses and springs:

$$m_p = \sum_{i=1}^{N} m_i \qquad (5.59)$$

$$k_p = \sum_{i=1}^{N} k_i. \qquad (5.60)$$

Fig. 5.18. N resistors in series.

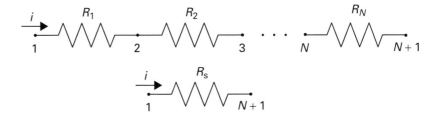

5.6 Series combination

Electrical elements connected in series share a common current. This is a direct consequence of Kirchoff's Current Law. Figure 5.18 shows N resistors in series. We can replace these resistors with a single equivalent resistor of resistance R_s that has the same current flow and the same voltage drop. Applying KVL, we have

$$v_{overall} = \sum_{i=1}^{N} v_{R_i},\qquad(5.61)$$

where $v_{overall}$ denotes the overall voltage drop across the equivalent resistor, and v_{R_i} represents the voltage drop across R_i. In terms of node voltages, Eq. (5.61) becomes

$$v_1 - v_{N+1} = (v_1 - v_2) + (v_2 - v_3) + \cdots + (v_N - v_{N+1}).\qquad(5.62)$$

Given that the current through all the resistors and their equivalent is i (because the resistors are in series), we apply the constitutive relationship for the resistors to rewrite Eq. (5.62) as

$$i R_s = i R_1 + i R_2 + \cdots + i R_N.\qquad(5.63)$$

By inspection, we note that the resistance for the equivalent resistor is simply

$$R_s = \sum_{i=1}^{N} R_i.\qquad(5.64)$$

Using a similar derivation, we can obtain the equivalent capacitance and inductance for capacitors and inductors in series as follows:

$$\frac{1}{C_s} = \sum_{i=1}^{N} \frac{1}{C_i}\qquad(5.65)$$

and

$$L_s = \sum_{i=1}^{N} L_i, \tag{5.66}$$

where C_i and L_i correspond to the capacitance and inductance of the ith capacitor and inductor in the series combinations, respectively.

Resistors and inductors in series are analogous to dampers and springs in series. They are combined exactly the same, as long as we keep track of the proportionality constant that appears in the constitutive equations. From our earlier derivations, recall for series combinations of dampers and springs, we have

$$\frac{1}{c_s} = \sum_{i=1}^{N} \frac{1}{c_i} \tag{5.67}$$

and

$$\frac{1}{k_s} = \sum_{i=1}^{N} \frac{1}{k_i}, \tag{5.68}$$

where Eq. (5.67) is the analog of Eq. (5.64), and Eq. (5.68) is the analog of Eq. (5.66). Note that some of the terms appear to add directly (the resistors) while others appear in reciprocals (the capacitors and inductors), depending on how the term appears in the original constitutive relationship.

Finally, in a mechanical system, masses are assumed to be rigid and referenced to a common fixed point. Thus, they are one-node elements and can only be combined in parallel because they are displaced by the same amount. Their electrical analogs, the capacitors, however, may be combined in series because they are two-node elements.

All of the parallel and series combination rules that we have previously defined are strictly valid for like-elements or elements of the same type, because we implicitly assumed that all the elements in the combination have the same constitutive relations. Thus, we can only combine springs or we can only combine capacitors. The concepts of parallel and series combinations, however, can be extended to elements of different type, as long as we obey the definitions of parallel and series. In Chapter 7, we will be able to combine different mechanical elements or different electrical elements using the concept of *impedance*.

5.7 Division of current in parallel combination

When multiple resistors act in parallel, each resistor has the same voltage drop but may carry a different fraction of the total current. Consider the system of Figure 5.17. From Eq. (5.50), we find

$$
\begin{aligned}
i_{R_j} &= \frac{v_1 - v_2}{R_i} \\
i_{\text{overall}} &= \frac{v_1 - v_2}{R_p}.
\end{aligned}
\tag{5.69}
$$

Taking the ratio of the current through the jth resistor to the total current, we get

$$
\frac{i_{R_j}}{i_{\text{overall}}} = \frac{\dfrac{1}{R_j}}{\dfrac{1}{R_p}} = \frac{\dfrac{1}{R_j}}{\displaystyle\sum_{i=1}^{N} \dfrac{1}{R_i}}.
\tag{5.70}
$$

This is called the *current divider equation* or the *current division rule*. Similar results can be extended to N capacitors in parallel and N inductors in parallel as follows:

$$
\frac{i_{C_j}}{i_{\text{overall}}} = \frac{C_j}{C_p} = \frac{C_j}{\displaystyle\sum_{i=1}^{N} C_i}
\tag{5.71}
$$

and

$$
\frac{i_{L_j}}{i_{\text{overall}}} = \frac{\dfrac{1}{L_j}}{\dfrac{1}{L_p}} = \frac{\dfrac{1}{L_j}}{\displaystyle\sum_{i=1}^{N} \dfrac{1}{L_i}}.
\tag{5.72}
$$

5.8 Division of voltage in series combination

When multiple resistors act in series, the same current will flow through each resistor but a different voltage drop may appear across each of the elements.

Consider the system of Figure 5.18. From Eqs. (5.61) and (5.63), we have

$$v_{R_j} = iR_j$$

$$v_{\text{overall}} = iR_s. \tag{5.73}$$

Manipulating the above equations, we find the voltage drop across a particular resistor R_j to be given by

$$v_{R_j} = iR_j = \frac{v_{\text{overall}}}{R_s} R_j = \frac{R_j}{\sum_{i=1}^{N} R_i} v_{\text{overall}}. \tag{5.74}$$

This is known as the *voltage divider equation* or the *voltage divider rule*. In words, Eq. (5.74) states that for N resistors in series, the ratio of the voltage drop across any resistor to the total voltage drop is simply the ratio of its resistance to the total resistance. Similar results can be extended to N capacitors in series and N inductors in series as follows:

$$\frac{d}{dt}(v_{C_j}) = \frac{\frac{1}{C_j}}{\sum_{i=1}^{N} \frac{1}{C_i}} \frac{d}{dt}(v_{\text{overall}}) \tag{5.75}$$

and

$$\int_{-\infty}^{t} v_{L_j} d\tau = \frac{L_j}{\sum_{i=1}^{N} L_i} \left(\int_{-\infty}^{t} v_{\text{overall}} d\tau \right), \tag{5.76}$$

where v_{C_j} and v_{L_j} denote the voltage drop across the jth capacitor and inductor, respectively.

5.9 Problems

5-1 Given R, L, C, and the input current $i(t)$ in Figure 5.19:
  (a) Find the governing equation for the voltage drop across the inductor, $v_L(t)$;
  (b) Find the governing equation for the voltage drop across the capacitor, $v_C(t)$.

5-2 Consider the circuit shown in Figure 5.20, where the inputs include the voltage source, $v_{\text{in}}(t)$, and the current source, $i_{\text{in}}(t)$.

Fig. 5.19. Figure for
Problem 5-1.

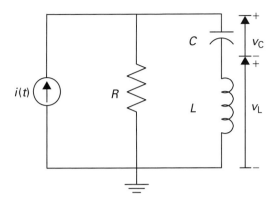

Fig. 5.20. Figure for
Problem 5-2.

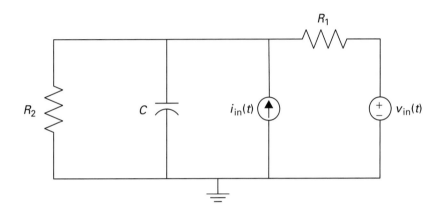

(a) Derive the governing equation for the voltage drop across the capacitor.

(b) Determine the governing equation for the current through the resistor R_1.

5-3 Consider the circuit shown in Figure 5.21 consisting of resistors. Given $v_{in} = v_1 - v_2$ and the resistor values, write the following current or voltage in terms of v_{in} and R. The current direction is assigned in the figure.

(a) The output voltage, v_{out}.

(b) The total current, i_1.

(c) The current, i_2, through the resistor with resistance $2R$.

5-4 Consider the circuit shown in Figure 5.22.

(a) Derive the governing equation for the voltage drop across the resistor, $v_R(t)$.

Fig. 5.21. Figure for
Problem 5-3.

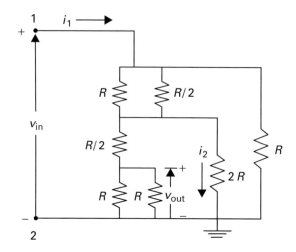

Fig. 5.22. Figure for
Problem 5-4.

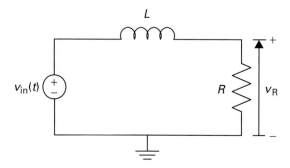

   (b) Derive the governing equation for the voltage drop across the inductor, $v_L(t)$.

5-5 Consider the circuit shown in Figure 5.23.
   (a) Find the total current $i(t)$ in terms of $v_{in}(t)$.
   (b) Find the voltage drop across R_2.
   (c) Find the current through R_1 in terms of $v_{in}(t)$.
   (d) It is desired to have a voltage drop across the 7 Ω resistor of 23 V.
   What $v_{in}(t)$ should be applied?

5-6 Consider the circuit shown in Figure 5.24, where $R_1 = 1\,\Omega$, $R_2 = 1\,\Omega$,
   $R_3 = 3\,\Omega$, $R_4 = 1.5\,\Omega$, $R_5 = 1\,\Omega$, $R_6 = \pi\,\Omega$, $v_{in}(t) = 3$ V and $i_{in}(t) =$
   2 A.
   (a) Find the voltage drop across R_5.

Fig. 5.23. Figure for
Problem 5-5.

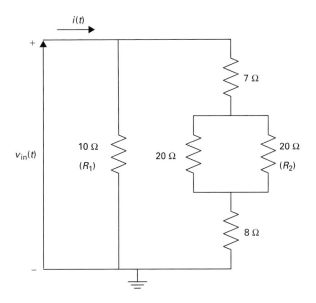

Fig. 5.24. Figure for
Problem 5-6.

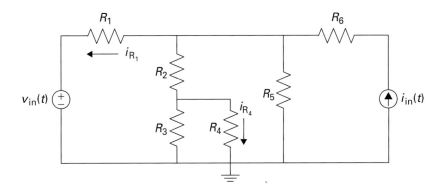

(b) Find the current through R_4 (in the direction shown).

(c) Find the voltage drop across R_2.

(d) Find the current through R_1 (in the direction shown).

(e) Find the voltage drop across the current source.

5-7 Consider the circuit in Figure 5.20. Determine the governing equation for the voltage drop across the capacitor using the principle of superposition. **Hint**: what does it mean to set the current source to zero and what does it mean to set the voltage source to zero?

5-8 Consider the circuit in Figure 5.24. Determine the voltage drop across R_5 using the principle of superposition. **Hint**: what does it mean to set the current source to zero and what does it mean to set the voltage source to zero?

6

Solution to differential equations

We have learned to derive the governing differential equations for mechanical and electrical systems. In a mechanical system, the input may be a specified force or a specified displacement. In an electrical system, the input may be a known voltage source or a known current source. In analyzing the behavior of systems, we are primarily interested in how the systems respond to certain inputs. This requires us to solve the governing differential equations.

In general, an nth order ordinary differential equation may be written as

$$a_n(t)\frac{d^n y}{dt^n} + a_{n-1}(t)\frac{d^{n-1} y}{dt^{n-1}} + \cdots + a_1(t)\frac{dy}{dt} + a_0(t)y = f(t). \qquad (6.1)$$

The above differential equation is said to be *linear* because y and its derivatives are raised to the first power only and they appear by themselves, that is they are not multiplied by other functions of y. Linear equations are convenient because we can apply the *principle of superposition*. If the coefficients $a_i(t)$ are constants, the system is said to be *time-invariant* or *autonomous*, and the governing equation is said to have constant coefficients. If $f(t) = 0$, the differential equation is said to be *homogeneous*. The ensuing response $y(t)$ is called the *free* or *natural* response, and it is caused by the initial conditions that are imparted on the system only, and not by any inputs. For an nth order differential equation, n initial conditions must be specified in order to uniquely determine its free response. If $f(t) \neq 0$, the differential equation is said to be *nonhomogeneous*, and the resulting response $y(t)$ is called the *forced* response. In general, the forced response depends on both the initial conditions and external excitations or inputs.

All of the mechanical and electrical elements are linear and time-invariant (LTI), so they produce LTI differential equations. Therefore, we will be exclusively interested in solving LTI differential equations in this text. We will first solve for the response of first- and second-order LTI ordinary differential equations to step inputs.

6.1 First-order ordinary differential equations

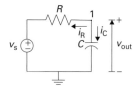

Fig. 6.1. A series *RC* circuit.

Figure 6.1 shows a resistor and capacitor in series with a voltage source. The governing equation for the voltage drop across the capacitor is computed by applying KCL at node 1. Summing the currents out of node 1, we obtain

$$i_R + i_C = 0. \tag{6.2}$$

Using the constitutive relationships for a resistor and a capacitor, we find

$$\frac{v_{out} - v_s}{R} + C\frac{d}{dt}(v_{out} - 0) = 0. \tag{6.3}$$

Upon rearranging, we have

$$C\frac{dv_{out}}{dt} + \frac{1}{R}v_{out} = \frac{1}{R}v_s. \tag{6.4}$$

Equation (6.4) constitutes the governing equation for the circuit of Figure 6.1. The output of the system is the voltage drop across the capacitor v_{out}, and the input is the voltage source v_s. For the system of Figure 6.1, the governing equation consists of a constant coefficient, first-order ordinary differential equation.

In another example, consider a rotating disk attached to a rotational damper, as shown in Figure 6.2. The disk is driven by twisting the right end of the torsional damper with an input angular displacement of $\phi_{in}(t)$, and by an external torque $T_{applied}$ applied to the disk. Summing the torques, we find

$$\sum T = I\ddot{\theta} = T_{applied} - c_t(\dot{\theta} - \dot{\phi}_{in}) \tag{6.5}$$

$$I\ddot{\theta} + c_t\dot{\theta} = T_{applied} + c_t\dot{\phi}_{in}. \tag{6.6}$$

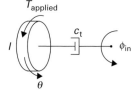

Fig. 6.2. Mechanical system consisting of a disk attached to a rotational damper.

This is a second-order, constant coefficient ordinary differential equation in θ, the angular displacement of the rotating disk. However, if we substitute the angular velocity $\omega = \dot{\theta}$, we can rewrite it as a first-order equation as follows:

$$I\dot{\omega} + c_t\omega = T_{applied} + c_t\dot{\phi}_{in}, \tag{6.7}$$

where the output is now the angular velocity ω, and the inputs are the applied torque T_{applied} and the input angular velocity $\dot{\phi}_{\text{in}}$.

Any differential equation can be manipulated into its *canonical*, *normal*, or *standard* form by making the coefficient of the leading derivative 1. For a first-order differential equation, its canonical form is given by

$$\frac{\mathrm{d}y}{\mathrm{d}t} + \frac{1}{\tau}y = \alpha f, \tag{6.8}$$

where y is the output and f depends on the input of the system, and α is typically a constant that depends on the system parameters. In this form, τ is called the *time constant* of the system, and it completely characterizes the response and behavior of the system. For Eq. (6.8) to be dimensionally correct, the time constant τ must have dimensions of time. The time constant gives a convenient means to describe the rate of exponential decay in the system. For the RC circuit of Figure 6.1, $y(t) = v_{\text{out}}$, $\tau = RC$, $\alpha = 1/\tau$ and $f = v_s$. For the rotating disk of Figure 6.2, $y(t) = \omega$, $\tau = I/c_t$, $\alpha = 1/I$ and $f = T_{\text{applied}} + c_t \dot{\phi}_{\text{in}}$. Note that the time constant τ depends only on the system parameters, and not on the inputs. It is convenient to write any first-order differential equation in its canonical form because we can easily extract τ. Moreover, we only need to solve the differential equation in canonical form just once, and express the solution in terms of τ. We can then find the response of any first-order system by substituting the appropriate value of τ. Expressing any first-order differential equation in its canonical form also allows us to compare the responses of seemingly different systems, and to make analogies between disparate physical systems. Finally, we note that in order to determine the response of a first-order system, we only need to specify one initial condition, $y(0) = y_0$.

6.2 Second-order ordinary differential equations

Figure 6.3 shows a mechanical system whose governing equation is given by a constant coefficient, second-order ordinary differential equation. Sketching the free body diagram and applying Newton's Second Law, we find that the equation of motion is given by

$$m\ddot{y} + ky = f(t). \tag{6.9}$$

Fig. 6.3. A spring–mass
system.

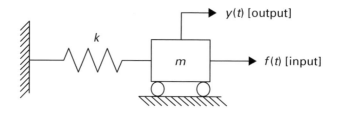

Fig. 6.4. A parallel RLC
circuit.

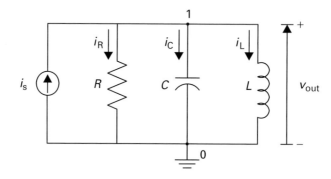

Figure 6.4 shows a RLC circuit driven by a current source. Applying KCL at node 1, we obtain

$$\sum_{\text{out of node 1}} i = -i_\text{s} + i_\text{L} + i_\text{C} + i_\text{R} = 0. \tag{6.10}$$

Using the constitutive relationships for the electrical elements, we can rewrite Eq. (6.10) in terms of the node voltages as follows:

$$C\frac{\text{d}}{\text{d}t}(v_1 - v_0) + \frac{1}{R}(v_1 - v_0) + \frac{1}{L}\int_{-\infty}^{t} (v_1 - v_0)\text{d}\tau = i_\text{s}. \tag{6.11}$$

Node 0 is connected to the ground, so $v_0 = 0$. By inspection, we note that

$$v_\text{out} = v_1 - v_0 = v_1. \tag{6.12}$$

Thus, taking the derivative of Eq. (6.11) and using Eq. (6.12), the governing equation becomes

$$C\frac{\text{d}^2 v_\text{out}}{\text{d}t^2} + \frac{1}{R}\frac{\text{d}v_\text{out}}{\text{d}t} + \frac{1}{L}v_\text{out} = \frac{\text{d}i_\text{s}}{\text{d}t}. \tag{6.13}$$

Any constant coefficient, second-order differential equation can be placed in canonical form by making the leading term 1. Thus, the canonical expression for any second-order differential equation is given by

$$\frac{d^2 y}{dt^2} + 2\zeta \omega_n \frac{dy}{dt} + \omega_n^2 y = \alpha f(t), \tag{6.14}$$

where y is the output and f depends on the input of the system, and α is generally a constant that depends on the system parameters. In the second-order canonical form given by Eq. (6.14), ζ is known as the *damping factor*, and ω_n is the *undamped natural frequency*, that is the frequency at which the system would oscillate during free response if there were no damping. Note that in order for Eq. (6.14) to be dimensionally correct, ζ must be unitless and ω_n must have units of 1/time or rad/s. When $\zeta = 0$, the system is said to be *undamped*, and no energy is dissipated when the system undergoes free response. When $\zeta > 0$, the system is known as *damped* and energy is lost during free response. Because damping is an energy dissipating mechanism, for real systems, $\zeta > 0$. These two canonical parameters, ζ and ω_n, completely characterize the response of the system, as τ did for a first-order system. They depend only on the system parameters, and are independent of the inputs. Finally, we note that in order to determine the response of a second-order system, we need to specify two initial conditions, $y(0) = y_0$ and $\dot{y}(0) = \dot{y}_0$.

In the case of the mechanical system of Figure 6.3, the governing equation in its canonical form is given by

$$\ddot{y} + \frac{k}{m} y = \frac{f(t)}{m}. \tag{6.15}$$

Comparing Eqs. (6.14) and (6.15), we see immediately that $\alpha = \frac{1}{m}$, $\zeta = 0$ (the system is undamped) and that the undamped natural frequency is given by

$$\omega_n = \sqrt{\frac{k}{m}}. \tag{6.16}$$

Note that the natural frequency is immediately apparent in the canonical form without having to solve the differential equation.

For the electrical system of Figure 6.4, its governing equation in the canonical form is as follows:

$$\frac{d^2 v_{\text{out}}}{dt^2} + \frac{1}{RC} \frac{dv_{\text{out}}}{dt} + \frac{1}{LC} v_{\text{out}} = \frac{1}{C} \frac{di_s}{dt}. \tag{6.17}$$

Comparing with Eq. (6.14), we immediately find that $\alpha = \frac{1}{C}$, $f(t) = \frac{di_s}{dt}$, and the expressions for the undamped natural frequency and the damping factor are given by

$$\omega_n = \sqrt{\frac{1}{LC}}$$

$$\zeta = \frac{1}{2\omega_n RC} = \frac{1}{2R\sqrt{\dfrac{C}{L}}}. \tag{6.18}$$

Having formulated the governing equations, we would now like to find the response of a system to any arbitrary input. The complete response is the sum of the homogeneous solution and the particular solution. The homogeneous solution is obtained by setting $f(t) = 0$, and it is independent of the inputs. The particular solution is obtained by satisfying the differential equation, and it depends on $f(t)$. We will be especially interested in the response of a system to a unit step input of the form

$$f(t) = u(t), \tag{6.19}$$

where $u(t)$ is called the *unit step function* or *Heavyside* function, defined as

$$u(t) = \begin{cases} 0 & t < 0 \\ 1 & t \geq 0. \end{cases} \tag{6.20}$$

The unit step function is often used to describe engineering applications that are either "on" or "off." It can also be used to model physical events that begin suddenly and thereafter are maintained. The response of a system to a unit step input is known as the *step response* or the *transient response*. The step input allows us to analyze the behavior of the system during the transient or the starting portion of the response.

We will also be interested in the response of the system to an exponential input of the form

$$f(t) = Fe^{j\omega t}, \tag{6.21}$$

where F denotes the excitation, forcing or input amplitude, and ω represents the excitation or forcing or input frequency. The response of a system to an exponential input is known as the *frequency response*. The frequency response allows us to analyze the behavior of the system at steady state. Knowing how a system responds to an exponential input also allows us to understand the behavior of the system to *any* arbitrary inputs. In the subsequent sections, we will analyze the response of the system to a step input. The response of the system to an exponential input will be covered in detail in Chapter 7.

6.3 Transient response

6.3.1 First-order systems

Let us first consider the unit step response of a canonical first-order system, whose governing equation has the form

$$\frac{dy}{dt} + \frac{1}{\tau}y = \alpha u(t). \tag{6.22}$$

To uniquely determine its response, we need to specify an initial condition $y(0) = y_0$. From the theory of differential equations, we recall that the complete solution to a nonhomogeneous differential equation can be written as the sum of the homogeneous solution, $y_h(t)$, and the particular solution, $y_p(t)$, as follows:

$$y(t) = y_h(t) + y_p(t). \tag{6.23}$$

The homogeneous solution satisfies the following homogeneous differential equation:

$$\frac{dy_h}{dt} + \frac{1}{\tau}y_h = 0. \tag{6.24}$$

Equation (6.24) requires the first derivative of y_h plus a constant $(1/\tau)$ multiplied by y_h to sum to zero. In order for Eq. (6.24) to hold, y_h and its first derivative must have the same functional form. Thus, we assume a homogeneous solution of the form

$$y_h(t) = Ae^{st}, \tag{6.25}$$

where A is some arbitrary constant, and s is a parameter that is to be determined. The presence of s ensures that the argument of the exponent is dimensionless, and it also allows us to satisfy the homogeneous equation. Substituting Eq. (6.25) into Eq. (6.24), we obtain

$$\left(s + \frac{1}{\tau}\right)Ae^{st} = 0. \tag{6.26}$$

Because $Ae^{st} \neq 0$ (otherwise we will have a trivial solution), Eq. (6.26) requires that

$$s + \frac{1}{\tau} = 0. \tag{6.27}$$

Equation (6.27) is known as the *characteristic equation*. Solving for s, we obtain $s = -1/\tau$. Thus, the homogeneous solution of any first-order system is given by the following exponential response:

$$y_{\mathrm{h}}(t) = Ae^{-\frac{t}{\tau}}. \tag{6.28}$$

While the homogeneous solution satisfies the homogeneous equation, the particular solution satisfies the differential equation

$$\frac{dy_{\mathrm{p}}}{dt} + \frac{1}{\tau} y_{\mathrm{p}} = \alpha u(t) = \alpha, \quad \text{for } t \geq 0. \tag{6.29}$$

For $t \geq 0$, the input is simply a constant. Thus, using the method of undetermined coefficients, we assume a particular solution of the form

$$y_{\mathrm{p}}(t) = y_{\mathrm{p}} = \text{constant}. \tag{6.30}$$

Substituting Eq. (6.30) into Eq. (6.29) and solving for y_{p}, we get

$$y_{\mathrm{p}} = \alpha\tau. \tag{6.31}$$

From Eq. (6.23), the complete response is then given by

$$y(t) = Ae^{-\frac{t}{\tau}} + \alpha\tau. \tag{6.32}$$

Applying the initial condition, we get

$$y(0) = y_0 = Ae^0 + \alpha\tau \tag{6.33}$$

from which we find

$$A = y_0 - \alpha\tau. \tag{6.34}$$

Hence the complete response is given by

$$y(t) = (y_0 - \alpha\tau)e^{-\frac{t}{\tau}} + \alpha\tau. \tag{6.35}$$

When a system is stable, τ must be positive and the response is a decaying exponential. This makes sense physically. For the systems of Figures 6.1 and 6.2, the system parameters are positive quantities, and hence the time constant τ must also be positive. For a stable first-order system subjected to a unit step input, the response approaches a *steady state* or *equilibrium* value as shown:

$$\lim_{t \to \infty} y(t) = \alpha \tau. \tag{6.36}$$

Physically, the steady state value is simply the response of the system as time approaches infinity, that is as we wait long enough. The exponential portion decays to zero at a rate depending on the time constant τ. If we wish to design a first-order system that has a fast response, we need to pick the system parameters such that the time constant τ is small.

6.3.2 Second-order systems

Consider now the unit step response of a second-order system whose governing equation has the form

$$\frac{d^2 y}{dt^2} + 2\zeta \omega_n \frac{dy}{dt} + \omega_n^2 y = \alpha u(t). \tag{6.37}$$

To uniquely determine its response, we need to specify the following initial conditions:

$$y(0) = y_0 \\ \dot{y}(0) = \dot{y}_0. \tag{6.38}$$

The complete response to Eq. (6.37) may be written as the sum of the homogeneous solution and the particular solution as follows:

$$y(t) = y_h(t) + y_p(t). \tag{6.39}$$

The homogeneous solution satisfies the following homogeneous differential equation with no input:

$$\frac{d^2 y_h}{dt^2} + 2\zeta \omega_n \frac{dy_h}{dt} + \omega_n^2 y_h = 0. \tag{6.40}$$

We guess a homogeneous solution of the form

$$y_h(t) = A e^{st}. \tag{6.41}$$

Substituting this guess into the differential equation yields

$$\left(s^2 + 2\zeta \omega_n s + \omega_n^2\right) A e^{st} = 0. \tag{6.42}$$

This clearly has the trivial solution $A e^{st} = 0$. The more interesting or the non-trivial solutions are found by solving the following characteristic equation:

$$s^2 + 2\zeta \omega_n s + \omega_n^2 = 0. \tag{6.43}$$

Table 6.1. Homogeneous solutions of Eq. (6.40) as a function of ζ

$\zeta = 0$	Undamped	$y_h(t) = A \cos \omega_n t + B \sin \omega_n t$
$0 < \zeta < 1$	Underdamped	$y_h(t) = e^{-\zeta \omega_n t}(A \cos \omega_d t + B \sin \omega_d t)$
$\zeta = 1$	Critically damped	$y_h(t) = e^{-\omega_n t}(A + Bt)$
$\zeta > 1$	Overdamped	$y_h(t) = e^{-\zeta \omega_n t}[A \cosh(\omega_n \sqrt{\zeta^2 - 1}\, t)$ $+ B \sinh(\omega_n \sqrt{\zeta^2 - 1}\, t)]$

Applying the quadratic formula, we obtain the two roots of s as follows:

$$s_{1,2} = \left(-\zeta \pm \sqrt{\zeta^2 - 1}\right)\omega_n. \tag{6.44}$$

Because there are two roots, we have the following solutions that satisfy Eq. (6.42):

$$(y_h)_1 = A_1 e^{s_1 t}$$
$$(y_h)_2 = A_2 e^{s_2 t}. \tag{6.45}$$

The system is linear, so superposition applies and the homogeneous solution is given by

$$y_h(t) = (y_h)_1 + (y_h)_2 = A_1 e^{s_1 t} + A_2 e^{s_2 t}. \tag{6.46}$$

On physical ground, the undamped natural frequency and the damping factor must both be positive. Thus, the value of ζ (≥ 0) dictates whether the roots s_1 and s_2 are purely imaginary, real or complex. In view of the above discussion, we have four distinct cases to consider. Not surprisingly, because the roots depend on ζ, the form of the homogeneous solution also depends on ζ. Table 6.1 shows the homogeneous solution as a function of ζ. For the underdamped case, ω_d is known as the *damped natural frequency*, and it is given by

$$\omega_d = \omega_n \sqrt{1 - \zeta^2}. \tag{6.47}$$

In the undamped case, the system oscillates forever at the *undamped natural frequency* ω_n during free response because there is no damping or energy loss. The undamped period of oscillation is given by $T = 2\pi / \omega_n$. In the underdamped case, the system oscillates at the *damped natural frequency* ω_d during free response. The damped period of oscillation is given by $T_d = 2\pi / \omega_d$. Except for the undamped case of $\zeta = 0$, at steady state all of the free responses tend to zero, that is,

$$\lim_{t \to \infty} y_h(t) = 0. \tag{6.48}$$

Consider a second-order spring–mass–damper system. Assume the system is released from rest. For the critically damped case, the free response due to initial displacement approaches the steady state value in the shortest time possible without any oscillation or overshoot. For the overdamped case, the free response approaches the steady state value more slowly and it also does not oscillate or overshoot.

The homogeneous solution is easy to obtain because it can be readily deduced from the value of ζ and Table 6.1. The particular solution is more difficult because it depends on the input. In the case of a unit step input, the particular solution must satisfy the following differential equation for $t \geq 0$:

$$\frac{d^2 y_p}{dt^2} + 2\zeta \omega_n \frac{dy_p}{dt} + \omega_n^2 y_p = \alpha. \tag{6.49}$$

Because the input is a constant, we assume the particular solution is also a constant of the form

$$y_p(t) = y_p = \text{constant}. \tag{6.50}$$

Substituting Eq. (6.50) into Eq. (6.49), we obtain an expression for the particular solution as

$$y_p = \frac{\alpha}{\omega_n^2}. \tag{6.51}$$

In summary, our complete solution is the sum of the homogeneous and particular solutions

$$y(t) = y_h(t) + \frac{\alpha}{\omega_n^2}. \tag{6.52}$$

The form of $y_h(t)$ depends on the value of ζ. To find the constant coefficients in the homogeneous solution (the A and B shown in Table 6.1), we must impose the initial conditions given in Eq. (6.38).

A step or constant input is important because it represents a physical event that begins suddenly and is applied continuously thereafter. The desired transient performance characteristics of a system of any order are often specified in terms of the response to a step input. For an nth order system, these *transient characteristics* are specified by assuming all of the initial conditions are identically zero, that is,

$$y(0) = \frac{dy}{dt}(0) = \frac{d^2 y}{dt^2}(0) = \cdots = \frac{dy^{n-1}}{dt^{n-1}}(0) = 0. \tag{6.53}$$

Fig. 6.5. Step responses of a second-order system as a function of ζ with $\omega_n = 1$ rad/s

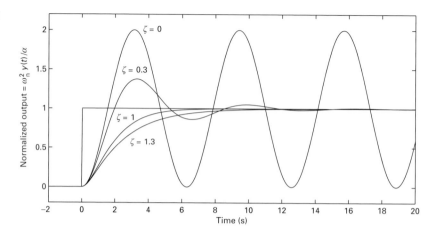

The step response (i.e., the response to a step input with all the initial conditions set to zero) of a second-order system is plotted in Figure 6.5 for each of the four cases of ζ with $\omega_n = 1$ rad/s. The MATLAB code used to generate the plots is also shown below for reference. Note that as time approaches infinity, the steady state value approaches $y_{ss} = y_p = \alpha/\omega_n^2$. Thus, for ease of plotting, the y-axis is normalized by this factor. Observe that in the undamped case, the output oscillates forever because no energy is dissipated. The underdamped solution oscillates but eventually settles to the steady state result. The damped frequency of oscillation ω_d is slightly lower than the natural frequency ω_n (see Eq. (6.47)). The critically damped solution approaches the steady state value from below, and never overshoots it. We say that for a critically damped system, the output approaches the steady state value in the quickest manner possible without any overshoot. The overdamped solution also approaches the steady state value from below, but much more slowly. Finally, we note that in all cases the output of the system never reaches more than twice the static deflection, that is $y(t)$ never exceeds $2\alpha/\omega_n^2$.

The MATLAB code for Figure 6.5 is as follows:

```
fs = 1e-3;
T = 20;
t = 0:fs:T;
omegan = 1;
% step input
t2 = -2:fs:T;
step = (t2>0);
```

```
hold off
plot(t2,step,'r')
hold on
% undamped
x1 = 1-cos(omegan*t);
plot(t,x1,'b')
% underdamped
zeta = 0.3
omegad = omegan*sqrt(1-zeta^2);
phi = atan(sqrt(1-zeta^2)/zeta);
x2 = 1-exp(-zeta*omegan*t).*...
  sin(omegad*t+phi)/sqrt(1-zeta^2);
plot(t,x2,'g');
% critically damped
x3 = 1+exp(-omegan*t).*(-1-omegan*t);
plot(t,x3,'c');
% overdamped
zeta = 1.3;
x4 = 1-exp(-zeta*omegan*t).*...
  (cosh(omegan*sqrt(zeta^2-1)*t)+...
(zeta/sqrt(zeta^2-1))* sinh(omegan*sqrt(zeta^2-...
  1)*t));
plot(t,x4,'m');
```

The desired performance characteristics of a system of any order may be evaluated in terms of *transient specifications*. The transient specifications are defined for a step input with zero initial conditions. We can describe the transient response of any system to a step input using the following transient specifications illustrated in Figure 6.6:

- t_d (delay time): time needed for the output to reach 50 percent of its steady state value for first time;
- t_r (rise time): time required for the output to rise from 0 percent to 100 percent of its steady state value;[1]
- t_p (peak time): time at which the output reaches its peak or maximum value;

---

[1] There is no universal definition of rise time. Some people refer to the 10–90 percent or 20–80 percent rise time. When reading a specification, be sure to check the definition.

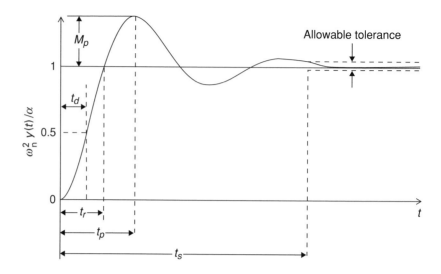

- M_p (maximum percent overshoot): maximum deviation of the output from its steady state value, defined as a percentage, and given by

$$M_p = \frac{y(t_p) - y_{ss}}{y_{ss}} \times 100\%; \qquad (6.54)$$

- t_s (settling time): time required for the output to settle within a specified percentage ϵ of the steady state value. Typically we use $\epsilon = 2$ percent or 5 percent.

The first three transient performance specifications dictate the speed of the response, while the last two govern the accuracy of the response. These definitions are valid for a system of any order. However, not all of these performance specifications may be defined for a given system. For instance, peak time and maximum percent overshoot are not defined for a first-order system, nor are they defined for a critically or an overdamped second-order system.

6.4 Transient specifications

Consider an underdamped second-order system with zero initial conditions whose governing equation is given by Eq. (6.37). Its complete response to a unit step input $f(t) = u(t)$ is given by

$$y(t) = \frac{\alpha}{\omega_n^2} \left[1 - \frac{e^{-\zeta \omega_n t}}{\sqrt{1 - \zeta^2}} \sin\left(\omega_d t + \phi\right) \right], \tag{6.55}$$

where

$$\omega_d = \omega_n \sqrt{1 - \zeta^2}$$

$$\phi = \tan^{-1}\left(\frac{\sqrt{1 - \zeta^2}}{\zeta}\right). \tag{6.56}$$

For a second-order system, the transient specifications can be expressed analytically in terms of the natural frequency and the damping factor, that is the canonical parameters. To fix ideas without any loss of generality, we will define the transient specifications in terms of ζ and ω_n for an underdamped second-order system.

The rise time t_r is found by solving for $y(t_r) = y_{ss}$. The peak time t_p is found by solving for the time at which $\dot{y} = 0$. To guarantee that a peak is found, we also need to make sure that $\ddot{y}(t_p) < 0$. The other transient specifications can also be derived directly from their definitions. The following equations show the transient performance specifications for underdamped, second-order systems:

$$t_r = \frac{1}{\omega_d} \left[\pi - \tan^{-1}\left(\frac{\sqrt{1 - \zeta^2}}{\zeta}\right) \right] \tag{6.57}$$

$$M_p = 100 e^{-\frac{\pi \zeta}{\sqrt{1 - \zeta^2}}} \% \tag{6.58}$$

$$t_p = \frac{\pi}{\omega_n \sqrt{1 - \zeta^2}} \tag{6.59}$$

$$t_s = \frac{-\ln\left(\epsilon \sqrt{1 - \zeta^2}\right)}{\zeta \omega_n}. \tag{6.60}$$

A second-order system is characterized by its canonical system parameters, ζ and ω_n. Thus, for a second-order system, we can at most satisfy two transient specifications exactly. If multiple transient specifications are given, then a trade-off must often be made. For first-order systems or other higher-order systems, their performance characteristics can be derived from the transient specification definitions. While analytical expressions may not always be possible to obtain for complicated systems, these transient performance specifications can always be measured experimentally if they exist.

Fig. 6.7. A spring–mass–
damper system.

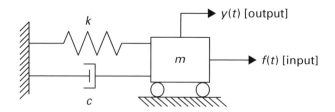

Example: A common application of these transient response parameters
is to solve for the component values that must satisfy a particular set of
transient response constraints. Consider the spring–mass–damper system of
Figure 6.7. A unit step force with amplitude $F_0 = 1$ N is applied to the mass.
The mass and the spring constant are given by $m = 1$ kg and $k = 4$ N/m,
respectively. We would like to have a maximum percent overshoot of no
more than 20 percent and a 100 percent rise time no greater than 3 seconds.
Let us give priority to the overshoot specification. What value of the viscous
damping coefficient c should we use for the damper?

Solution: Sketching the free body diagram and applying Newton's Second
Law, we obtain the equation of motion for the mass as follows:

$$m\ddot{y} + c\dot{y} + ky = f(t) = F_0 u(t) = F_0, \quad \text{for } t \geq 0. \tag{6.61}$$

In terms of the given system parameters, Eq. (6.61) becomes

$$\ddot{y} + c\dot{y} + 4y = 1. \tag{6.62}$$

The maximum percent overshoot is given by Eq. (6.58). Thus, the maximum
overshoot constraint leads to

$$100\, e^{-\frac{\pi \zeta}{\sqrt{1-\zeta^2}}}\, \% \leq 20\%. \tag{6.63}$$

Solving for ζ, we obtain $\zeta \geq 0.456$. Because the system oscillates (it overshoots), it is inherently underdamped. Thus, the damping factor must be in
the range

$$1 > \zeta \geq 0.456. \tag{6.64}$$

The 100 percent rise time is given by Eq. (6.57), and the rise time constraint
yields

$$\frac{1}{\omega_d}\left[\pi - \tan^{-1}\left(\frac{\sqrt{1-\zeta^2}}{\zeta}\right)\right] \leq 3. \tag{6.65}$$

For the given set of system parameters, the undamped natural frequency is $\omega_n = \sqrt{k/m} = 2 \, \text{rad/s}$. Because $\omega_d = \omega_n\sqrt{1 - \zeta^2}$, Eq. (6.65) can be rewritten in terms of ζ only as follows:

$$\frac{1}{2\sqrt{1 - \zeta^2}} \left[\pi - \tan^{-1}\left(\frac{\sqrt{1 - \zeta^2}}{\zeta} \right) \right] \leq 3. \tag{6.66}$$

Equation (6.66) does not yield an analytical solution for ζ. However, ζ can easily be solved using either a graphical or a numerical means. Solving for the damping factor, we find

$$0 < \zeta \leq 0.895. \tag{6.67}$$

The lower bound is implicit because the system is damped. Combining Eqs. (6.67) and (6.64), we find

$$0.895 \geq \zeta \geq 0.456. \tag{6.68}$$

Any value of ζ in this range would satisfy both of the constraints. However, because the priority was given to the overshoot constraint, that is minimizing the overshoot, we must select the largest possible value for ζ. Thus, we choose $\zeta = 0.895$, which then yields the desired damping coefficient c as follows:

$$\frac{c}{m} = 2\zeta\omega_n \quad \Longrightarrow \quad c = 2\zeta\omega_n m = 3.580 \, \frac{\text{N s}}{\text{m}}. \tag{6.69}$$

6.5 State space formulation

We now describe yet another approach to analyzing the dynamics of engineering systems whereby we investigate the *states* of the system. The approach is commonly known as the state space formulation. While this text is restricted to linear time-invariant systems, the state-space approach is applicable to nonlinear and time-varying systems, making the method more versatile. A key feature of the state space approach is that it reduces a complex problem into a set of first-order differential equations, whose matrix solution can be easily obtained analytically. Formulating a problem in terms of state variables (to be defined shortly) also allows it to be easily solved numerically using high-speed digital computers. Additionally, the state variable method provides a unified approach that is used extensively in modern control theory.

Consider a linear, time-invariant system described by the following nth-order differential equation:

$$a_n \frac{d^n y(t)}{dt^n} + \cdots + a_2 \frac{d^2 y(t)}{dt^2} + a_1 \frac{dy(t)}{dt} + a_0 y(t) = f(t), \tag{6.70}$$

where a_n is a nonzero constant, $y(t)$ represents the output variable, and $f(t)$ depends on the input or forcing function. We will extend what we do to include the general case of multiple inputs. For now, however, we restrict our analysis to only a single forcing term. Our objective is to represent Eq. (6.70) by n state equations. The state variables for a given system are not uniquely assigned. Typically, they consist of the output $y(t)$ and its derivatives as follows:

$$
\begin{aligned}
y_1(t) &= y(t) \\
y_2(t) &= \frac{dy(t)}{dt} \\
&\;\;\vdots \\
y_i(t) &= \frac{d^{i-1} y(t)}{dt^{i-1}} \\
&\;\;\vdots \\
y_n(t) &= \frac{d^{n-1} y(t)}{dt^{n-1}}.
\end{aligned}
\tag{6.71}
$$

These variables describe the current state or condition of the system. Note that for the nth-order system, we have n independent variables, $y_i(t)$, where $i = 1, \ldots, n$. Taking the first derivative of all these variables with respect to time, we obtain

$$
\begin{aligned}
\dot{y}_1(t) &= y_2(t) \\
\dot{y}_2(t) &= y_3(t) \\
&\;\;\vdots \\
\dot{y}_i(t) &= y_{i+1}(t) \\
&\;\;\vdots \\
\dot{y}_n(t) &= -\frac{a_0}{a_n} y_1(t) - \frac{a_1}{a_n} y_2(t) - \cdots - \frac{a_{n-1}}{a_n} y_n(t) + \frac{1}{a_n} f(t),
\end{aligned}
\tag{6.72}
$$

where the last equation is obtained from Eq. (6.70) by dividing through by the coefficient a_n. The set of n variables, $y_i(t)$, can also be written as a column vector given by

$$\mathbf{y}(t) = \begin{bmatrix} y_1(t) \\ y_2(t) \\ \vdots \\ y_i(t) \\ \vdots \\ y_n(t) \end{bmatrix} \qquad (6.73)$$

The elements of the vector $\mathbf{y}(t)$ represent the states of the nth-order system, and they are known as the *state variables* of the system. They constitute the smallest set of variables required such that knowing these variables at some prescribed time $t = t_0$ and knowing the input function for $t \geq t_0$ completely characterizes the response of the system for any time $t \geq t_0$. In terms of the *state vector* of Eq. (6.73), Eq. (6.70) can be written compactly in matrix form as

$$\dot{\mathbf{y}}(t) = [A]\mathbf{y}(t) + \mathbf{b}f(t), \qquad (6.74)$$

where

$$[A] = \begin{bmatrix} 0 & 1 & 0 & \cdots & \cdots & 0 \\ 0 & 0 & 1 & \cdots & \cdots & 0 \\ \vdots & \vdots & \vdots & \ddots & \vdots & \vdots \\ 0 & 0 & 0 & \cdots & 1 & 0 \\ \vdots & \vdots & \vdots & \vdots & \vdots & \ddots \\ -\dfrac{a_0}{a_n} & -\dfrac{a_1}{a_n} & \cdots & \cdots & \cdots & -\dfrac{a_{n-1}}{a_n} \end{bmatrix}, \quad \mathbf{b} = \begin{bmatrix} 0 \\ 0 \\ \vdots \\ 0 \\ \vdots \\ \dfrac{1}{a_n} \end{bmatrix}. \qquad (6.75)$$

Matrix $[A]$ is a real $n \times n$ constant matrix and it is known as the *state matrix* of the system. In general, $[A]$ is nonsymmetric. Vector $\mathbf{b}$ is a constant column vector and $f(t)$ is a scalar input function. Note that what was originally an nth-order differential equation is now reduced to a set of n first-order differential equations that can be conveniently expressed in compact matrix form.

In spite of the above organized approach to assigning state variables, it is important to bear in mind that the selection of state variables is by no means unique. Moreover, the state variables do not need to be physically measurable or even physically meaningful. In fact, the ability to choose the state variables freely is an advantage of this particular approach of solving complex problems. Typically, however, the choice of state variables is suggested by the system's governing equations. Whenever possible, state variables are selected to coincide with easily measurable quantities, because this allows us to track the states of the system for the purpose of designing control systems.

In the following sections, we will use the state variable approach to obtain the general solution of a linear time-invariant system. We will first analyze the free response and then the forced response. We will see that there are many similarities between the solution of Eq. (6.74) and the solution of a scalar first-order differential equation given by Eq. (6.8). The state variable approach of solving for the complete response utilizes many of the important concepts that were introduced earlier. Specifically, we will apply the principle of superposition, the orthogonality relations, and the results of Section 6.1.

6.5.1 Free response using the state equations

For the free response where $f(t) = 0$, Eq. (6.74) reduces to

$$\dot{\mathbf{y}}(t) = [A]\mathbf{y}(t). \tag{6.76}$$

The matrix differential equation of Eq. (6.76) can be solved using the same techniques that we employed to solve for the free response of a scalar equation. Note that a vector $\mathbf{y}(t)$ is a solution to Eq. (6.75) provided its first time-derivative equals the product of a constant matrix and itself. This can occur if and only if $\mathbf{y}(t)$ and its first time-derivative have the same functional dependency on t. Thus, we seek a solution to Eq. (6.76) having the following exponential form:

$$\mathbf{y}(t) = e^{\lambda t}\mathbf{u}, \tag{6.77}$$

where $\mathbf{u}$ is an unknown constant column vector and λ is an unknown scalar quantity that must satisfy Eq. (6.76). Note the similarity to Eq. (6.25). The parameters λ and $\mathbf{u}$ are formally known as the *eigenvalue* and *eigenvector* of matrix $[A]$, respectively, and together they constitute an *eigensolution* of the system.

To find the eigensolutions of the system, we first calculate the eigenvalues of matrix $[A]$. Substituting Eq. (6.77) into Eq. (6.76) yields

$$\lambda e^{\lambda t}\mathbf{u} = [A]e^{\lambda t}\mathbf{u}. \tag{6.78}$$

Because exponentials can never vanish, Eq. (6.78) can also be expressed as

$$[A]\mathbf{u} = \lambda\mathbf{u}. \tag{6.79}$$

Equation (6.79) is known as the *eigenvalue problem*. We now introduce an identity matrix, $[I]$, defined as a square matrix whose diagonal elements are all ones and whose off-diagonal terms are all zero. Then Eq. (6.79) can be rewritten as

$$\{[A] - \lambda[I]\}\, \mathbf{u} = \mathbf{0}. \tag{6.80}$$

In order to have a nontrivial solution for $\mathbf{u}$ in Eq. (6.80), the determinant of the coefficient matrix must be identically zero[2]:

$$\det\{[A] - \lambda[I]\} = |[A] - \lambda[I]| = 0, \tag{6.81}$$

where both $\det[\phi]$ and $|\phi|$ can be used to denote the determinant of the matrix $[\phi]$.

Expanding the characteristic determinant of Eq. (6.81), we obtain an nth-order polynomial in λ as follows:

$$\lambda^n + \frac{a_{n-1}}{a_n}\lambda^{n-1} + \cdots + \frac{a_1}{a_n}\lambda^1 + \frac{a_0}{a_n} = 0. \tag{6.82}$$

Equation (6.82) is also known as the characteristic equation of matrix $[A]$. Assuming the eigenvalues to be distinct,[3] Eq. (6.82) yields n eigenvalues. For each eigenvalue λ_i, the corresponding eigenvector $\mathbf{u}_i$ is obtained by solving Eq. (6.80). The eigenvalues and their corresponding eigenvectors are unique for a given system except for the magnitude of the eigenvectors. This can be clearly illustrated by noting that if $\mathbf{u}_i$ is a solution to Eq. (6.80), then so is $\alpha\mathbf{u}_i$, where α is any arbitrary constant, because Eq. (6.80) is homogeneous. While the magnitude of the elements of each eigenvector is not unique, the ratio between any two components of the eigenvector remains the same.

Because the system of Eq. (6.70) is linear, the principle of superposition applies. Assuming the n eigenvalues are distinct, then the free response of the state vector $\mathbf{y}(t)$ at any time can be expressed as a linear combination of the eigenvectors as follows:

$$\mathbf{y}(t) = \sum_{i=1}^{n} \alpha_i \mathbf{u}_i e^{\lambda_i t}, \tag{6.83}$$

---

[2] See *Linear Algebra and Its Applications* by G. Strang, Academic Press, New York, 1976.

[3] For the case of repeated eigenvalues, the solution scheme becomes more complicated but still tractable; see *Mathematical Methods in Physics and Engineering* by J. W. Dettman, McGraw-Hill, New York, 1969.

where the constants α_i are generally complex and are uniquely determined by enforcing the initial conditions of the system. Specifically, at $t = 0$, Eq. (6.83) reduces to a set of algebraic equations given by

$$\mathbf{y}(0) = \mathbf{y}_0 = \sum_{i=1}^{n} \alpha_i \mathbf{u}_i. \tag{6.84}$$

For a given set of specified initial conditions, Eq. (6.84) leads to a set of n coupled algebraic equations that allows us to solve for the n unknown coefficients α_i. While conceptually simple, the solution of α_i does not lend itself easily to manual calculation because the algebra involved may be complicated and lengthy. Fortunately, the resultant set of algebraic equations can be solved easily using computationally efficient numerical techniques such as Gaussian eliminations.

6.5.2 Forced response using state equations

Consider now the forced response of Eq. (6.74) where $f(t) \neq 0$. Not surprisingly, the solution to the forced response case is substantially more complicated. To this end, we will harness the orthogonality properties of the eigenvectors to reduce the set of n first-order coupled differential equations into a set of n first-order uncoupled differential equations. Once uncoupled, each equation has the form given by Eq. (6.8), whose solution can be easily obtained.

In general, the state matrix $[A]$ is nonsymmetric. To show the orthogonality properties of the eigenvectors, let us consider the eigenvalue problem associated with $[A]^T$ and write it as

$$[A]^T \mathbf{v} = \lambda \mathbf{v}, \tag{6.85}$$

where λ and $\mathbf{v}$ are the eigenvalue and eigenvector of $[A]^T$. For simplicity the eigenvalues are assumed to be distinct. Equation (6.85) is known as the *adjoint* eigenvalue problem. From the elementary theory in linear algebra, we know that because the determinants of $[A]$ and $[A]^T$ are identical, $[A]$ and $[A]^T$ must share the same characteristic equation. Therefore, the eigenvalues of $[A]^T$ are identical to those of $[A]$. The eigenvectors of $[A]^T$, on the other hand, are different from those of $[A]$. To prove the orthogonality relations of the eigenvectors, let us consider the ith eigensolution of Eq. (6.79) and the jth eigensolution of Eq. (6.85), where $i \neq j$. Thus, these eigensolutions must satisfy

$$[A]\mathbf{u}_i = \lambda_i \mathbf{u}_i \tag{6.86}$$

$$[A]^T \mathbf{v}_j = \lambda_j \mathbf{v}_j. \tag{6.87}$$

Multiplying Eq. (6.86) by $\mathbf{v}_j^T$ and multiplying Eq. (6.87) by $\mathbf{u}_i^T$, we obtain

$$\mathbf{v}_j^T [A]\mathbf{u}_i = \lambda_i \mathbf{v}_j^T \mathbf{u}_i \tag{6.88}$$

$$\mathbf{u}_i^T [A]^T \mathbf{v}_j = \lambda_j \mathbf{u}_i^T \mathbf{v}_j. \tag{6.89}$$

Taking the transpose of Eq. (6.89) and subtracting the resulting expression from Eq. (6.86), we get

$$(\lambda_i - \lambda_j)\mathbf{v}_j^T \mathbf{u}_i = 0. \tag{6.90}$$

Because we have assumed that $i \neq j$ from the outset, $\lambda_i \neq \lambda_j$, we must have

$$\mathbf{v}_j^T \mathbf{u}_i = 0 \quad \text{for} \quad i, j = 1, 2, \ldots, n. \tag{6.91}$$

In words, we say that the eigenvectors of $[A]$ and the eigenvectors of $[A]^T$ corresponding to distinct eigenvalues are orthogonal. Substituting Eq. (6.91) into Eq. (6.86), we conclude that for $\lambda_i \neq \lambda_j$,

$$\mathbf{v}_j^T [A]\mathbf{u}_i = 0 \quad \text{for} \quad i, j = 1, 2, \ldots, n. \tag{6.92}$$

We say that the eigenvectors $\mathbf{u}_i$ and $\mathbf{v}_j$ are also orthogonal with respect to matrix $[A]$. When $i = j$, the products $\mathbf{v}_i^T \mathbf{u}_i$ and $\mathbf{v}_i^T [A]\mathbf{u}_i$ are not zero. It is convenient to normalize the two sets of eigenvectors by letting

$$\mathbf{v}_j^T \mathbf{u}_i = \delta_i^j \quad i, j = 1, 2, \ldots, n, \tag{6.93}$$

where δ_i^j is the *Kronecker delta*, defined as

$$\delta_i^j = \begin{cases} 1 & \text{when } i = j \\ 0 & \text{when } i \neq j. \end{cases} \tag{6.94}$$

Then from Eq. (6.86), we conclude that

$$\mathbf{v}_j^T [A]\mathbf{u}_i = \lambda_i \delta_i^j \quad i, j = 1, 2, \ldots, n. \tag{6.95}$$

The orthogonality conditions of Eqs. (6.93) and (6.95) can be expressed compactly in matrix form as follows:

$$[V]^T [U] = [I] \tag{6.96}$$

$$[V]^T [A][U] = [\Lambda], \tag{6.97}$$

where the column vectors of matrices $[V]$ and $[U]$ consist of the eigenvectors $\mathbf{v}_i$ and $\mathbf{u}_i$, respectively, as shown:

$$
\begin{aligned}
[V] &= [\mathbf{v}_1 \ \ \mathbf{v}_2 \ \ \cdots \ \ \mathbf{v}_i \ \ \cdots \ \ \mathbf{v}_n] \\
[U] &= [\mathbf{u}_1 \ \ \mathbf{u}_2 \ \ \cdots \ \ \mathbf{u}_i \ \ \cdots \ \ \mathbf{u}_n]
\end{aligned}
\tag{6.98}
$$

and $[\Lambda]$ is a diagonal matrix whose ith element corresponds to the ith eigenvalue λ_i.

To solve for the forced response of Eq. (6.74), we introduce the following linear transformation:

$$
\mathbf{y}(t) = [U]\mathbf{z}(t).
\tag{6.99}
$$

Substituting Eq. (6.99) into Eq. (6.74), we obtain

$$
[U]\dot{\mathbf{z}}(t) = [A][U]\mathbf{z}(t) + \mathbf{b}f(t).
\tag{6.100}
$$

Multiplying Eq. (6.100) by $[V]^T$ and noting the orthogonality conditions of Eqs. (6.96) and (6.97), Eq. (6.100) becomes

$$
[I]\dot{\mathbf{z}}(t) = [\Lambda]\mathbf{z}(t) + \mathbf{N}(t),
\tag{6.101}
$$

where the column vector $\mathbf{N}(t)$ is given by

$$
\mathbf{N}(t) = [V]^T \mathbf{b}f(t).
\tag{6.102}
$$

Because matrices $[I]$ and $[\Lambda]$ are both diagonal, Eq. (6.102) represents a set of n uncoupled first-order equations of the form

$$
\dot{z}_i(t) = \lambda_i z_i(t) + N_i(t), \quad \text{for } i = 1, 2, \ldots, n,
\tag{6.103}
$$

where $N_i(t)$ represents the ith component of $\mathbf{N}(t)$. Equation (6.103) has the same structure as the first-order differential equation of Eq. (6.8). Hence, the solution of Eq. (6.103) can be obtained by the methods outlined in Section 6.1. In order to determine the complete response of Eq. (6.103), the initial conditions $z_i(0)$ must be specified. From Eq. (6.99), we note that the initial conditions in the new coordinates are related to the initial conditions in the original state variables as follows:

$$
\mathbf{y}(0) = [U]\mathbf{z}(0).
\tag{6.104}
$$

Multiplying Eq. (6.104) by $[V]^T$, we obtain the initial conditions in the decoupled coordinate system:

$$
\mathbf{z}(0) = [V]^T \mathbf{y}(0).
\tag{6.105}
$$

Knowing the initial conditions in the decoupled coordinates and the $N_i(t)$, we can readily determine the responses in the decoupled coordinates $z_i(t)$. By applying the linear transformation of Eq. (6.99), we can find the desired complete forced response of Eq. (6.74). Finally, when the state matrix $[A]$ is symmetric, the problem is greatly simplified because one does not need to solve the adjoint eigenvalue problem. For a symmetric state matrix, the previous results are still applicable by simply setting $[V] = [U]$.

The above derivations can be extended to analyze a system with m inputs, l outputs and n state variables. All the coefficients in the governing equations describing the behavior of the linear time-invariant system are constants. The matrix state and output equations are then given by the following matrix equations:

$$\dot{\mathbf{y}}(t) = [A]\mathbf{y}(t) + [B]\mathbf{u}(t) \tag{6.106}$$

and

$$\mathbf{x}(t) = [C]\mathbf{y}(t) + [D]\mathbf{u}(t). \tag{6.107}$$

The column vectors $\mathbf{y}(t)$, $\mathbf{u}(t)$ and $\mathbf{x}(t)$ correspond to the state vector, the input vector and the output vector, and they are of length n, m and l, respectively. Matrices $[A]$, $[B]$, $[C]$ and $[D]$ all have constant elements, and they are of size $n \times n$, $n \times m$, $l \times n$ and $l \times m$, respectively. Equation (6.106) is first solved for the state vector $\mathbf{y}(t)$. This result is then used to determine the desired output $\mathbf{x}(t)$ in Eq. (6.107). The response of a system with multiple inputs and multiple outputs is beyond the scope of this text and will not be covered. Interested readers are encouraged to refer to any text on control systems.[4]

Example: Consider the circuit shown in Figure 6.8. Determine the currents $i_1(t)$ and $i_2(t)$, assuming that all the voltages and currents are zero at $t = 0$, the instant when the switch is closed.

Solution: Because the unknowns are the currents, the governing equations for the circuit are obtained by applying KVL. The left loop of the circuit yields

$$L\frac{di_1}{dt} + R_1(i_1 - i_2) = v_{in}(t), \tag{6.108}$$

---

[4] For example, see *Automatic Control Systems* by B. C. Kuo, Prentice-Hall, New Jersey, 1991, or *Modern Control Engineering* by K. Ogata, Prentice Hall, New Jersey, 2002.

Fig. 6.8. An electrical
circuit with a switch.

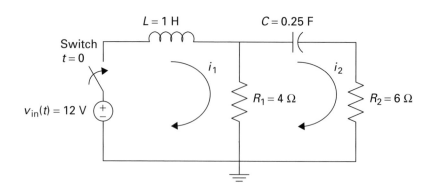

where $i_1(t)$ and $i_2(t)$ are the current in the left and right loops, respectively. The voltage $R_1(i_1 - i_2)$ denotes the voltage drop across the resistor R_1 because $i_1(t)$ and $i_2(t)$ flow through the resistor in opposite directions. Applying KVL for the right loop, we obtain

$$R_2 i_2 + R_1(i_2 - i_1) + \frac{1}{C} \int_{-\infty}^{t} i_2(\tau)d\tau = 0. \tag{6.109}$$

Upon differentiation, Eq. (6.109) becomes

$$(R_1 + R_2)\frac{di_2}{dt} - R_1\frac{di_1}{dt} + \frac{1}{C}i_2 = 0. \tag{6.110}$$

Equations (6.108) and (6.110) constitute the governing equations for the system of Figure 6.8. Rearranging Eqs. (6.108) and (6.110), we obtain

$$\frac{di_1}{dt} = -\frac{R_1}{L}(i_1 - i_2) + \frac{1}{L}v_{in}(t) \tag{6.111}$$

and

$$\frac{di_2}{dt} = \frac{R_1}{R_1 + R_2}\frac{di_1}{dt} - \frac{1}{C(R_1 + R_2)}i_2. \tag{6.112}$$

Substituting Eq. (6.111) into Eq. (6.112), we obtain

$$\frac{di_2}{dt} = -\frac{R_1^2}{R_1 + R_2}i_1 + \left[\frac{R_1^2}{L(R_1 + R_2)} - \frac{1}{C(R_1 + R_2)}\right]i_2$$

$$+ \frac{R_1}{L(R_1 + R_2)}v_{in}(t). \tag{6.113}$$

Note that the right-hand sides of Eqs. (6.111) and (6.113) are now in terms of the currents $i_1(t)$, $i_2(t)$ and the input $v_{in}(t)$, thus allowing Eqs. (6.111) and (6.113) to be expressed compactly in the form of Eq. (6.74).

For the given set of system parameters and for $v_{in}(t) = 12$ V, we can express the governing equations for the system in state vector form as follows:

$$\frac{d}{dt}\begin{bmatrix} i_1 \\ i_2 \end{bmatrix} = \begin{bmatrix} -4.0 & 4.0 \\ -1.6 & 1.2 \end{bmatrix}\begin{bmatrix} i_1 \\ i_2 \end{bmatrix} + \begin{bmatrix} 12 \\ 4.8 \end{bmatrix}, \tag{6.114}$$

where the state matrix is simply

$$[A] = \begin{bmatrix} -4.0 & 4.0 \\ -1.6 & 1.2 \end{bmatrix}. \tag{6.115}$$

Using the MATLAB command eig, the eigenvalues of the state matrix are found to be

$$\begin{aligned} \lambda_1 &= -0.8 \\ \lambda_2 &= -2.0. \end{aligned} \tag{6.116}$$

The eigenvector[5] associated with λ_1 is calculated from Eq. (6.80) with $\lambda = \lambda_1 = -0.8$ and by setting $\mathbf{u} = \mathbf{u_1} = [u_{11} \quad u_{12}]^T$, yielding

$$\{[A] - \lambda_1[I]\}\,\mathbf{u}_1 = \mathbf{0} \tag{6.117}$$

or

$$\begin{bmatrix} -3.2 & 4.0 \\ -1.6 & 2.0 \end{bmatrix}\begin{bmatrix} u_{11} \\ u_{12} \end{bmatrix} = \begin{bmatrix} 0 \\ 0 \end{bmatrix}. \tag{6.118}$$

Solving for the elements u_{11} and u_{12} (the subscript ij denotes the ith eigenvector and its jth component) that satisfies Eq. (6.113), we obtain[6]

$$\mathbf{u}_1 = \begin{bmatrix} 1.0 \\ 0.8 \end{bmatrix}. \tag{6.119}$$

Similarly, substituting $\lambda = \lambda_2$ into Eq. (6.80) and solving for $\mathbf{u}_2$, we get

$$\mathbf{u}_2 = \begin{bmatrix} 1.0 \\ 0.5 \end{bmatrix}. \tag{6.120}$$

In order to use the results of this section, we also need to solve the adjoint eigenvalue problem of Eq. (6.85). The eigenvalues of $[A]^T$ are identical to those of $[A]$, and the corresponding eigenvectors are obtained by solving

$$\{[A]^T - \lambda[I]\}\mathbf{v} = \mathbf{0}. \tag{6.121}$$

---

[5] The MATLAB command eig also gives the complete set of eigenvectors of the system. Here, we will show how to find the eigenvectors by hand.

[6] MATLAB will yield a different eigenvector for $\mathbf{u}_1$. However, we should bear in mind that the eigenvalues are unique for a system, but their eigenvectors may be arbitrary up to a constant.

After some algebra, we find

$$\mathbf{v}_1 = \begin{bmatrix} -0.5 \\ 1.0 \end{bmatrix} \tag{6.122}$$

and

$$\mathbf{v}_2 = \begin{bmatrix} 0.8 \\ -1.0 \end{bmatrix}. \tag{6.123}$$

Before applying the results of this section, we recall that the eigenvectors must be normalized according to Eq. (6.93). By inspection, we note that eigenvectors corresponding to distinct eigenvalues are indeed orthogonal as predicted by Eq. (6.93). Specifically,

$$\mathbf{v}_1^T \mathbf{u}_2 = \mathbf{v}_2^T \mathbf{u}_1 = 0. \tag{6.124}$$

For identical eigenvalues, we normalize $\mathbf{u}_i$ and $\mathbf{v}_i$ using

$$\begin{aligned} \hat{\mathbf{u}}_i &= \frac{1}{\sqrt{\mathbf{v}_i^T \mathbf{u}_i}} \mathbf{u}_i \\ \hat{\mathbf{v}}_i &= \frac{1}{\sqrt{\mathbf{v}_i^T \mathbf{u}_i}} \mathbf{v}_i, \end{aligned} \tag{6.125}$$

where the (^) is used to denote a normalized eigenvector. For the sake of convenience, the "hat" notation is subsequently dropped in favor of renaming the normalized eigenvector as simply $\mathbf{u}_i$ and $\mathbf{v}_i$. Using Eq. (6.125), we find the normalized eigenvectors of $[A]$ to be given by

$$\mathbf{u}_1 = \begin{bmatrix} 1.8257 \\ 1.4606 \end{bmatrix} \text{ and } \mathbf{u}_2 = \begin{bmatrix} 1.8257 \\ 0.9129 \end{bmatrix} \tag{6.126}$$

and the normalized eigenvectors of $[A]^T$ are

$$\mathbf{v}_1 = \begin{bmatrix} -0.9129 \\ 1.8257 \end{bmatrix} \text{ and } \mathbf{v}_2 = \begin{bmatrix} 1.4606 \\ -1.8257 \end{bmatrix}. \tag{6.127}$$

Finally, the column vector $\mathbf{N}(t)$ (see Eq. (6.102)) is given by

$$\mathbf{N}(t) = [V]^T \begin{bmatrix} 12 \\ 4.8 \end{bmatrix} = \begin{bmatrix} -0.9129 & 1.8257 \\ 1.4606 & -1.8257 \end{bmatrix} \begin{bmatrix} 12 \\ 4.8 \end{bmatrix}$$

$$= \begin{bmatrix} -2.1914 \\ 8.7638 \end{bmatrix}. \tag{6.128}$$

Using the approach outlined in this section, the coupled governing equations of the system in matrix form (see Eq. (6.114)) can be reduced to the following set of uncoupled first-order differential equations:

$$\dot{z}_1 = -0.8z_1 - 2.1914$$
$$\dot{z}_2 = -2.0z_2 + 8.7638. \tag{6.129}$$

Because the initial currents are zero as specified in the problem statement, Eq. (6.105) also yields zero initial conditions in the new coordinates, $z_1(0) = z_2(0) = 0$. Solving Eq. (6.129) using the procedure outlined in Section 6.3.1, we find the complete response of the system in the decoupled coordinates as follows

$$z_1(t) = 2.7393 \left(e^{-0.8t} - 1 \right)$$
$$z_2(t) = -4.3819 \left(e^{-2.0t} - 1 \right). \tag{6.130}$$

Finally, by applying the linear transformation of Eq. (6.99), we obtain the complete response for the currents in the circuit,

$$\mathbf{y}(t) = \begin{bmatrix} i_1(t) \\ i_2(t) \end{bmatrix} = [U]\mathbf{z}(t) = \begin{bmatrix} 1.8257 & 1.8257 \\ 1.4606 & 0.9129 \end{bmatrix} \begin{bmatrix} z_1(t) \\ z_2(t) \end{bmatrix} \tag{6.131}$$

from which we find

$$i_1(t) = 5e^{-0.8t} - 8e^{-2.0t} + 3$$
$$i_2(t) = 4e^{-0.8t} - 4e^{-2.0t}. \tag{6.132}$$

Note that in the limit as t approaches zero, the current i_1 has a steady state value of 3 A and the current i_2 has a steady state value of zero. Do the results make sense? Can you explain it physically?

6.6 Problems

6-1 Consider a canonical first-order system subjected to a unit step input, whose governing equation is given by

$$\dot{x} + \frac{1}{\tau}x = \alpha f(t) = \alpha u(t).$$

The system is initially at rest (zero initial condition).

(a) Determine the delay time, t_d, of the system. The delay time is defined as the time required for the output to reach 50 percent of its steady state value for the very first time.

(b) Determine the rise time, t_r, of the system. The rise time is defined as the time required for the output to rise from, say, 10 percent to 90 percent of its steady state value.

Fig. 6.9. Figure for
Problem 6-2.

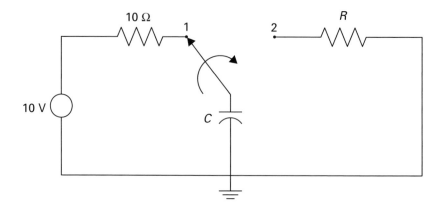

Fig. 6.10. Figure for
Problem 6-3.

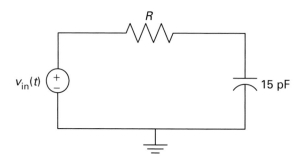

6-2 Consider the circuit shown in Figure 6.9.

  (a) At $t = 0$, the switch is moved to position 1. If the capacitor voltage
  charges up to 95 percent of its steady-state value in 15 μs, what is
  the value of the capacitance? Assume that the capacitor is initially
  discharged.

  (b) After flipping to position 2, what should R be if the capacitor
  voltage discharges to 63 percent of its initial voltage (immediately
  after flipping the switch to position 2) in 1 μs? Use the capacitance
  value from part **(a)**.

  (c) Plot the voltage drop across the capacitor if the switch is initially
  in position 1 for 10 μs and then flips to position 2, where it remains
  for a long period of time.

6-3 For the RC circuit shown in Figure 6.10, find the largest possible value
of R if the capacitor voltage must be charged up to 95 percent of its
steady-state value in less than 15 μs. Assume that $v_{in}(t)$ is a unit step
input, and that there is zero voltage drop across the capacitor initially.

Fig. 6.11. Figure for
Problem 6-4.

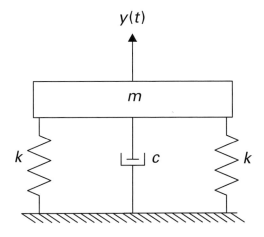

$y(t)$

m

k c k

Fig. 6.12. Figure for
Problem 6-5.

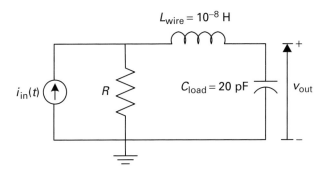

$L_{wire} = 10^{-8}$ H

$i_{in}(t)$ R $C_{load} = 20$ pF v_{out} +
                                                              −

6-4 Consider a 20 kg machine mounted on springs and a damper as shown in Figure 6.11. The system parameters are $k = 4 \times 10^3$ N/m and $c = 130$ N s/m. A velocity of 100 mm/s is imparted to the system initially at rest.

(a) Determine the free response for the vertical displacement, $y(t)$, of the system.

(b) For a given application, the desired performance characteristics can be specified by the system's transient response to a unit step input with zero initial conditions. For $t_p = 1$ s and $t_s = 10$ s (for $\epsilon = 0.02$), determine the required spring stiffness, k, and the viscous damping coefficient, c, if $m = 20$ kg.

6-5 The system shown in Figure 6.12 represents the output circuit of a transistor amplifier driving a capacitive load through an inductive wire. What range of R should be chosen to ensure that there is no more

Fig. 6.13. Figure for
Problem 6-6.

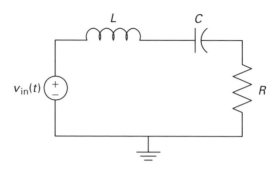

than 10 percent overshoot in response to a step change in the current
delivered by the current source? Assume $v_{out}(0) = \dot{v}_{out}(0) = 0$.

6-6 Consider the circuit shown in Figure 6.13, where the voltage drop
across the capacitor is the output and $R = 1\,k\Omega$. For a certain application the desired performance characteristics can be specified by its
transient response to a unit step input. Determine the required capacitance C (in farads) and inductance L (in henries) for $t_p = 1\,\mu s$ and
$t_s = 10\,\mu s$ (for $\epsilon = 0.02$).

6-7 Consider the RLC circuit in Figure 6.14, where $i_s(t)$ represents the
input and the current through the inductor, $i_L(t)$, represents the output.
(a) Determine the governing equation for the system.
(b) For a step input,

$$i_s(t) = I_0 u(t) = I_0 = \text{constant}, \quad \text{for } t > 0$$

and for

$$i_L(0) = \frac{di_L}{dt}(0) = 0,$$

the output $i_L(t)$ of the system, normalized to I_0, is shown in the
graph. From this particular response, determine the values of L
and C, assuming $R = 20\,\Omega$. Note that the time-axis is scaled by
1×10^{-4}.

6-8 Consider the circuit shown in Figure 6.15, where the switch S is initially at position 1. The circuit is driven by a constant voltage source,
v_s.
(a) Find the voltage drop across the capacitor when the switch is at
position 1. Assume the capacitor is initially discharged.

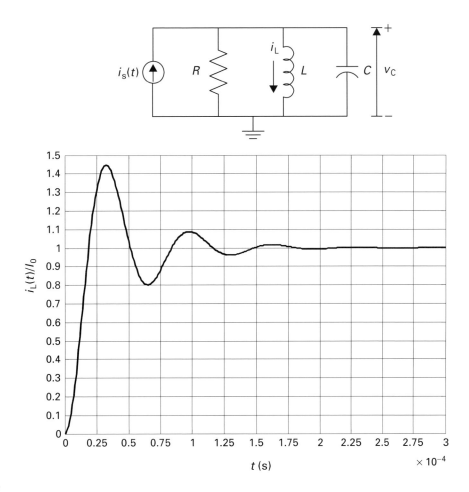

Fig. 6.14. Figure for
Problem 6-7.

Fig. 6.15. Figure for
Problem 6-8.

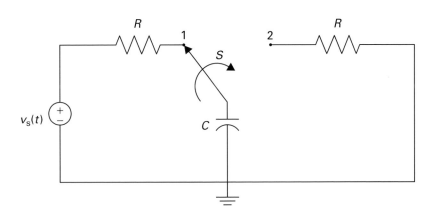

(b) After the steady state is reached, S is switched to position 2. When the switch has been in position 2 for 0.2 s, it is desired that the voltage drop across the capacitor be at 90 percent of its steady state value when the switch is at 1. For $R = 20\,\Omega$, determine the required capacitance, C, needed to achieve this objective.

6-9 Determine the canonical parameters for the system of Problem 4-2.

6-10 Determine the canonical parameters for the system of Problem 4-7.

6-11 Consider the system of Problem 4-8.
   (a) Determine the canonical parameters of the system when the output is $y(t)$.
   (b) Determine the canonical parameters of the system when the output is $\theta(t)$.
   (c) Please explain your results.

6-12 Consider the example problem in Figure 6.8. For $L = 2 \times 10^{-3}$ H, $C = 1 \times 10^{-3}$ F, $R_1 = 4\,\Omega$, $R_2 = 10\,\Omega$ and $v_{in}(t) = 5$ V, determine and plot the currents $i_1(t)$ and $i_2(t)$ for the following cases:
   (a) If the initial currents are $i_1(0) = 15$ A and $i_2(0) = 5$ A.
   (b) If the initial currents are $i_1(0) = 0$ and $i_2(0) = 8$ A.
   (c) If the system parameters are changed to $L = 2.5 \times 10^{-2}$ H, $C = 5 \times 10^{-3}$ F, $R_1 = 20\,\Omega$, $R_2 = 12\,\Omega$, $v_{in}(t) = 7$ V, and the initial currents are zero.

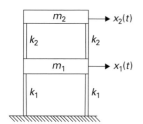

Fig. 6.16. Figure for Problem 6-13.

6-13 Consider a scaled model of a two-story building, as shown in Figure 6.16, in which the floors are modeled as lumped masses, and the supporting columns are assumed to be massless with stiffnesses k_1 and k_2. The assumption of massless columns is reasonable because the masses of the columns are much less than the masses of the floors. Assume $m_1 = 10$ kg, $m_2 = 8$ kg, $k_1 = 7500$ N/m and $k_2 = 5000$ N/m.
   (a) Determine the governing equations for the system.
   (b) Using the state space approach, determine the horizontal displacements, $x_1(t)$ and $x_2(t)$, of the system, if the initial conditions are

$$x_1(0) = x_2(0) = 0, \ \dot{x}_1(0) = 0.02 \text{ m} \ \text{ and } \ \dot{x}_2(0) = 0.015 \text{ m},$$

   where the initial velocities are caused by an impulse-type movement of the Earth, which is often used to model an earthquake.
   (c) Plot the responses of $x_1(t)$ and $x_2(t)$.

7 Input–output relationship using frequency response

Sinusoidal inputs are often encountered. For example, unbalanced rotating machinery like a poorly loaded washing machine produces sinusoidal variations. Reciprocating engines and electrical circuits with alternating current or voltage sources also experience sinusoidal excitation. Inputs encountered in the real world, however, are seldom harmonic (sinusoid with one excitation frequency) but periodic. Periodic excitation implies the existence of several, sometimes even an infinite number of, different frequencies. Fourier showed that any periodic or nonperiodic function can be expressed in terms of sines and cosines (or complex exponentials). Thus, the response to pure harmonic inputs gives us the foundation which we can use to generalize the periodic or nonperiodic excitations. Knowing the response to a harmonic input, we can apply the principle of superposition to obtain the response to periodic and nonperiodic excitations.

To fix ideas without any loss of generality, let us consider a simple spring–mass–damper system excited by a complex exponential with an input amplitude of F and an excitation frequency of ω,

$$m\ddot{y} + c\dot{y} + kx = f(t) = Fe^{j\omega t}. \tag{7.1}$$

In canonical form, Eq. (7.1) becomes

$$\ddot{y} + 2\zeta\omega_n\dot{y} + \omega_n^2 y = \frac{f(t)}{m} = \frac{F}{m}e^{j\omega t}. \tag{7.2}$$

The complete solution is again the sum of the homogeneous and the particular solutions. If there is any damping in the system ($\zeta > 0$) and if we wait sufficiently long, the homogeneous solution will eventually approach zero and die out. Therefore, we see that the steady state solution corresponds to the particular solution, if the system is damped. Because the input is

Fig. 7.1. Input–output
relationship using the
frequency response
function.

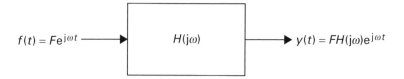

a complex exponential, we guess that the output must also be a complex
exponential of the same frequency

$$y(t) = Y e^{j\omega t}, \tag{7.3}$$

where the output amplitude or phasor Y is complex-valued with an
unknown amplitude and phase. By substituting this particular solution into
the differential equation, we find

$$\left(-\omega^2 + 2\zeta\omega_n\omega j + \omega_n^2\right) Y e^{j\omega t} = \frac{F}{m} e^{j\omega t}. \tag{7.4}$$

Because $e^{j\omega t}$ never vanishes, it can be canceled and Eq. (7.4) reduces to

$$\left(-\omega^2 + 2\zeta\omega_n\omega j + \omega_n^2\right) Y = \frac{F}{m}. \tag{7.5}$$

The ratio of the output amplitude to the input amplitude is given by Y/F.
This ratio is known as the *frequency response function*, and it is denoted by
$H(j\omega)$. Thus, the frequency response for the system of Eq. (7.1) is given by

$$H(j\omega) = \frac{Y}{F} = \frac{\dfrac{1}{m}}{\omega_n^2 - \omega^2 + 2\zeta\omega_n\omega j}. \tag{7.6}$$

The frequency response function is widely used to represent the input–
output relations of LTI systems. Notice that the frequency response is purely
algebraic. It is much easier working with algebraic equations than differential equations so the frequency response is very convenient for calculations.
Given the frequency response and the input amplitude, it is easy to formulate
the output amplitude as

$$Y = FH(j\omega). \tag{7.7}$$

Thus, the desired output is simply

$$y(t) = Y e^{j\omega t} = FH(j\omega)e^{j\omega t}. \tag{7.8}$$

This input–output relationship is illustrated in the block diagram in
Figure 7.1. The input $f(t)$ is a complex exponential. The system is

described by its frequency response function $H(j\omega)$. Thus, the output $y(t)$ is also a complex exponential (of the same frequency as the input) weighted by the frequency response. Alternatively, we say the output of the system is simply the input multiplied by the frequency response. When the input is more complicated, we can break it into a weighted sum of complex exponentials using the Fourier series that we studied earlier. The output is then a new weighted sum of complex exponentials.

7.1 Frequency response of linear, time-invariant systems

The *frequency response function* of a system is defined as the ratio of the output amplitude to the input amplitude when the input is a complex exponential. To find the frequency response function of a system, we assume the input, $f(t)$, to be a complex exponential of the form

$$f(t) = F e^{j\omega t}, \tag{7.9}$$

where F is the input amplitude. Using the method of undetermined coefficients, we know that the output, $y(t)$, must be of the form

$$y(t) = Y e^{j\omega t}. \tag{7.10}$$

In general, Y and F may be complex numbers. The frequency response function is defined as

$$H(j\omega) = \frac{Y}{F} = \frac{\text{output amplitude}}{\text{input amplitude}}. \tag{7.11}$$

This approach gives us the particular solution $y_p(t)$. Because the homogeneous solution always decays to zero in a stable system, the initial conditions become irrelevant if we wait long enough, and in this case, the particular solution defines the steady state output, which we will simply designate as $y(t)$. Thus, $y(t)$ is sometimes called the *steady state sinusoidal* response, and it is given by

$$y(t) = F H(j\omega) e^{j\omega t}. \tag{7.12}$$

Example: Consider the mass attached to a spring and damper shown in Figure 7.2, driven by a moving base with a displacement input of $y_{in}(t)$ connected to the damper. Determine the frequency response function of the system.

Fig. 7.2. A spring–mass–damper system under base excitation.

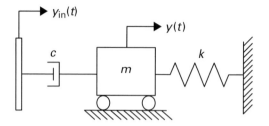

Solution: Sketching the free body diagram and applying Newton's Second Law, we can determine the equation of motion as

$$m\ddot{y} + c\dot{y} + ky = c\dot{y}_{\text{in}}. \tag{7.13}$$

Dividing through by m, we manipulate the governing equation to its canonical form

$$\ddot{y} + 2\zeta\omega_{\text{n}}\dot{y} + \omega_{\text{n}}^2 x = 2\zeta\omega_{\text{n}}\dot{y}_{\text{in}}. \tag{7.14}$$

To find the frequency response function, we set the input $y_{\text{in}}(t)$ to

$$y_{\text{in}}(t) = Y_{\text{in}}e^{j\omega t}. \tag{7.15}$$

We expect the output to be of the form

$$y(t) = Ye^{j\omega t}. \tag{7.16}$$

Substituting Eqs. (7.15) and (7.16) into Eq. (7.14), we obtain

$$\left(-\omega^2 + 2\zeta\omega_{\text{n}}j\omega + \omega_{\text{n}}^2\right)Ye^{j\omega t} = 2\zeta\omega_{\text{n}}j\omega Y_{\text{in}}e^{j\omega t}. \tag{7.17}$$

Because $e^{j\omega t}$ never vanishes, we can safely divide it out of both sides:

$$\left(-\omega^2 + 2\zeta\omega_{\text{n}}j\omega + \omega_{\text{n}}^2\right)Y = 2\zeta\omega_{\text{n}}j\omega Y_{\text{in}}. \tag{7.18}$$

Taking the ratio of the output amplitude to the input amplitude gives

$$H(j\omega) = \frac{Y}{Y_{\text{in}}} = \frac{2\zeta\omega_{\text{n}}\omega j}{-\omega^2 + 2\zeta\omega_{\text{n}}\omega j + \omega_{\text{n}}^2}. \tag{7.19}$$

Equation (7.19) corresponds to the frequency response function for the system of Figure 7.2.

In summary, we can easily find the frequency response function from the governing equation by substituting complex exponentials for the input and output, and performing some algebraic manipulations. Once we know the frequency response, we can easily find the output amplitude and the output when the input is a complex exponential. This input–output relationship is

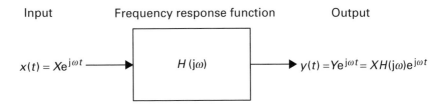

Input   Frequency response function   Output

$x(t) = Xe^{j\omega t}$ ⟶ $H(j\omega)$ ⟶ $y(t) = Ye^{j\omega t} = XH(j\omega)e^{j\omega t}$

Fig. 7.3. Input–output relationship using the frequency response function.

shown in block diagram form in Figure 7.3, where $x(t) = Xe^{j\omega t}$ is the input, then $Y = XH(j\omega)$ is the output amplitude, and $y(t) = Ye^{j\omega t} = XH(j\omega)e^{j\omega t}$ is the output.

The above formulation can also be extended to the case where the input is a constant. For a constant input, the excitation frequency is $\omega = 0$. Assume the constant input has an amplitude of X_0, then the output is also a constant given by $X_0 H(j0)$. This is consistent with the results of Chapter 6, where we noted that for a constant input, the particular solution is also a constant. Thus, for a constant input, the output is simply the input scaled by the frequency response function evaluated at $\omega = 0$.

7.2 Frequency response to a periodic input and any arbitrary input

Consider a stable LTI system with frequency response function $H(j\omega)$. Assume that the input is of the form

$$x(t) = \sum_{k=-\infty}^{\infty} X_k e^{jk\omega_0 t}. \tag{7.20}$$

Because the system is linear, we can apply the principle of superposition. Thus, the output to the input of Eq. (7.20) is the sum of the responses to each part of the input. In other words, the steady state solution is given by

$$y(t) = \sum_{k=-\infty}^{\infty} X_k H(jk\omega_0) e^{jk\omega_0 t}. \tag{7.21}$$

Note that in order to obtain the output of the system, $y(t)$, all we need is the frequency response function of the system, $H(j\omega)$, and the amplitudes X_k of the input at each frequency. We can use the Fourier series to find the amplitudes and frequencies of any periodic input. Even for a non-periodic or a random input, we can make it periodic by repeating it after some time interval of interest. In an introductory differential equations course, we could only find the particular solution when the input was of a simple form where we could guess the appropriate output. Using Fourier series and frequency response, we can find the output for *any* input! Thus, as long as we know the spectrum of the input signal (from signal processing) and the frequency response function of the system (from modeling and its governing equation), we can determine the response of the system to *any* input.

7.3 Bode plots

The frequency response function $H(j\omega)$ is often a complex function of frequency ω, which can be expressed in polar form as

$$H(j\omega) = |H(j\omega)|\, e^{j\angle H(j\omega)}, \tag{7.22}$$

where $|H(j\omega)|$ and $\angle H(j\omega)$ are known as the *gain* (or *magnitude*) and the *phase* of the frequency response function, respectively. Because the output depends on the frequency response function, it is imperative that we fully understand how $|H(j\omega)|$ and $\angle H(j\omega)$ vary with frequency. We often plot the gain and phase with ω on a logarithmic scale. By using the logarithmic scale, we can conveniently span many orders of magnitude on a single plot. Moreover, we will see that certain asymptotic behaviors of $H(j\omega)$ result in straight-line approximations, which allow for ease of plotting and interpretation of the data. Finally, using a log scale, the mathematical operations of multiplication and division are reduced to simple addition and subtraction, respectively. Such a graph is called a *Bode plot*.

Even though we have computer tools such as MATLAB to help us draw the Bode plots, it is still important for us to be able to sketch the Bode plots by hand. Whenever we get a result from a computer, it is important to check that the result is correct. Computers apply the "garbage-in, garbage-out" rule. Moreover, sketching the Bode plots by hand allows us to quickly predict the behavior of the system in limiting cases (that is, at very high and very low frequencies). Finally, knowing the characteristic features of $H(j\omega)$ associated with a system, we can properly identify and accurately model the system.

So far we have focused on modeling a physical system from the constitutive relationships. In the real world, many systems are too complicated to model analytically. However, we can still describe the system by applying

Fig. 7.4. Bode plot axes.

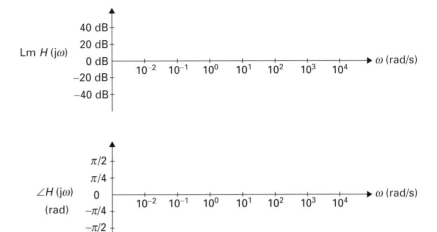

certain inputs (for example, sinusoids of various frequencies) and measuring the outputs. From these input–output measurements, we can deduce the frequency response function of the system experimentally, from which we can then construct a mathematical model of the system. This procedure is called *system identification*.

Some definitions are needed before we can plot the frequency response function, or the Bode plot, by hand. We will use the following definitions extensively.

- Log magnitude (Lm): the logarithm of the gain, expressed in decibels (dB). It is defined as

$$\text{Lm } H(j\omega) = 20 \log_{10} |H(j\omega)| .$$

- Decade above: the frequency range from ω to $10\,\omega$.
- Decade below: the frequency range from ω to $\omega/10$.

The definition of decade is useful for plotting purposes. Using a logarithmic scale, any frequency ratio can be represented by the same distance. For example, the distance from $\omega = 1$ rad/s to $\omega = 10$ rad/s is identical to the distance from $\omega = 100$ rad/s to $\omega = 1000$ rad/s.

A *Bode plot* or a *Bode diagram* is a graph of Lm $H(j\omega)$ and $\angle H(j\omega)$ plotted as a function of ω on a logarithmic scale. Figure 7.4 shows the axes for a Bode plot. We will use a logarithmic scale for ω, and a linear scale for Lm $H(j\omega)$ (but in dB) and $\angle H(j\omega)$ (in radians or degrees).

7.3.1 Constant and first-order factors

Consider a canonical first-order system whose governing equation is given by

$$\frac{dy}{dt} + \frac{1}{\tau} y = x(t). \tag{7.23}$$

The frequency response function is found by assuming the input and output are complex exponentials, from which we obtain

$$H(j\omega) = \frac{Y}{X} = \frac{1}{j\omega + \frac{1}{\tau}} = \frac{\tau}{j\omega\tau + 1}. \tag{7.24}$$

The frequency response function of Eq. (7.24) shows that the numerator and the denominator consist of a constant factor (τ) and a first-order factor ($j\omega\tau + 1$), respectively. Because multiplication and division become addition and subtraction on a log scale, we will consider each factor separately, then add or subtract them to obtain the desired Bode plot.

Constant factors

We begin with the constant factor τ. The constant is often called the *gain factor*. Assuming the system is stable ($\tau > 0$), we can rewrite the constant factor in polar form as

$$\tau = \tau e^{j0}. \tag{7.25}$$

In this form, the gain is simply τ and the phase is always zero. Thus, the log magnitude and the phase of the constant factor are

$$\text{Lm } \tau = 20 \log_{10} \tau \tag{7.26}$$

$$\angle\tau = 0.$$

If $\tau > 1$, its log magnitude is positive. If $0 < \tau < 1$, its log magnitude is negative. Changing τ merely moves the curve up or down, because the gain and phase of a constant factor are independent of ω. The gain and phase of a constant factor are plotted in Figure 7.5. These two graphs constitute the Bode plot.

First-order factors in the denominator

The first-order factor in the denominator is more interesting. It may be rewritten in polar form as

Fig. 7.5. Bode plot of a positive constant factor.

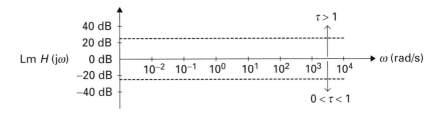

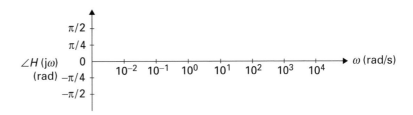

$$\frac{1}{j\omega\tau + 1} = \frac{1}{\sqrt{(\omega\tau)^2 + 1} \; e^{j\tan^{-1}(\omega\tau)}} = \frac{1}{\sqrt{(\omega\tau)^2 + 1}} \; e^{-j\tan^{-1}(\omega\tau)}. \qquad (7.27)$$

In this form, we readily determine the log magnitude and phase of this first-order factor in the denominator as follows:

$$\text{Lm} \; \frac{1}{j\omega\tau + 1} = 20\log_{10} \frac{1}{\sqrt{(\omega\tau)^2 + 1}} = -20\log_{10} \sqrt{(\omega\tau)^2 + 1}$$

$$= -10\log_{10}[(\omega\tau)^2 + 1] \qquad (7.28)$$

and

$$\angle \frac{1}{j\omega\tau + 1} = -\tan^{-1}(\omega\tau). \qquad (7.29)$$

The asymptotic behavior of the log magnitude can be obtained by considering the limiting cases of $(\omega\tau) \ll 1$ and $(\omega\tau) \gg 1$. The following shows the resulting approximations:

• For $(\omega\tau) \ll 1$, we can approximate Eq. (7.28) as

$$\text{Lm} \; \frac{1}{j\omega\tau + 1} \approx -10\log_{10} 1 = 0 \, \text{dB}. \qquad (7.30)$$

• For $(\omega\tau) \gg 1$, we can approximate Eq. (7.28) as

$$\text{Lm} \; \frac{1}{j\omega\tau + 1} \approx -20\log_{10}(\omega\tau) \, \text{dB}. \qquad (7.31)$$

Equations (7.30) and (7.31) are known as the low frequency and high frequency asymptotes, respectively, of a first-order factor in the denominator.

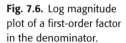

Fig. 7.6. Log magnitude plot of a first-order factor in the denominator.

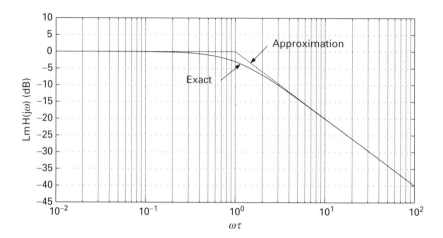

By definition, the *corner frequency* ω_c is the frequency at which the low and the high frequency asymptotic behaviors intersect. It is obtained by setting the two limits of Eqs. (7.30) and (7.31) equal and solving for the frequency:

$$0 = -20\log_{10}(\omega_c \tau) \implies \omega_c = \frac{1}{\tau}. \tag{7.32}$$

Equations (7.30) and (7.31) are approximations, and they predict a value of 0 dB at the corner frequency. It is interesting to compare this approximation with the exact log magnitude at the corner frequency. At ω_c, the exact log magnitude can be obtained from Eq. (7.28), and it is

$$\text{Lm}\,\frac{1}{j\omega_c \tau + 1} = 20\log_{10}\frac{1}{\sqrt{(\omega_c \tau)^2 + 1}}$$

$$= 20\log_{10}\frac{1}{\sqrt{2}} \approx -3\,\text{dB}. \tag{7.33}$$

Therefore, the corner frequency is sometimes called the "minus 3 dB point." Near ω_c there will be significant error in the straight-line approximation of the Bode plot. Thus, if we wish to obtain a more accurate sketch, we need to modify the approximation near ω_c.

The log magnitude of the first-order factor in the denominator is shown in Figure 7.6. At frequencies well below the corner frequency, that is, $\omega\tau \ll 1$, the log magnitude is zero. At frequencies well above the corner frequency, that is $\omega\tau \gg 1$, the log magnitude drops off with frequency at a slope of -20 dB/decade. Alternatively, for every factor of 10 increase in ω or $\omega\tau$, the log magnitude drops by a 20 dB (again, this is strictly valid for high frequencies). We also call this a roll-off of -20 dB/decade. At low and high frequencies, that is at $\omega\tau \ll 1$ and $\omega\tau \gg 1$, the asymptotic approximations are very good. At the corner frequency, the approximation predicts 0 dB while we know the exact log magnitude is approximately -3 dB. Therefore we expect the exact plot will roll-off as shown by the lower line rather than making a sharp bend at the corner (see Figure 7.6).

We have analyzed the asymptotic behaviors for the log magnitude in the previous discussion. We now focus our attention on plotting the phase of Eq. (7.29). The asymptotic behavior of the phase can be obtained by considering the following three limiting cases:

- For $\omega\tau = 1$ (at the corner frequency, $\omega_c = 1/\tau$), the phase is

$$\angle\frac{1}{\mathrm{j}\omega\tau + 1} = -\tan^{-1}(\omega\tau) = -\frac{\pi}{4} = -45°. \tag{7.34}$$

- For $\omega\tau = 1/10$ (a decade below the corner frequency), the phase is

$$\angle\frac{1}{\mathrm{j}\omega\tau + 1} = -\tan^{-1}(\omega\tau) \approx -5.71°. \tag{7.35}$$

At lower frequencies, the arctangent becomes even smaller and the phase approaches zero. We can approximate the phase as zero for all frequencies $\omega\tau \leq 1/10$.

- For $\omega\tau = 10$ (a decade above the corner frequency), the phase is

$$\angle\frac{1}{\mathrm{j}\omega\tau + 1} = -\tan^{-1}(\omega\tau) \approx -84.29°. \tag{7.36}$$

At higher frequencies, the arctangent becomes even larger and the phase approaches $-90°$. We can approximate the phase as $-90°$ ($-\pi/2$ radians) for all frequencies $\omega\tau \geq 10$.

Now we can complete our Bode plot for a first-order factor in the denominator. Recall that its log magnitude plot has a magnitude of zero for frequencies below $\omega\tau$. It drops off with a slope of approximately -20 dB/decade for frequencies above $\omega\tau$. The approximation is worst at the corner frequency $\omega\tau = 1$, where the exact log magnitude is approximately -3 dB, not zero. Nevertheless, this is still a pretty good approximation. In the phase plot, we know the phase is exactly $-\pi/4$ at $\omega\tau = 1$. More than one decade below ($\omega\tau \leq 1/10$), the phase is approximately zero. More than one decade above ($\omega\tau \geq 10$), the phase is approximately $-\pi/2$. Interpolating these points, we see that the phase drops off at about $-\pi/4$ per decade between these extremes.

Fig. 7.7. Bode plot of a
first-order factor in the
denominator.

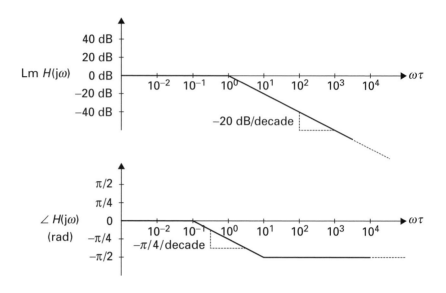

Figure 7.7 shows the Bode plot for a first-order factor in the denominator as a function of the normalized frequency $\omega\tau$, making the results independent of τ. It is made using the straight-line approximation. The exact Bode plot varies slightly from the straight-line approximation. The plot matches very well for the limiting cases of $\omega\tau \ll 1$ and $\omega\tau \gg 1$. It is worst at the corner frequency $\omega\tau = 1$. At low frequencies, its magnitude or gain is one, so its log magnitude is zero. At high frequencies, its magnitude becomes much less than one and its log magnitude therefore becomes negative. In other words, this first-order factor *attenuates* high frequency inputs. Thus, we call a system whose frequency response function is given by Eq. (7.27) a *low-pass filter*. Finally, in order to use the previously derived results, it is important to note that the denominator must be of the form $1 + j\omega\tau$, where the constant term is unity.

First-order factors in the numerator

Consider now a first-order factor in the numerator, whose frequency response function can be expressed as

$$H(j\omega) = j\omega\tau + 1 = \sqrt{(\omega\tau)^2 + 1}\ e^{j\tan^{-1}(\omega\tau)}. \tag{7.37}$$

We could repeat the analysis with limiting cases as we have done previously, but we know that the log of the reciprocal of a function is the negative of the log of the function. Therefore we expect the log magnitude will be the same except that it increases rather than decreases with frequency, as shown

Fig. 7.8. Bode plot of a
first-order factor in the
numerator.

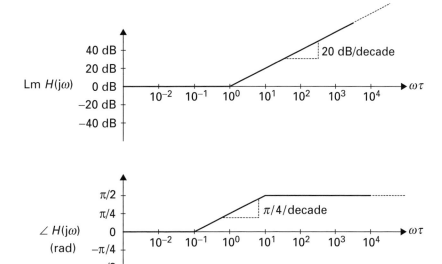

in Figure 7.8. Similarly, the phase will have the opposite sign. The reader
is encouraged to verify these asymptotic results by considering the various
limiting cases.

7.3.2 Multiple first-order factors

Multiplication of functions becomes the addition of their logarithms. Thus,
computing the log magnitude of frequency response functions with products becomes an easy task. For example, consider the frequency response
function shown below, which consists of two first-order factors in the
denominator:

$$H(j\omega) = \frac{1}{(1 + j\omega\tau_1)(1 + j\omega\tau_2)}. \tag{7.38}$$

For simplicity, we assume $\tau_1 = 1$ s and $\tau_1 = 0.1$ s (or in general any positive
real number).

Figure 7.9(a) plots the log magnitude of the first-order factor with the
τ_1 term. The corner frequency is at $\omega_c = 1/\tau_1 = 10^0$ rad/s. Figure 7.9(b)
plots the log magnitude of the first-order factor with the τ_2 term. The corner
frequency is at $\omega_c = 1/\tau_2 = 10^1$ rad/s. In each case the log magnitude is
zero below the corner frequency and drops off at -20 dB/decade above
the corner frequency. The log magnitude of the overall function is the sum

Fig. 7.9. Log magnitude plot of the product of two first-order factors in the denominator (see Eq. (7.38)), for $\tau_1 = 1$ s and $\tau_2 = 0.1$ s.

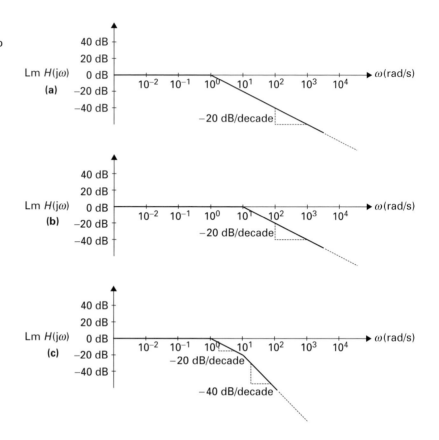

of the two log magnitudes, shown in Figure 7.9(c). At frequencies below 10^0 rad/s, the sum is zero. At frequencies between 10^0 rad/s and 10^1 rad/s, the sum drops at -20 dB/decade. At frequencies above 10^1 rad/s, the sum drops at -40 dB/decade.

The phase plot can be constructed in a similar fashion. For the first-order factor with the τ_1 term, the phase drops from 0 to $-\pi/2$ radians between one decade below and one decade above the corner frequency of 10^0 rad/s (Figure 7.10(a)). For the first-order factor with the τ_2 term, the phase drops from 0 to $-\pi/2$ radians between one decade below and one decade above the corner frequency of 10^1 rad/s (see Figure 7.10(b)). The slope is $-\pi/4$ rad/decade in each case. The phase of the overall function is the sum of the two phases. At frequencies below 10^{-1} rad/s it is zero. At frequencies between 10^{-1} rad/s and 10^0 rad/s it drops at $-\pi/4$ rad/decade. At frequencies between 10^0 rad/s and 10^1 rad/s it drops at $-\pi/2$ rad/decade. At frequencies between 10^1 rad/s and 10^2 rad/s it drops at $-\pi/4$ rad/decade.

Fig. 7.10. Phase plot of the product of two first-order factors in the denominator (see Eq. (7.38)) for $\tau_1 = 1$ s and $\tau_2 = 0.1$ s.

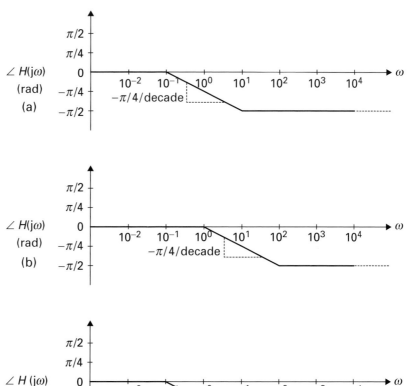

At frequencies above 10^2 rad/s it flattens out at $-\pi$ radians. These plots are shown in Figure 7.10(c).

Figure 7.11 compares the Bode plot that we obtain with the straight-line approximation with the exact values calculated with MATLAB. Observe that the plots match very well. The worst error is -3 dB at the corner frequency of the log magnitude plot.

7.3.3 Second-order factors

A general second-order system can be described with the following canonical expression:

Fig. 7.11. Straight-line
approximation versus the
exact Bode plot for the
system of Eq. (7.38) for
$\tau_1 = 1$ s and $\tau_2 = 0.1$ s.

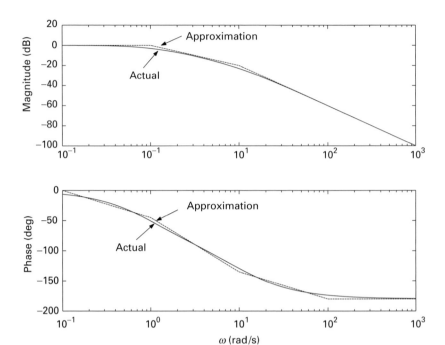

Fig. 7.11. Straight-line approximation versus the exact Bode plot for the system of Eq. (7.38) for $\tau_1 = 1$ s and $\tau_2 = 0.1$ s.

$$\ddot{y} + 2\zeta\omega_n\dot{y} + \omega_n^2 y = x. \tag{7.39}$$

The frequency response function is found by substituting

$$\begin{aligned} x(t) &= X\mathrm{e}^{\mathrm{j}\omega t} \\ y(t) &= Y\mathrm{e}^{\mathrm{j}\omega t} \end{aligned} \tag{7.40}$$

into Eq. (7.39) and taking the ratio of the output amplitude to the input amplitude:

$$\frac{Y}{X} = \frac{1}{-\omega^2 + 2\zeta\omega_n\omega\mathrm{j} + \omega_n^2} = \frac{\left(\dfrac{1}{\omega_n}\right)^2}{1 - \left(\dfrac{\omega}{\omega_n}\right)^2 + 2\zeta\dfrac{\omega}{\omega_n}\mathrm{j}}. \tag{7.41}$$

The quantity in the numerator is a constant. It contributes a gain but does not vary with frequency. Therefore we will ignore it for now and concentrate on the second-order or quadratic factor in the denominator:

$$H(\mathrm{j}\omega) = \frac{1}{1 - \left(\dfrac{\omega}{\omega_n}\right)^2 + 2\zeta\dfrac{\omega}{\omega_n}\mathrm{j}}. \tag{7.42}$$

For factors that are critically damped or overdamped where $\zeta \geq 1$, the denominator can be factored into two first-order terms. After some algebraic manipulations, we factor the second-order denominator polynomial as the product of two first-order polynomials as shown:

$$1 - \left(\frac{\omega}{\omega_n}\right)^2 + 2\zeta \frac{\omega}{\omega_n}j = (1 + j\omega\tau_1)(1 + j\omega\tau_2), \tag{7.43}$$

where τ_1 and τ_2 are

$$\begin{aligned} \tau_1 &= \frac{\zeta - \sqrt{\zeta^2 - 1}}{\omega_n} \\ \tau_2 &= \frac{\zeta + \sqrt{\zeta^2 - 1}}{\omega_n}. \end{aligned} \tag{7.44}$$

It is apparent that these time constants are positive real numbers because $\zeta \geq 1$. Therefore, the Bode plot can be constructed exactly as was done in Section 7.3.2.

The undamped and underdamped case of $0 \leq \zeta < 1$ is more interesting. We cannot simply factor the quadratic term into two first-order factors because the resulting time constants would be complex. For any value of ζ, the frequency response function of Eq. (7.42) can be rewritten in polar form as

$$H(j\omega) = \frac{1}{1 - \left(\frac{\omega}{\omega_n}\right)^2 + 2\zeta \frac{\omega}{\omega_n}j} = |H(j\omega)|\, e^{j\angle H(j\omega)}, \tag{7.45}$$

where the gain is

$$|H(j\omega)| = \frac{1}{\sqrt{\left[1 - \left(\frac{\omega}{\omega_n}\right)^2\right]^2 + \left[2\zeta \frac{\omega}{\omega_n}\right]^2}} \tag{7.46}$$

and the phase is

$$\angle H(j\omega) = -\tan^{-1}\left(\frac{2\zeta \dfrac{\omega}{\omega_n}}{1 - \left(\dfrac{\omega}{\omega_n}\right)^2}\right). \tag{7.47}$$

We will now construct the approximate Bode plot for the frequency response function of Eq. (7.45) for the undamped and the underdamped cases. The log magnitude of Eq. (7.45) is

$$\text{Lm } H(j\omega) = 20 \log_{10} \cfrac{1}{\sqrt{\left[1 - \left(\dfrac{\omega}{\omega_n}\right)^2\right]^2 + \left[2\zeta\dfrac{\omega}{\omega_n}\right]^2}}$$
$$= -10 \log_{10} \left\{\left[1 - \left(\dfrac{\omega}{\omega_n}\right)^2\right]^2 + \left[2\zeta\dfrac{\omega}{\omega_n}\right]^2\right\}. \tag{7.48}$$

Observe that the second form eliminates the square root and moves the function to the numerator. We can again consider the limiting cases:

- For $\omega/\omega_n \ll 1$, we can approximate Eq. (7.48) as

$$\text{Lm } H(j\omega) \approx -10 \log_{10}(1) = 0 \, \text{dB}. \tag{7.49}$$

- For $\omega/\omega_n \gg 1$, we can approximate Eq. (7.48) as

$$\text{Lm } H(j\omega) \approx -10 \log_{10} \left(\dfrac{\omega}{\omega_n}\right)^4 = -40 \log_{10} \left(\dfrac{\omega}{\omega_n}\right) \, \text{dB}. \tag{7.50}$$

Equations (7.49) and (7.50) are known as the low and high frequency asymptotes, respectively, of a second-order factor in the denominator. To determine the corner frequency, ω_c, we compute the frequency at which these two limiting cases intersect by setting them equal. For this second-order factor, we find the corner frequency is $\omega_c = \omega_n$. Recall that the straight-line approximation is worst at the corner frequency. Let us compute the exact value of the frequency response function at this frequency. Substituting $\omega = \omega_c = \omega_n$ into Eq. (7.48), we obtain

$$\text{Lm } H(j\omega) = -10 \log_{10} \left\{\left[1 - \left(\dfrac{\omega_n}{\omega_n}\right)^2\right]^2 + \left[2\zeta\dfrac{\omega_n}{\omega_n}\right]^2\right\}$$
$$= -20 \log_{10}(2\zeta) \, \text{dB}. \tag{7.51}$$

The straight-line approximation predicts that the log magnitude should be zero at the corner frequency. If ζ is close to $1/2$, this is nearly correct. However, if ζ is very small, that is close to zero, the log magnitude may become very large, and the straight-line yields a very poor approximation. The exact log magnitude exhibits a peak near, but not at, the corner frequency. The *exact* log magnitude value (in decibels) at the corner or the natural frequency is $-20 \log_{10}(2\zeta)$.

The frequency at which the peak occurs is known as the *resonant frequency*, and the condition is called *resonance*. The resonant frequency can be found by maximizing the gain of the frequency response function. For a system whose frequency response function is given by Eq. (7.45), its resonant frequency is obtained as follows

$$\frac{d}{d\omega}\left(\frac{1}{\sqrt{\left[1 - \left(\frac{\omega}{\omega_n}\right)^2\right]^2 + \left[2\zeta\left(\frac{\omega}{\omega_n}\right)\right]^2}}\right) = 0. \tag{7.52}$$

After some algebra, we find that the resonant frequency, ω_r, occurs at

$$\omega_r = \omega_n\sqrt{1 - 2\zeta^2}. \tag{7.53}$$

The resonant frequency is a physical quantity. As such, it exists only for a damping factor of $\zeta < \sqrt{0.5}$ (obtained by setting the term inside the radical of Eq. (7.53) greater than zero). Above this value, there will be no resonance, and the exact log magnitude will not exhibit any peak. For a damping factor in the range of $0 \leq \zeta < \sqrt{0.5}$, the lower the damping factor the higher the resonant peak. Note that for a very lightly damped case, that is for $\zeta \ll 1$, the resonant frequency and the natural frequency are nearly identical, that is $\omega_r \approx \omega_n$. In fact, for $\zeta = 0$, the resonant frequency is exactly the same as the natural frequency, and the resonant peak is at infinity.

It is worth emphasizing that the resonant frequency expression of Eq. (7.53) is strictly valid for a second-order system whose frequency response function is given by Eq. (7.45). For higher-order systems, we may have multiple resonant frequencies. For such systems, the resonant frequencies are defined by the local maxima in the gain of the frequency response function.

We now focus our attention on plotting the phase of Eq. (7.45), which is given by

$$\angle\frac{1}{1 - \left(\frac{\omega}{\omega_n}\right)^2 + 2\zeta\frac{\omega}{\omega_n}j} = -\tan^{-1}\left(\frac{2\zeta\frac{\omega}{\omega_n}}{1 - \left(\frac{\omega}{\omega_n}\right)^2}\right). \tag{7.54}$$

The asymptotic behavior of the phase can be obtained by considering the following three limiting cases:

- For $\omega/\omega_n = 1$ (at the corner frequency $\omega_c = \omega_n$), the phase is exactly

$$\angle \frac{1}{1 - \left(\dfrac{\omega}{\omega_n}\right)^2 + 2\zeta \dfrac{\omega}{\omega_n}j} = -\frac{\pi}{2} = 90°. \tag{7.55}$$

- For $\omega/\omega_n \leq 1/10$ (a decade or more *below* the corner frequency), the phase is approximately

$$\angle \frac{1}{1 - \left(\dfrac{\omega}{\omega_n}\right)^2 + 2\zeta \dfrac{\omega}{\omega_n}j} \approx -\tan^{-1}\left(\frac{0}{1}\right) = 0°. \tag{7.56}$$

- For $\omega/\omega_n \geq 10$ (a decade or more *above* the corner frequency), the phase is approximately

$$\angle \frac{1}{1 - \left(\dfrac{\omega}{\omega_n}\right)^2 + 2\zeta \dfrac{\omega}{\omega_n}j} \approx -\tan^{-1}\left(\frac{2\zeta\dfrac{\omega}{\omega_n}}{-\left(\dfrac{\omega}{\omega_n}\right)^2}\right)$$

$$= -\tan^{-1}\left(\frac{2\zeta}{-\dfrac{\omega}{\omega_n}}\right) \approx -\pi = -180°. \tag{7.57}$$

Figure 7.12 shows a Bode plot for the second-order factor using the straight-line approximation. For ease of plotting, ω is non-dimensionalized by ω_n. The log magnitude is zero below the corner frequency. Above the corner frequency it drops off at a slope of -40 dB/decade. This is twice as steep as in a first-order factor. The phase is exactly $-\pi/2$ at the corner frequency (or $\omega/\omega_n = 1 = 10^0$). It is zero at one decade below ω_c (or $\omega/\omega_n = 0.1 = 10^{-1}$) and $-\pi$ at one decade above ω_c (or $\omega/\omega_n = 10 = 10^1$). The straight-line approximation works fairly well for $0.5 < \zeta < 1$. For small values of ζ, the log magnitude displays a spike near the corner frequency. Finally, in order to use the previously derived results, it is important to note that the frequency response function must be identical to Eq. (7.42), where the constant term in the denominator is unity.

Figure 7.13 shows the Bode plots for various values of ζ. Observe how the resonant peak increases and the phase drops more steeply than the straight-line approximation predicts for small ζ. The second-order

Fig. 7.12. Bode plot of an undamped or underdamped second-order factor in the denominator.

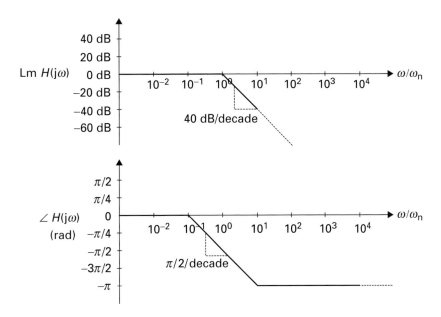

Fig. 7.13. Exact Bode plot of an underdamped second-order factor in the denominator.

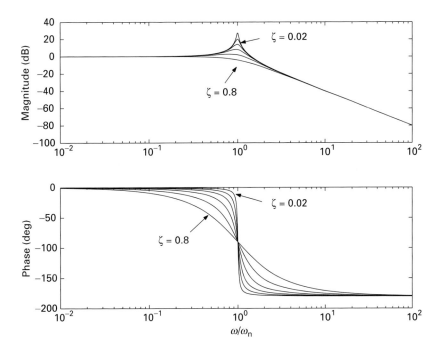

factor is usually called a *low-pass filter* because it passes low frequencies but greatly attenuates higher frequencies. If ζ is small, it enhances frequencies around resonance. Therefore, it is sometimes also called a *band-pass filter*. From Eq. (7.53), note that the resonant peak actually falls slightly below the corner frequency, exactly as we had anticipated. Moreover, note that for $\zeta > \sqrt{0.5}$, the exact log magnitude does not exhibit any peak.

7.4 Impedance

The concept of frequency response function is not only useful for determining the input–output relationship for a system, it is also useful for analyzing the input–output relationship for any element. The concept of impedance has its origin in the field of electricity. As often happens in science or engineering, the concepts and ideas developed in one branch often inspire new approaches and theories in another branch, and indeed this was the case for mechanical impedance. Utilizing the concept of electrical or mechanical impedance, we can apply parallel and series combination rules with elements of different type; we can apply the division rules (current and voltage division rules for electrical systems, and force and relative displacement division rules for mechanical systems) with elements of different type; we can derive the frequency response function of a system without having to first obtain the governing equation in the time-domain. Finally, impedance allows us to gain physical understanding regarding commonly occurring combinations of elements.

The impedance of an electrical element is given by the frequency response function of its constitutive equation, where the voltage drop across the element is defined as the output, and the current through the element is defined as the input. Recall that in the time-domain, the constitutive or current–voltage relationships for electrical elements are given by

$$i_C = C\frac{\mathrm{d}}{\mathrm{d}t}(v_a - v_b) = C\frac{\mathrm{d}v_C}{\mathrm{d}t}$$

$$i_R = \frac{1}{R}(v_a - v_b) = \frac{1}{R}v_R \qquad (7.58)$$

$$i_L = \frac{1}{L}\int_{-\infty}^{t}(v_a - v_b)\mathrm{d}\tau = \frac{1}{L}\int_{-\infty}^{t}v_L\mathrm{d}\tau.$$

The combination rules and division rules obtained in Chapter 5 are strictly valid for elements of the *same* type. Because the constitutive relations for the various elements involve different orders, we can only apply the combination rules or the division rules when the elements are identical.

Because impedances correspond to the frequency response functions, we can reduce the current–voltage relationships to purely algebraic relationships that depend only on frequency. Thus, impedances yield frequency-domain relations, as opposed to the time-domain relations of Eq. (7.58). Since the frequency response function or impedance of any electrical element is purely algebraic, electrical impedance plays the role of a generalized frequency dependent resistance. Alternatively, we may view impedances as complex resistances. Thus, we can readily combine the impedances and apply the division rules in terms of impedances for elements of different type. Impedance is designated by the letter Z.

In summary, the electrical impedance for each element corresponds to its frequency response function, where the voltage across the element is defined as the output, and the current through the element is defined as the input. In the following, we will derive the impedance for each of the three electrical elements.

7.4.1 Inductor

The constitutive relationship for an inductor is given by

$$i_L = \int_{-\infty}^{t} (v_a - v_b)d\tau = \frac{1}{L} \int_{-\infty}^{t} v_L(\tau)d\tau, \tag{7.59}$$

where v_L represents the voltage drop across the inductor. Its impedance is found by substituting

$$i_L(t) = I_L e^{j\omega t}$$
$$v_L(t) = V_L e^{j\omega t} \tag{7.60}$$

and taking the ratio of the output amplitude to the input amplitude

$$Z_L = \frac{V_L}{I_L} = j\omega L. \tag{7.61}$$

Note that at low frequencies ($\omega \to 0$), the impedance of an inductor approaches zero ($Z_L \to 0$). We say that at low frequencies, an inductor behaves like a *short circuit* or a wire with zero resistance. At high frequencies ($\omega \to \infty$), the impedance of an inductor approaches infinity ($Z_L \to \infty$). Thus, at high frequencies an inductor behaves like an *open circuit* or a complex resistor with infinite resistance. Note that the magnitude of the impedance is proportional to frequency, that is,

$$|Z_L| = \omega L \propto \omega. \tag{7.62}$$

Finally, from Eq. (7.61), we note that the phase of the impedance is $\angle Z_L = 90°$, which implies that the voltage *leads* the current by 90°, or alternatively, the current *lags* the voltage by 90°.

7.4.2 Resistor

The constitutive relationship for a resistor is given by

$$i_R(t) = \frac{1}{R}(v_a - v_b) = \frac{1}{R}v_R, \tag{7.63}$$

where v_R represents the voltage drop across the resistor. Its impedance is found by substituting

$$\begin{aligned} i_R(t) &= I_R e^{j\omega t} \\ v_R(t) &= V_R e^{j\omega t} \end{aligned} \tag{7.64}$$

and taking the ratio of the output amplitude to the input amplitude

$$Z_R = \frac{V_R}{I_R} = R. \tag{7.65}$$

Note that the impedance of a resistor is independent of frequency, ω. Because the impedance is real, the current and the voltage are in phase.

7.4.3 Capacitor

The constitutive relationship for a capacitor is given by

$$i_C(t) = C\frac{d}{dt}(v_a - v_b) = C\frac{dv_C}{dt}, \tag{7.66}$$

where v_C represents the voltage drop across the capacitor. Its impedance is found by setting

$$\begin{aligned} i_C(t) &= I_C e^{j\omega t} \\ v_C(t) &= V_C e^{j\omega t} \end{aligned} \tag{7.67}$$

and taking the ratio of the output amplitude to the input amplitude

$$Z_C = \frac{V_C}{I_C} = \frac{1}{j\omega C}. \tag{7.68}$$

Note that at low frequencies ($\omega \to 0$), the impedance of a capacitor approaches infinity ($Z_C \to \infty$). Thus, we say at low frequencies a capacitor behaves like an open circuit. At high frequencies ($\omega \to \infty$), the impedance of a capacitor approaches zero ($Z_C \to 0$). Thus, at high frequencies a capacitor behaves like a short circuit. Note that the magnitude of the impedance is inversely proportional to frequency, that is

$$|Z_C| = \frac{1}{\omega C} \propto \frac{1}{\omega}. \tag{7.69}$$

Finally, from Eq. (7.68), we note that the phase of the impedance is $\angle Z_C = -90°$, which implies that the voltage *lags* the current by 90°, or alternatively, the current *leads* the voltage by 90°.

7.5 Combination and division rules using impedance

Because the impedance of any electrical element is purely *algebraic*, electrical impedance plays the role of a generalized *frequency dependent resistance*. Once we replace all of the electrical elements by their corresponding impedances, we can treat the circuit as consisting of a collection of "resistors," thus allowing us to treat the different elements on equal footing. Hence, using impedances, we can combine different types of electrical elements in the same manner as we do resistors.

Specifically, when N elements are in parallel, we can combine them into a single element whose equivalent impedance is given by

$$\frac{1}{Z_p} = \sum_{i=1}^{N} \frac{1}{Z_i}, \tag{7.70}$$

where Z_i denotes the impedance of the ith element. Similarly, when N elements are in series, we can combine them into a single element whose equivalent impedance is given by

$$Z_s = \sum_{i=1}^{N} Z_i. \tag{7.71}$$

The concept of impedance also allows us to apply the current division and voltage division rules for elements of different type. For N elements in parallel (see Figure 7.14(a)), the ratio of the current through the ith element to the total current is given by

$$I_i = \frac{\dfrac{1}{Z_i}}{\dfrac{1}{Z_p}} I_{\text{overall}}. \tag{7.72}$$

Fig. 7.14. (a) Parallel and (b) series impedance combinations.

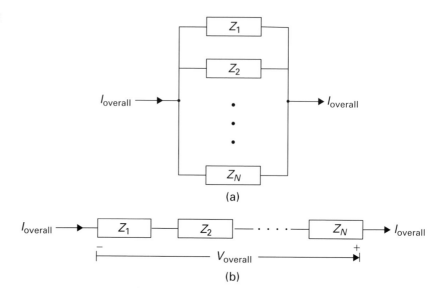

(a)

(b)

For N elements in series (see Figure 7.14(b)), the ratio of the voltage drop across the ith element to the total voltage drop is simply the ratio of its impedance to the total impedance as follows:

$$V_i = \frac{Z_i}{Z_s} V_{\text{overall}}. \tag{7.73}$$

Equation (7.73) is the familiar voltage divider rule.

7.5.1 Examples

We will now consider various examples and demonstrate the utility of the impedance approach of solving circuit problems.

Example 1: Consider the circuit of Figure 7.15. Our objective is to find the frequency response of the system, where the voltage drop across the capacitor is the output and the voltage source is the input.

Solution: The frequency response of the system can be obtained directly by first obtaining the governing equation, assuming complex input and output, and then taking the ratio of the output amplitude to the input amplitude. Alternatively, we can determine the frequency response function by using impedances. We will compute the frequency response function using both approaches for this example problem.

Fig. 7.15. Electric circuit for Example 1.

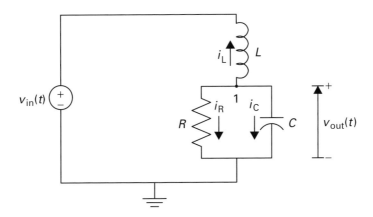

Applying KCL at node 1, we obtain

$$\sum_{\text{out of note 1}} i = 0. \tag{7.74}$$

In terms of the currents through the various elements, we have

$$i_L + i_R + i_C = 0. \tag{7.75}$$

Substituting the constitutive equations for the elements, we get

$$\frac{1}{L} \int_{-\infty}^{t} (v_{\text{out}} - v_{\text{in}}) d\tau + \frac{1}{R} v_{\text{out}} + C \frac{dv_{\text{out}}}{dt} = 0. \tag{7.76}$$

Taking the derivative of Eq. (7.76) with respect to time and rearranging, we obtain the following governing equation for the system of Figure 7.15:

$$C \frac{d^2 v_{\text{out}}}{dt^2} + \frac{1}{R} \frac{dv_{\text{out}}}{dt} + \frac{1}{L} v_{\text{out}} = \frac{1}{L} v_{\text{in}}. \tag{7.77}$$

To find the desired frequency response function, we set

$$v_{\text{in}}(t) = V_{\text{in}} e^{j\omega t}$$
$$v_{\text{out}}(t) = V_{\text{out}} e^{j\omega t}. \tag{7.78}$$

Substituting Eq. (7.78) into Eq. (7.77) and taking the ratio of the output amplitude to the input amplitude, we find

$$H(j\omega) = \frac{V_{\text{out}}}{V_{\text{in}}} = \frac{\dfrac{1}{L}}{-C\omega^2 + j\dfrac{\omega}{R} + \dfrac{1}{L}} = \frac{R}{-LCR\omega^2 + j\omega L + R}. \tag{7.79}$$

Equation (7.79) constitutes the frequency response function for the system of Figure 7.15.

Instead of solving the problem in the time-domain and then introducing complex exponentials to transform the problem into the frequency-domain,

Fig. 7.16. Circuit of Figure 7.15 in terms of impedances.

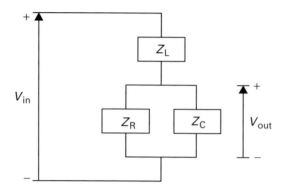

we can solve for the frequency response function directly in the frequency-domain using impedances. Figure 7.16 shows the circuit of Figure 7.15, with all the elements replaced with their corresponding impedances. Thus, it is implicitly understood that we are now in the frequency-domain, and the input and output can all be replaced with their respective amplitudes, denoted by V_{in} and V_{out}. By inspection, we note that Z_R and Z_C are in parallel, and their equivalent is given by

$$\frac{1}{Z_{eq}} = \frac{1}{Z_R} + \frac{1}{Z_C}. \tag{7.80}$$

Solving for Z_{eq}, we get

$$Z_{eq} = \frac{Z_R Z_C}{Z_R + Z_C} = \frac{\dfrac{R}{C\omega j}}{R + \dfrac{1}{C\omega j}} = \frac{R}{CR\omega j + 1}. \tag{7.81}$$

We note that Z_L and Z_{eq} are in series. To find the voltage drop across Z_{eq}, we apply the voltage division rule as follows:

$$V_{out} = \frac{Z_{eq}}{Z_L + Z_{eq}} V_{in} \tag{7.82}$$

or

$$V_{out} = \frac{\dfrac{R}{j\omega CR + 1}}{j\omega L + \dfrac{R}{j\omega CR + 1}} V_{in}. \tag{7.83}$$

Manipulating Eq. (7.83), we obtain the desired frequency response function:

$$H(j\omega) = \frac{V_{out}}{V_{in}} = \frac{R}{-LCR\omega^2 + j\omega L + R} \tag{7.84}$$

which is identical to Eq. (7.79).

Fig. 7.17. General
frequency response
function.

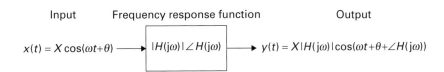

The frequency response function allows us to readily determine the steady state response of a stable system when its input is a complex exponential (see Figure 7.3). It is a relatively straightforward exercise to extend the result of Figure 7.3 to an input, $x(t)$, that is a general sinusoid given by

$$x(t) = X \cos(\omega t + \theta). \tag{7.85}$$

If the frequency response function of a stable system is denoted by $H(j\omega)$, then its steady state sinusoidal response is given by

$$y(t) = X \, |H(j\omega)| \cos(\omega t + \theta + \angle H(j\omega)), \tag{7.86}$$

where $|H(j\omega)|$ and $\angle H(j\omega)$ correspond to the gain and phase of the frequency response function. The result of Eq. (7.86) is one of the most important results in the study of LTI systems. Stated in words, we say that

For a stable, linear time-invariant system having a frequency response function of $H(j\omega)$, its steady state response to an input sinusoid will be a sinusoid of the same frequency having an amplitude that is scaled by $|H(j\omega)|$, and a phase shift of $\angle H(j\omega)$ relative to the input.

Figure 7.17 represents this important input–output relationship using the frequency response function. Finally, if the input is

$$x(t) = X_1 \sin(\omega_1 t + \theta_1) + X_2 \cos(\omega_2 t + \theta_2) \tag{7.87}$$

then its steady state sinusoidal response can be obtained using superposition as follows:

$$y(t) = X_1 \, |H(j\omega_1)| \sin(\omega_1 t + \theta_1 + \angle H(j\omega_1)) \tag{7.88}$$
$$+ X_2 \, |H(j\omega_2)| \cos(\omega_2 t + \theta_2 + \angle H(j\omega_2)).$$

Note that the gain and phase associated with each input are evaluated at the excitation frequency of that particular input.

Fig. 7.18. Electric circuit for Example 2.

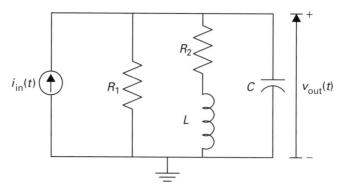

Fig. 7.19. Circuit of Figure 7.18 in terms of impedances.

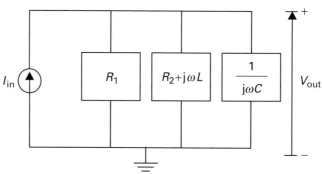

Example 2: Consider Figure 7.18. Determine the steady state voltage drop across the capacitor when the current source is $i_{in}(t) = 10\cos(1000t)$ A. The system parameters are $R_1 = R_2 = 10\,\Omega$, $L = 10\,\text{mH}$, and $C = 1000\,\mu\text{F}$.

Solution: Using impedances, the circuit of Figure 7.18 becomes the circuit of Figure 7.19. Because R_2 and L are in series, they can be combined to give $R_2 + j\omega L$ (see Figure 7.19). Thus, we can simplify Figure 7.19 to Figure 7.20. The equivalent impedance for the circuit can be obtained by applying the parallel combination rule as follows:

$$\frac{1}{Z_{eq}} = \frac{1}{R_1} + \frac{1}{R_2 + j\omega L} + j\omega C. \tag{7.89}$$

But $R_1 = R_2 = R$, so

$$Z_{eq} = \frac{R^2 + j\omega RL}{-RLC\omega^2 + (L + CR^2)j\omega + 2R}. \tag{7.90}$$

The frequency response function of the system is simply the equivalent impedance of the circuit, that is

$$H(j\omega) = \frac{V_{out}}{I_{in}} = Z_{eq}. \tag{7.91}$$

Fig. 7.20. Simplified circuit of Figure 7.19.

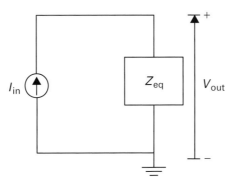

For an excitation frequency of $\omega = 1000 \, \text{rad/s}$, the frequency response function becomes

$$H(j\omega)|_{\omega=1000} = \frac{100 + 100j}{-80 + 110j} = \frac{100\sqrt{2}e^{j\frac{\pi}{4}}}{10\sqrt{185}e^{j\tan^{-1}(\frac{110}{-80})}} \tag{7.92}$$

$$= 1.0389e^{j(\frac{\pi}{4}-2.1996)} = 1.0389e^{j(-1.4142)}.$$

As a side note, in determining the phase of the denominator polynomial $(-80 + 110j)$, we have to be careful to properly identify the quadrant in which the complex vector lies. By inspection, we note the vector $(-80 + 110j)$ lies in the second-quadrant of the complex plane, thus its phase is given by 2.1996 rad. At steady state, the output amplitude is simply the input amplitude multiplied by the gain of the frequency response function, and the output picks up the phase of the frequency response function relative to the input. Thus, the steady state voltage drop across the capacitor is simply

$$(v_{\text{out}})_{\text{ss}} = I_s \, |H(j\omega)| \cos{(\omega t + \angle H(j\omega))}$$

$$= 10(1.0389)\cos{(1000t - 1.4142)} \text{ V} \tag{7.93}$$

$$= 10.398\cos(1000t - 1.4142) \text{ V}.$$

Example 3: Consider the circuit of Figure 7.21. Determine the steady state voltage drop across the inductor when the voltage source is $v_{\text{in}}(t) = 500\cos(10t)$ V. The system parameters are $R = R_1 = R_2 = 10\,\Omega, L = 1$ H, and $C = 0.01$ F.

Solution: Replacing all of the electrical elements by their impedances, we obtain the circuit of Figure 7.22. By inspection, we note that the resistor R_2

Fig. 7.21. Electric circuit for Example 3.

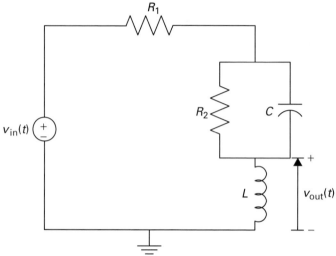

Fig. 7.22. Circuit of Figure 7.21 in terms of impedances.

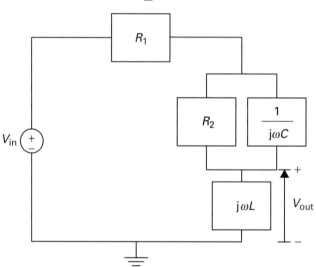

and the capacitor are in parallel, and their equivalent impedance is given by

$$\frac{1}{Z_p} = \frac{1}{R_2} + j\omega C \tag{7.94}$$

or

$$Z_p = \frac{R_2}{1 + j\omega R_2 C} \tag{7.95}$$

which is in series with the resistor R_1 and the inductor L. To find the voltage drop across the inductor, we apply the voltage division rule as follows:

$$V_{out} = \frac{Z_L}{Z_{R_1} + Z_p + Z_L} V_{in} = \frac{j\omega L}{R_1 + \dfrac{R_2}{1 + j\omega R_2 C} + j\omega L} V_{in}. \tag{7.96}$$

Fig. 7.23. Figure for Problem 7-1.

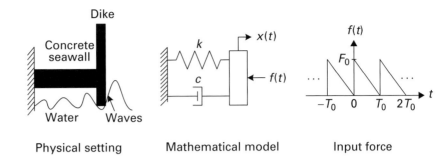

Dike

Concrete seawall

Water   Waves

Physical setting

Mathematical model

Input force

But $R = R_1 = R_2$, so we readily obtain the frequency response function of the system

$$H(j\omega) = \frac{V_{\text{out}}}{V_{\text{in}}} = \frac{j\omega L - LRC\omega^2}{R + j\omega C R^2 + R + j\omega L - LRC\omega^2}. \qquad (7.97)$$

In terms of the given system parameters, Eq. (7.97) reduces to

$$H(j\omega) = \frac{10j - 10}{20j + 10} = \frac{\sqrt{10^2 + 10^2}}{\sqrt{20^2 + 10^2}} e^{j\left[\tan^{-1}\left(\frac{10}{-10}\right) - \tan^{-1}\left(\frac{20}{10}\right)\right]} \qquad (7.98)$$
$$= 0.6325 e^{j(1.248)}.$$

Thus, the steady state voltage drop across the inductor is

$$(v_{\text{out}})_{\text{ss}} = V_{\text{in}} |H(j\omega)| \cos(\omega t + \angle H(j\omega))$$
$$= 500(0.6325) \cos(10t + 1.248) \text{ V} \qquad (7.99)$$
$$= 316.25 \cos(10t + 1.248) \text{ V}.$$

The impedance method offers a convenient approach for formulating the frequency response function of a system, and for determining its steady state response. The approach is purely algebraic, and the combination and division rules can be easily applied even if the elements are of different type. Because impedance is formulated in the frequency domain, it also allows us to develop a feel for how elements or combinations of elements will behave as a function of the excitation frequency.

7.6 Problems

7-1 Waves impact a sea wall. It is desired to calculate the resulting vibration. Figure 7.23 illustrates the physical system and a suitable lumped-element mathematical model. The impact from the waves can be

Fig. 7.24. Figure for
Problem 7-3.

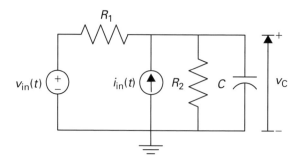

described by a periodic force $f(t)$ as shown. Calculate the response
of the sea wall–dike system to such a load. Please leave your solution
in terms of ζ, ω_n and other relevant system and input parameters.

7-2 We have shown that the response of a linear time-invariant or LTI
system to a complex exponential input of the form

$$f(t) = Fe^{j\omega t}$$

is given by

$$y(t) = F \cdot H(j\omega)e^{j\omega t},$$

where $H(j\omega)$ represents the frequency response function of the LTI
system. Now consider a real input of the form

$$f(t) = F_0 \cos(\omega t + \theta),$$

where F_0 and θ denote the input amplitude and the input phase (both
real constants). Show that the response of a LTI system, whose system
parameters are all real, to this real input $f(t)$ is given by

$$y(t) = F_0 |H(j\omega)| \cos(\omega t + \theta + \angle H(j\omega)),$$

where $|H(j\omega)|$ and $\angle H(j\omega)$ are the magnitude and phase of the fre-
quency response function, respectively.

7-3 Consider the circuit shown in Figure 7.24 with a voltage input, $v_{in}(t)$,
and a current input, $i_{in}(t)$. The voltage drop across the capacitor, $v_C(t)$,
is the output of the system.

(a) Using the concept of impedance, find the frequency response function $H(j\omega)$ of the system when $i_{in}(t) = 0$. In this case, $v_{in}(t)$ is the
input and $v_C(t)$ is the output.

Fig. 7.25. Figure for
Problem 7-4.

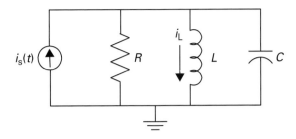

Fig. 7.26. Figure for
Problem 7-5.

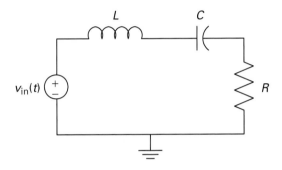

(**b**) Assuming $R_1 = R_2 = 20\,\text{k}\Omega$ and $C = 2\,\mu\text{F}$, sketch the straight
line approximation of the Bode plot for the frequency response
function of part (**a**). Identify all of the relevant break points and
the corresponding frequencies.

(**c**) Repeat part (**a**) when $v_{in}(t) = 0$. In this case, $i_{in}(t)$ is the input and
$v_C(t)$ is the output.

7-4 Consider the circuit shown in Figure 7.25, where $i_s(t)$ is the input and
$i_L(t)$ is the output.

(**a**) Using the concept of impedance, find the frequency response function $H(j\omega)$ of the system.

(**b**) If $R = 20\,\Omega$, $L = 0.1\,\text{mH}$ and $C = 1\,\mu\text{F}$, sketch the straight line
approximation of the Bode plot. Identify and plot the exact value
of $\text{Lm}\,(H(j\omega))$ at the corner frequency.

7-5 Consider the circuit in Figure 7.26.

(**a**) Determine the frequency response function of the system, where
$v_{in}(t)$ is the input and the voltage drop across the capacitor, $v_C(t)$,
is the output.

(**b**) If $R = 1\,\text{k}\Omega$, $L = 1.276\,\text{mH}$, $C = 78.21\,\text{pF}$, and $v_{in}(t) = [3 + \cos(\omega_n t)]\,\text{V}$, where ω_n is the undamped natural frequency

Fig. 7.27. Figure for
Problem 7-6.

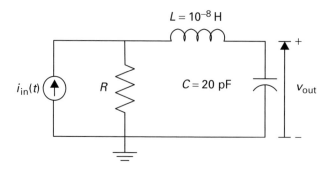

of the system, what is the steady state voltage drop across the
capacitor, $v_C(t)$?

7-6 Consider the circuit in Figure 7.27, where $R = 40\,\Omega$.

    (a) Determine the frequency response function of the system, where
the current source, $i_{in}(t)$, constitutes the input, and the voltage drop
across the capacitor, $v_{out}(t)$, is the output.

    (b) Sketch the Bode plot of the system using the straight line approx-
imations. Identify all of the relevant break points and the corresponding frequencies on the Bode plot.

    (c) Sketch the exact Bode plot.

    (d) If you were given the exact Bode plot of part **(c)** and a value
for R, describe how you would determine the values for L and
C, assuming they are not specified. This constitutes an inverse
problem that is often encountered in engineering applications.

7-7 The schematic for the harmonic filter in the NorCal 40A amateur
band transceiver is shown in Figure 7.28. It is designed to reduce the
harmonics of the 7 MHz carrier produced by the power amplifier. These
spurious emissions (or *spurs* for short) occur at harmonic frequencies
of 14, 21, and 28 MHz and can interfere with other radio services.

    (a) Show that the frequency response of the filter is given by

$$H(j\omega) = \frac{R}{RL_1L_2C_2C_3(j\omega)^4 + L_1L_2C_2(j\omega)^3 + R(L_1C_2 + L_1C_3 + L_2C_3)(j\omega)^2 + (L_1 + L_2)(j\omega) + R}.$$

    (b) Use MATLAB to plot a Bode diagram showing the magnitude
and phase of the frequency response from 10^3 to 10^8 Hz. What is
the magnitude (gain) at 7, 14, 21, and 28 MHz?

    (c) Determine an expression for the impedance at the input of the filter
and plot a Bode diagram showing its magnitude and phase from
10^3 to 10^8 Hz.

Fig. 7.28. Figure for
Problem 7-7.

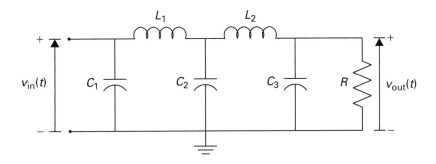

$$L_1 = L_2 = 1.3\ \mu\text{H},\ C_1 = C_3 = 330\ \text{pF},\ C_2 = 820\ \text{pF},\ R = 50\ \Omega$$

(d) Determine the steady-state response for

$$v_{\text{in}}(t) = \{(10\cos[2\pi(7 \times 10^6)t] + 10\sin[2\pi(14 \times 10^6)t]\}\ \text{V}.$$

7-8 Consider the following frequency response function:

$$H(j\omega) = \frac{4 + 4j\omega}{j\omega(200 + 20j\omega)}.$$

Sketch the Bode plot (magnitude and phase) using straight line approximations. **Hint**: The $j\omega$ factor produces a straight line without any
approximation.

7-9 Impedances are useful for finding the frequency response function
of a system directly without having to first determine the governing
equation. Using the concept of impedance, determine the frequency
response functions, V_C / V_{in} and V_C / I_{in}, for Problem 5-2, where V_C is
the amplitude of the voltage drop across the capacitor, and V_{in} and
I_{in} are the amplitudes for the voltage source and the current source,
respectively. Compare your results with the frequency response functions determined by first deriving the governing equations. **Hint**: A
system with N inputs and one output will have N frequency response
functions. To find the ith frequency response function, we set all of
the inputs to zero except for the ith input.

7-10 Use impedances to determine the frequency response functions,
$H(j\omega)$, for the circuit shown in Figure 7.29. The input is $v_{\text{in}}(t)$ in
all cases.

Fig. 7.29. Figure for
Problem 7-10.

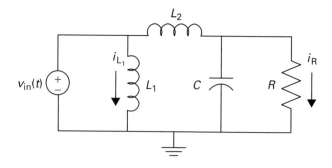

Fig. 7.30. Figure for
Problem 7-11.

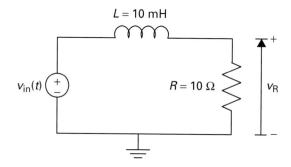

(a) If V_{L_2} is the output amplitude, find $H(j\omega)$. Why is $H(j\omega)$ independent of L_1?

(b) If I_{L_1} is the output amplitude, find $H(j\omega)$. Why is $H(j\omega)$ independent of L_2, R and C? Is this still true if the voltage source input were changed to a current source input?

(c) If I_R is the output amplitude, find $H(j\omega)$. How will $H(j\omega)$ change if the current through the resistor, I_R, is drawn as going upward rather than downward?

7-11 Consider the circuit shown in Figure 7.30.

(a) Derive the governing equation for the voltage drop across the resistor, $v_R(t)$.

(b) Show that the frequency response function is given by

$$H(j\omega) = \frac{1}{1 + j\omega L/R}.$$

(c) Determine the steady state output when the input is

$$v_{in}(t) = [10\cos(t) + 10\cos(1000t) + 10\cos(10\,000t)] \text{ V}.$$

Fig. 7.31. Figure for Problem 7-12.

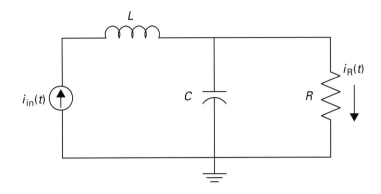

Fig. 7.32. Figure for Problem 7-13.

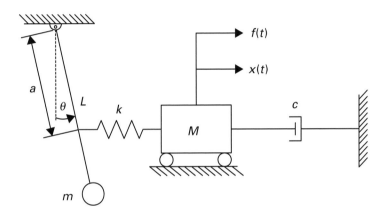

(d) Based on your results, is this circuit a high-pass or a low-pass filter? Please explain.

7-12 Consider the circuit shown in Figure 7.31.

(a) Find the steady state current through the resistor, $i_R(t)$, for an input current source of $i_{in}(t) = i_1 \sin(10t) + i_2 \cos(5t)$, where i_1 and i_2 are constants.

(b) For $i_2 = 0$ and $i_{in}(t) = i_1 \sin(10t)$, it is desired that at steady state the amplitude of $i_R(t)$ be greater than 25 percent of $i_{in}(t)$ and the phase shift be less than $-\pi/4$. What values of R will accomplish this for $L = 1\,\mathrm{H}$, $C = 1\,\mu\mathrm{F}$ and $i_1 = 1\,\mathrm{A}$?

7-13 A spring–mass–damper system is connected to a pendulum of length L, having a bob of mass m, as shown in Figure 7.32.

(a) Assuming small displacements, determine the governing equations that describe the motion of the outputs, $x(t)$ and $\theta(t)$, of the system.

Fig. 7.33. Figure for
Problem 7-14.

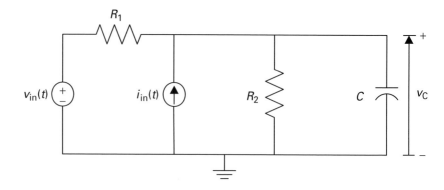

(b) Determine the frequency response function of the system, where
the output is the angular displacement, $\theta(t)$, of the pendulum and
the input is the applied force, $f(t)$.

(c) Determine the frequency response function of the system, where
the output is the horizontal displacement, $x(t)$, of the cart and the
input is the applied force, $f(t)$.

(d) Compare the denominator of the frequency response functions in
parts **(a)** and **(b)**. What conclusions can you draw? Please explain
your answer.

7-14 Inputs to the circuit shown in Figure 7.33 are the voltage source, $v_{in}(t)$,
and the current source, $i_{in}(t)$. The output is the voltage drop across the
capacitor, $v_C(t)$.

(a) Derive the governing equation for the circuit and leave the solution
in canonical form. Express the canonical parameter in terms of R_1,
R_2 and C.

(b) Determine the frequency response function of the circuit if:

    i) $i_{in}(t) = 0$;

    ii) $v_{in}(t) = 0$.

(c) If $R_1 = R_2 = 20 \times 10^3\,\Omega$ and $C = 2 \times 10^{-6}\,\mathrm{F}$, determine the
steady state output voltage $v_C(t)$ when $i_{in}(t) = 1 \times 10^{-3}\,\mathrm{A}$ and
$v_{in}(t) = 2\cos(50t + \pi/4)\,\mathrm{V}$.

7-15 Given the circuit parameters C_1, C_2, C_3 and C_4, determine the frequency response function $H(j\omega)$ of the circuit shown in Figure 7.34.

7-16 Given R and C of the circuit shown in Figure 7.35, determine the
frequency response function $H(j\omega)$.

Fig. 7.34. Figure for
Problem 7-15.

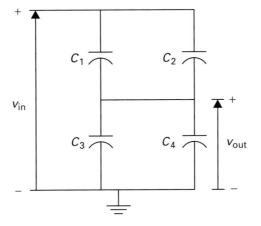

Fig. 7.35. Figure for
Problem 7-16.

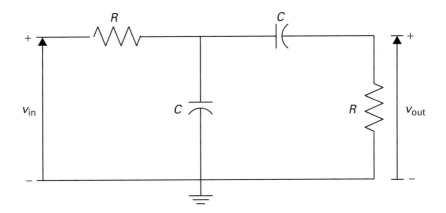

7-17 Consider the circuit shown in Figure 7.36. At *high frequencies*, the
circuit can be simplified using only three elements. Sketch the simplified circuit and determine the frequency response function for the
simplified circuit, for $\omega \gg 1$.

7-18 Consider the circuit shown in Figure 7.37, where $R_1 = 5\,\Omega$, $L =$
$1.5\,\text{H}$, $C = 10\,\text{mF}$, $R_2 = 10\,\Omega$ and $i_{in}(t) = [3 + 10\cos(10t)]$ A.
  Let the voltage across the capacitor, $v_1(t)$, be the output of the
system.
(a) Determine the frequency response function, $H(j\omega)$, of the system.
(b) Determine the steady-state voltage $v_1(t)$.
(c) Sketch the Bode plot for $H(j\omega)$ using straight line approximations.
  Clearly identify all the relevant frequencies and slopes on the plots.
(d) Sketch the exact Bode plot.

Fig. 7.36. Figure for
Problem 7-17.

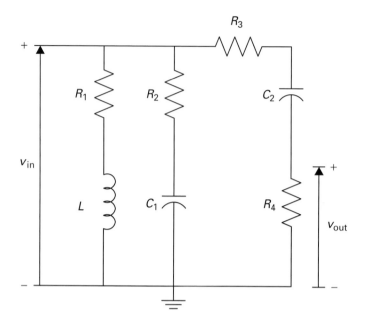

Fig. 7.37. Figure for
Problem 7-18.

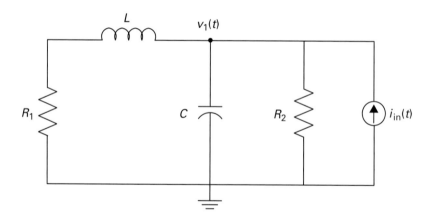

7-19 Consider a system whose frequency response function (gain and phase) is shown in Figure 7.38(a). Determine an expression for the steady state output $y(t)$ for an input $x(t)$ given by the pulse train shown in Figure 7.38(b).

7-20 Consider the RC circuit shown in Figure 7.39(a), whose unit step response is shown in Figure 7.39(b). The resistance of the circuit is $R = 20\,\Omega$.

(a) Determine the capacitance, C.

Fig. 7.38. Figure for
Problem 7-19.

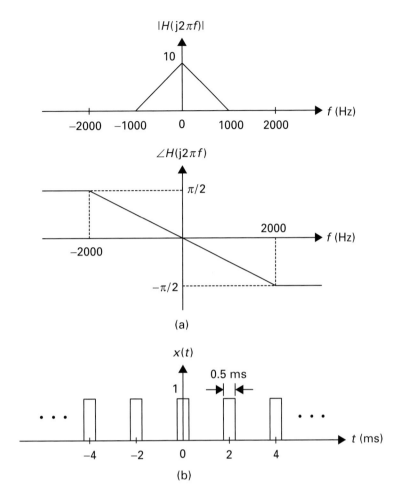

(a)

(b)

(b) Sketch the Bode plot for the circuit using straight-line approximations.

7-21 Consider a second-order system with the following frequency response
function:

$$H(j\omega) = \cfrac{1}{1 - \left(\cfrac{\omega}{\omega_n}\right)^2 + 2\zeta\cfrac{\omega}{\omega_n}j}.$$

(a) Show that its resonant frequency, ω_r, is given by $\omega_r = \omega_n\sqrt{1 - 2\zeta^2}$.

(b) By definition, the frequencies ω_1 and ω_2 at which the gain of $H(j\omega)$
falls to $1/\sqrt{2}$ of its peak amplitude are known as the *half-power* or

Fig. 7.39. Figure for
Problem 7-20.

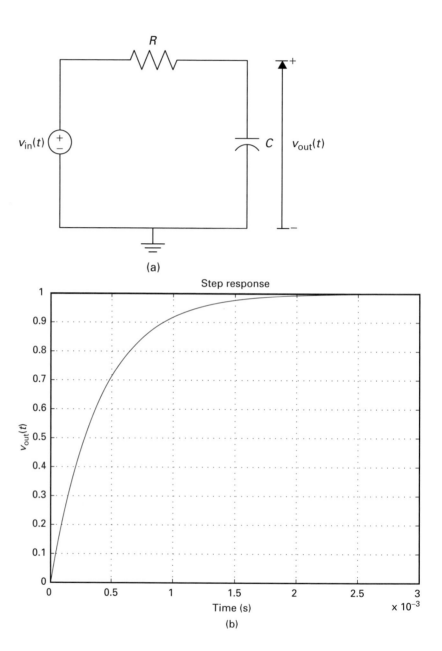

(a)

(b)

3 dB points, and the frequency range from ω_1 to ω_2 is referred to as the *bandwidth* of the system. Show that for light damping, the bandwidth for the given $H(j\omega)$ is

$$\Delta\omega = \omega_2 - \omega_1 \approx 2\zeta\omega_n.$$

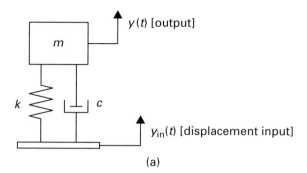

y(t) [output]

m

k   c

$y_{in}(t)$ [displacement input]

(a)

Fig. 7.40. Figure for
Problem 7-22.

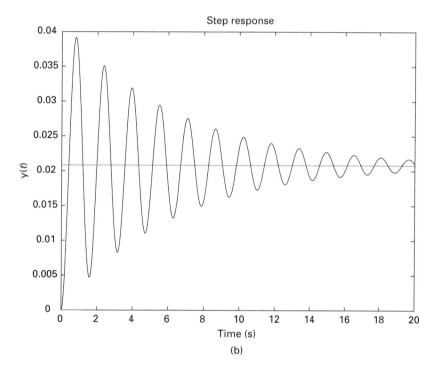

(b)

Bandwidth is important because it can be used as a quick way of
estimating the damping factor ζ.

7-22 Consider the base excitation system shown in Figure 7.40(a), where
      $m = 3\,\text{kg}$. The input and output of the system are $y_{in}(t)$ and $y(t)$,
      respectively.

      (a) When the base is held fixed, the response of the system to a unit
      step force input applied to m is shown in Figure 7.40(b). Determine
      the system parameters c and k.

(b) Using straight-line approximations, sketch the Bode plot of the system when the base is allowed to move.

(c) Sketch the exact Bode plot and determine the resonant frequency of the system.

7-23 In order to perform microgravity experiments in space, an active suspension system for the experimental enclosure is constructed. However, experiments require power, cooling and data transfer, which necessitate connections to the space platform. Our objective is to determine the stiffness and damping of the umbilical that provides the aforementioned necessities to the experiments (see Figure 7.41(a)), where $x(t)$ represents the displacement of the umbilical at its right end. In order to compute the umbilical stiffness and damping, it is necessary to simulate a weightless environment, which can be mimicked by hanging the experimental box as shown in Figure 7.41(b). A mathematical model of the experimental setup is shown in Figure 7.41(c).

(a) Assuming small displacements, determine the governing equation for the system.

(b) For an input of $x(t) = x_0 \sin(\omega t)$, determine the steady state response of the system.

(c) Figure 7.41(d) illustrates the experimentally determined Bode plot of the system, where the input and output of the system are $x(t)$ and $\theta(t)$, respectively. For an excitation frequency of $\omega = 4\,\text{rad/s}$ and an input amplitude of x_0, determine the steady state sinusoidal response of the system.

(d) Set up, but do not solve, the equation that you would need to use to determine the resonant frequency, ω_r, of the system.

(e) For a particular umbilical chosen, the damping is so small that for all practical purposes $\omega_r \approx \omega_n$. For $L = 1.3\,\text{m}$ and $m = 14.5\,\text{kg}$, determine the stiffness, k, and damping coefficient, c, of the umbilical.

7-24 The impedance of a mechanical element is given by the frequency response function of its constitutive equation, where the force through the element is defined as the output, and the relative velocity between the endpoints is defined as the input. This is analogous to the reciprocal of electrical impedance, in the same way that the damping coefficient is analogous to the reciprocal of the electrical resistance. Recall that in

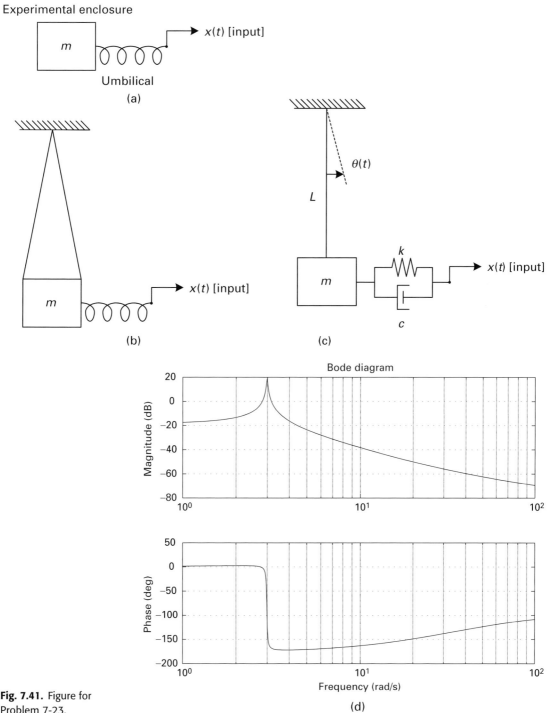

Fig. 7.41. Figure for Problem 7-23.

Section 5.2, we can write the constitutive equations for the mechanical elements in terms of velocity, $u(t)$.

(a) Show that the impedance for a spring element is

$$Z_s = \frac{F_s}{U_s} = \frac{k}{j\omega},$$

where F_s and U_s are the force amplitude and relative velocity amplitude for the spring element, respectively.

(b) Show that the impedance for a damper element is

$$Z_d = \frac{F_d}{U_d} = c,$$

where F_d and U_d are the force amplitude and relative velocity amplitude for the damper element, respectively.

(c) Show that the impedance for a mass element is

$$Z_m = \frac{F_m}{U_m} = j\omega m,$$

where F_m and U_m are the force amplitude and velocity amplitude for the mass element, respectively.

(d) The electrical impedance plays the role of a generalized frequency dependent resistance. What role does the mechanical impedance play?

(e) Let the two endpoints of a mechanical element be given by x_a and x_b (or x_{near} and x_{far}). If the relative displacement for the mechanical element is given by a complex exponential,

$$x_a - x_b = (X_a - X_b)e^{j\omega t} = X e^{j\omega t},$$

show that the ratio of the relative displacement amplitude, X, to the relative velocity amplitude, U, can be expressed as

$$\frac{X}{U} = \frac{1}{j\omega}.$$

7-25 Consider the mechanical system shown in Figure 7.42. Using the concept of mechanical impedance (see Problem 7-24), determine the steady state response of the mass when it is driven by an applied force given by $f(t) = F \sin(\omega t + \theta)$.

7-26 Consider the system shown in Figure 7.43, where $x(t)$ and $f(t)$ are the output and input, respectively.

Fig. 7.42. Figure for
Problem 7-25.

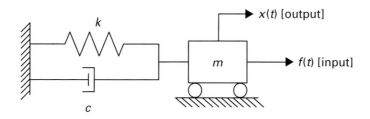

Fig. 7.43. Figure for
Problem 7-26.

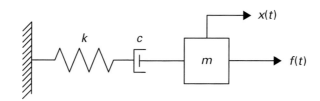

(a) Determine its frequency response function using the time-domain
approach, that is first obtain the governing equation, then assume
complex input and output, and finally take the ratio of the output
amplitude to the input amplitude.

(b) Using the concept of mechanical impedance (see Problem 7-24),
determine the frequency response function of the system. **Hint**:
Recall that the impedance for a mechanical element is given by
its frequency response function, where the force through the element is defined as the output, and the relative velocity between the
endpoints is defined as the input.

8 Digital signal processing

In the first part of this text, we focused on signals. Specifically, we looked at how to represent signals, how to transform them, and how to construct them from building blocks. We examined transformations between discrete-time and continuous-time signals. We often represented signals as weighted sums of complex exponentials where the weights revealed the frequency content of the signal. The second part of the text focused on systems. In particular, we learned how to model mechanical and electrical systems, how to represent these systems with their governing equations, and how to characterize them through their frequency response functions. We found that the same complex exponentials, combined with the frequency response, allow us to find the response of any LTI system to arbitrary inputs. In the last part of the text we will combine the signals and systems with *digital signal processing* (DSP) and look at applications. DSP includes the mathematics, the algorithms and the techniques used to manipulate signals after they have been converted to digital form. DSP can be used to accomplish a wide variety of objectives, including filtering, image enhancement, speech recognition, data compression for storage and transmission, and simulation of continuous-time systems, to name just a few.

We have already seen some examples. Suppose we would like to determine the response of a continuous-time LTI system to various inputs. The first step is to express the input signal $x(t)$ as a weighted sum of complex exponential building blocks. Assuming $x(t)$ is periodic,[1] we can write it as

$$x(t) = x(t + T_0) \approx \sum_{k=-N}^{N} c_k e^{jk\omega_0 t}, \tag{8.1}$$

---

[1] If the input signal $x(t)$ is not periodic, we can make it periodic by repeating it after the interval of interest, T_0.

where $\omega_0 = 2\pi/T_0$. For simple signals we can integrate

$$c_k = \frac{1}{T_0} \int_{T_0} x(t) e^{-jk\omega_0 t} \, dt \tag{8.2}$$

to obtain the coefficients of Eq. (8.1). For complicated signals, Eq. (8.2) may be difficult if not impossible to integrate, and it may not even have a closed-form solution. In such cases, it is often more convenient to solve for the coefficients numerically using the FFT.

Once we know the coefficients, we must determine the frequency response function (FRF) of the system, usually denoted as $H(j\omega)$. The FRF is defined as the ratio of the steady state complex output amplitude to the complex input amplitude when the input consists of a complex exponential. If the input is not a complex exponential, we cannot use the frequency response directly. Fortunately, we can represent almost any input as a weighted sum of complex exponentials. Then, for LTI systems we can apply the frequency response to each individual exponential, and sum the results to obtain the desired output of the system to the complicated input.

The FRF for a simple system can be derived from the governing equation once a suitable mathematical model of the physical system has been constructed. For a complicated system, on the other hand, the basic principles governing the behavior of the system may not be fully understood or even known, and the relevant physical parameters may not be readily available. For such a system, we may not be able to create any mathematical model based on the application of basic physical principles. Thus, we will not be able to obtain an expression for the governing equation, and we will not be able to find the FRF analytically. Fortunately, we can still obtain the FRF experimentally by applying sinusoidal inputs of various frequencies and measuring the amplitude and phase of the output. Knowing the characteristic features of the FRF associated with the system, we can then construct a suitable mathematical model to describe the system based on the experimental data, that is the measured FRF results.

Knowing how to express the input signal as a weighted sum of complex exponentials and knowing the frequency response function, $H(j\omega)$, of the system, we can readily determine the periodic output signal by summing the weighted complex exponentials as follows:

$$y(t) = y(t + T_0) \approx \sum_{k=-N}^{N} c_k H(jk\omega_0) e^{jk\omega_0 t} = \sum_{k=-N}^{N} d_k e^{jk\omega_0 t}, \tag{8.3}$$

where d_k represent the coefficients of the output complex exponentials. By inspection, note that the output coefficients and the input coefficients are related as

$$d_k = c_k H(\mathrm{j}k\omega_0). \tag{8.4}$$

In other words, using analysis or the FFT we express the input signal as the sum of complex exponentials. Then using the frequency response function $H(\mathrm{j}\omega)$ specific to the system, we compute the output of the system to each of these complex exponentials individually. Finally, using superposition we combine the resulting modified complex exponentials to get the desired output.

We have used this procedure many times already throughout this text. To make the process easier, we will develop some useful MATLAB utilities. We will then examine ways to construct the frequency response functions to achieve certain specifications. We will also look at governing equations and FRFs for discrete-time systems. We will explore the similarities and differences between continuous and discrete-time FRFs. Finally, we will look at more applications of signals and systems.

As a warning, these techniques are dependent on the system being linear. Certain nonlinear systems may be approximated as linear over a limited operating range of interest. Others are highly nonlinear and are much harder to solve.

8.1 More frequency response

The Fourier series and the FRF are extremely important. Let us look at it from a graphical and a block diagram perspective to gain a better understanding of the input–output relationships. Suppose our input signal $x(t)$ has the spectrum shown in Figure 8.1(a) and our system has the FRF shown in Figure 8.1(b). We know that if the input signal $x(t)$ is real, the positive and negative coefficients must be complex conjugates

$$c_{-k} = c_k^*. \tag{8.5}$$

For systems that produce a real output signal for a real input signal, the positive and negative frequency responses must also be complex conjugates, that is

$$H(-\mathrm{j}\omega) = H^*(\mathrm{j}\omega). \tag{8.6}$$

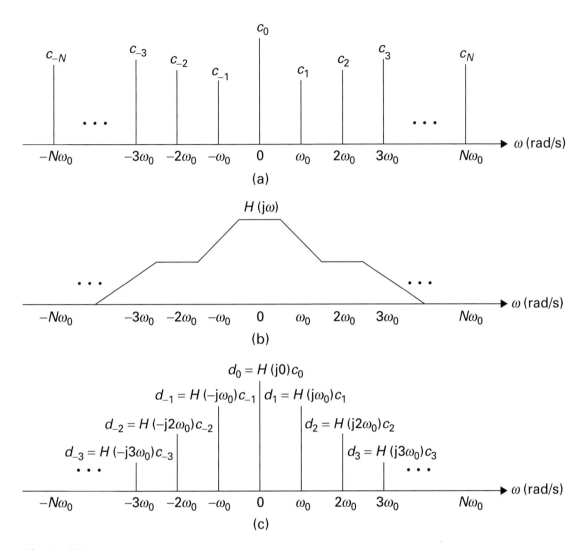

Fig. 8.1. (a) Input spectrum, (b) frequency response function and (c) output spectrum.

The output spectrum d_k is the product of the input spectrum and the frequency response, shown in Figure 8.1(c). The output $y(t)$ is reconstructed from its spectrum by summing the weighted complex exponentials. We can also view this process in block diagram form. For simple signals, we can obtain the coefficients by integrating, as shown in Figure 8.2. Then the coefficients are multiplied by the FRF and the output is reconstructed by summing the weighted complex exponentials.

For complicated signals where the integral is difficult if not impossible to evaluate, we address the problem numerically, as shown in Figure 8.3. The input $x(t)$ must be sampled with a continuous-to-discrete converter to give

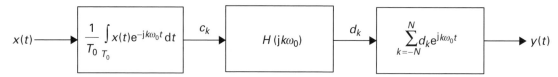

Fig. 8.2. Computing response using Fourier series and FRF.

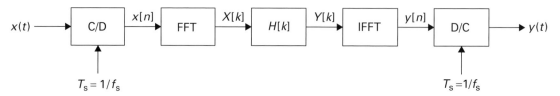

Fig. 8.3. Computing response using FFT.

a sequence $x[n]$ that we can process numerically. We use the fast Fourier transform (FFT) on these samples to find the input spectrum, then multiply it by the frequency response of the system to get the output spectrum. The inverse fast Fourier transform (IFFT) or the `sumcos` function (see Laboratory Exercise 1) returns the sampled output signal. If we need a continuous output, we may regenerate it with a discrete-to-continuous converter. The signals in this diagram are defined as follows:

$$
\begin{aligned}
x[n] &= x(nT_s) \\
X[k] &= \tfrac{1}{N}\mathrm{fft}(x) \approx c_k \\
H[k] &= H(jk\omega_0) \\
Y[k] &= X[k]H[k] \\
y[n] &= N\mathrm{ifft}(Y) \\
y[n] &\approx y(nT_s).
\end{aligned}
\tag{8.7}
$$

8.2 Notation clarification

Incidentally, we have used X as $X[k]$ and X_k. They are different and should not be confused. In Chapter 2 we showed that we could represent a cosine either as the sum of two complex continuous-time exponentials or as the real part of a single complex continuous-time exponential as follows:

$$
A\cos(\omega_0 t) = \frac{A}{2}e^{jk\omega_0 t} + \frac{A}{2}e^{-jk\omega_0 t} = \mathrm{Re}\{Ae^{jk\omega_0 t}\},
\tag{8.8}
$$

where index $k = 1$. The coefficients on the complex exponentials are denoted with the c_k. In this case, $c_1 = A/2$ and $c_{-1} = A/2$. The coefficients on the real part of a single complex exponential are given by X_k. In this case, $X_1 = A$. In general, $X_0 = c_0$ and for $k = 0$ $X_k = 2c_k$ for real functions. On the other hand, $X[k] \approx c_k$ represents the coefficient on the complex exponential returned by the FFT. The second use is the most important one; unfortunately both get the same letter because of a shortage of letters in the alphabet! The relationship between $X[k]$ and X_k can be best summarized as follows: if the signal $x(t)$ is real, then

$$x(t) = \sum_{k=-\infty}^{\infty} c_k e^{jk\omega_0 t}$$

$$= c_0 + 2\sum_{k=1}^{\infty} [a_k \cos(k\omega_0 t) + b_k \sin(k\omega_0 t)] \tag{8.9}$$

$$= \mathrm{Re}\left\{\sum_{k=0}^{\infty} X_k e^{jk\omega_0 t}\right\}.$$

8.3 Utilities

The procedure discussed previously is repeated so often that it is handy to define some MATLAB utilities, fdomain and tdomain, to assist with the computation.

[X, f] = fdomain(x,fs)
x: input signal x[n]
fs: rate at which f[n] was sampled
X: vector of coefficients c_k (ordered from negative to positive)
f: vector of frequencies kf_0 corresponding to c_k

The most challenging part of using the FFT is getting the frequency indices correct. The MATLAB code fdomain does this for us automatically, and works for vectors with both even and odd numbers of elements. If N is even, we can create a vector of angular frequencies $\omega[n]$ using the relationship

$$\omega[n] = \begin{cases} \left[\frac{n-1}{N}\right] 2\pi f_s & 1 \leq n \leq \frac{N}{2} \\ \left[\frac{n-1}{N} - 1\right] 2\pi f_s & \frac{N}{2}+1 \leq n \leq N \end{cases} \tag{8.10}$$

which generates the frequency vector

$$\omega[n] = \frac{2\pi f_s}{N} \left[\underbrace{0, 1, 2, \ldots, \frac{N}{2}-1}_{\text{positive}}, \underbrace{-\frac{N}{2}, -\frac{N}{2}+1, \ldots, -2, -1}_{\text{negative}} \right]. \tag{8.11}$$

If N is odd, we find a similar relationship

$$\omega[n] = \begin{cases} \left[\frac{n-1}{N}\right] 2\pi f_s & 1 \leq n \leq \frac{N+1}{2} \\ \left[\frac{n-1}{N} - 1\right] 2\pi f_s & \frac{N+1}{2} + 1 \leq n \leq N. \end{cases} \tag{8.12}$$

In this case, the frequency vector becomes

$$\omega[n] = \frac{2\pi f_s}{N} \left[\underbrace{0, 1, 2, \ldots, \frac{N-1}{2}}_{\text{positive}}, \underbrace{-\frac{N-1}{2}, -\frac{N-1}{2}+1, \ldots, -2, -1}_{\text{negative}}\right]. \tag{8.13}$$

If we plot the spectrum in this order, we obtain a shape in which the negative frequencies appear on the right instead of the left. It would be more intuitive and conventional to arrange the frequencies in ascending order as follows: (assuming N even):

$$\omega_s[n] = \frac{2\pi f_s}{N} \left[\underbrace{-\frac{N}{2}, -\frac{N}{2}+1, \ldots, -2, -1}_{\text{negative}}, \underbrace{0, 1, 2, \ldots, \frac{N}{2}-1}_{\text{positive}}\right]. \tag{8.14}$$

Similarly, for N odd, we would like to order the frequencies as follows:

$$\omega_s[n] = \frac{2\pi f_s}{N} \left[\underbrace{-\frac{N-1}{2}, -\frac{N-1}{2}+1, \ldots, -2, -1}_{\text{negative}}, \underbrace{0, 1, 2, \ldots, \frac{N-1}{2}}_{\text{positive}}\right]. \tag{8.15}$$

This reordering is so common that MATLAB has a built-in function `fftshift` that accomplishes this automatically. Specifically, the following command `ws = fftshift(w)` moves the right side of the vector (corresponding to the negative frequencies) back to the left. Fdomain in Section 8.4 takes care of this.

8.4 DSP example and discrete-time FRF

We will now apply the various techniques that we have developed. In particular, we will use the ability to decompose a signal into its constituent frequencies and the ability to determine how a continuous-time system responds to various frequencies. We will then use the frequency response approach to determine the transient response of a system to a step input. Finally, we will extend the notion of frequency response function to discrete-time systems.

Consider the electrical system shown in Figure 8.4, where the voltage source, $x(t)$, is the input and the voltage drop across the capacitor,

Fig. 8.4. Second-order
electrical system.

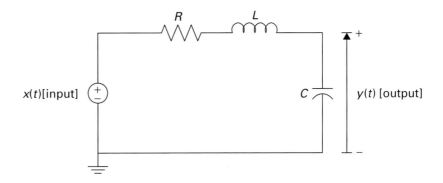

$y(t)$, is the output. Applying KVL, we find that its governing equation is
given by

$$LC\frac{d^2y}{dt^2} + RC\frac{dy}{dt} + y = x \tag{8.16}$$

or in canonical form

$$\ddot{y} + 2\xi\omega_n\dot{y} + \omega_n^2 y = \omega_n^2 x. \tag{8.17}$$

To find the frequency response function of the system, we set

$$\begin{aligned} x &= Xe^{j\omega t} \\ y &= Ye^{j\omega t}. \end{aligned} \tag{8.18}$$

Substituting Eq. (8.18) into Eq. (8.17), we find

$$H(j\omega) = \frac{Y}{X} = \frac{\omega_n^2}{\omega_n^2 - \omega^2 + 2\xi\omega_n\omega j} = \frac{1}{1 - \left(\dfrac{\omega}{\omega_n}\right)^2 + 2\xi\dfrac{\omega}{\omega_n}j}. \tag{8.19}$$

We can solve for the step response[2] of the system either analytically or
numerically. The step response for the system of Figure 8.4 is obtained by
setting $x(t) = u(t)$.

Analytically, we can solve for the step response by using the theory
of differential equations, as we did in Chapter 6. The analytical approach
requires us to determine the homogeneous and particular solutions, then sum
the two expressions and impose the initial conditions to uniquely determine
the unknown constants in the homogeneous solution. If we assume the

---

[2] The step response of a system is defined as the response of the system to a step input with all of
the initial conditions set to zero.

Digital signal processing

Fig. 8.5. Pulse train input.

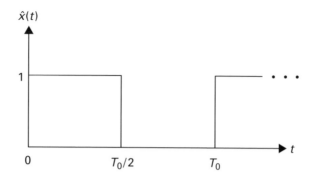

system to be underdamped, the exact step response of the system is given by

$$y(t) = 1 - \frac{e^{-\zeta\omega_n t}}{\sqrt{1 - \zeta^2}} \sin\left[\omega_n\sqrt{1 - \zeta^2}\, t + \tan^{-1}\left(\frac{\sqrt{1 - \zeta^2}}{\zeta}\right)\right]. \qquad (8.20)$$

Alternatively, we can solve for the step response numerically by using the frequency response approach. To do this, we first need to use the Fourier series to find the frequencies contained in the step input. However, the Fourier series assumes that the input is periodic, which $x(t) = u(t)$ is clearly not. Fortunately, we can easily make the input periodic by repeating our step function in time over some interval T_0. This gives us the pulse train shown in Figure 8.5, where $\hat{x}(t)$ represents the input pulse train. An important question to ask is how long ought the interval T_0 to be so that the response of the system to a step input and a pulse train of equal amplitude becomes nearly identical over $0 \le t < T_0/2$?

If the system of Figure 8.4 is underdamped, we know its response to the rising edge of the input of Figure 8.5 should first rise, then overshoot, and eventually oscillate about the final value, as shown in Figure 8.6. The response to the falling edge should drop, overshoot again, and then oscillate. Thus, based on the results of Figure 8.6, we conclude that the half-period $T_0/2$ must be longer than the settling time so that most of the oscillations are damped out before the next edge is encountered. Recall that the settling time for an underdamped second-order system is given by

$$t_s = \frac{-\ln\left(\epsilon\sqrt{1 - \zeta^2}\right)}{\zeta\omega_n}, \qquad (8.21)$$

where ϵ is a parameter that dictates how close we want the output to settle. A common value is given by the 2 percent settling time, defined as $\epsilon = 0.02$,

Fig. 8.6. Pulse train response for the underdamped second-order system of Figure 8.4.

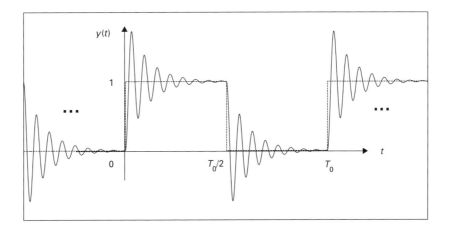

which yields a response that settles to within 2 percent of the final or steady state value. If $\zeta = 0.1$ and $\omega_n = 10$ rad/s, we find the 2 percent settling time is $t_s = 3.92$ s. Let us arbitrarily choose $T_0 = 10$ s, so that the half-period $T_0/2$ is greater than the 2 percent settling time.

Because we are solving the problem numerically, we also need to select a sampling time. In particular, we need to choose enough samples so that we can accurately describe the response of the system. The input pulse train has components at high frequencies because of the sharp edge rate. These components will alias down to the lower frequencies. Thus, our sampling rate should be high enough so that aliasing is insignificant. Because we have quite a bit of computing power at our disposal, let us choose $f_s = 1000$ Hz. This will give us $N = f_s T_0 = 10\,000$ samples, which allows us to represent the signal as

$$\hat{x}(t) = \hat{x}(t + T_0) \approx \sum_{k=-5000}^{4999} c_k e^{jk\omega_0 t}. \tag{8.22}$$

We can now write a MATLAB code to do all of the numerical computations. The code (steprespdemo.m) is given below. In step 1, we create the desired pulse train input of period T_0. Notice how the logical command "<" is used to make a pulse. In step 2, we compute the spectrum of the input using the fdomain utility discussed previously. In step 3, we find the frequency response and multiply it by the spectrum. Recall that the frequency response is given in terms of $\omega = 2\pi f$. In step 4, we compute the output using the tdomain utility. The fdomain and tdomain utilities are also listed. They compute the FFT and IFFT, respectively. The output of

Fig. 8.7. Pulse train
response found with
Fourier technique.

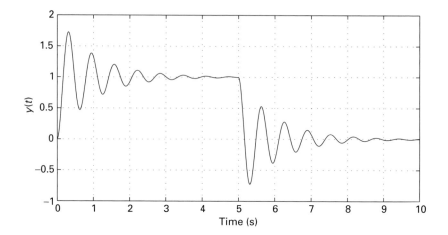

Fig. 8.8. Pulse train
response obtained with
Fourier technique (solid
line) overlaid with the
exact step response.

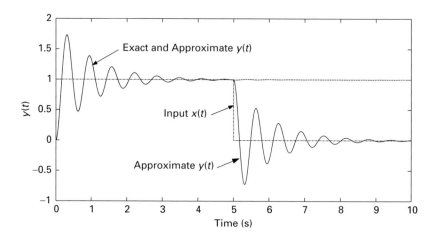

the MATLAB code is shown in Figure 8.7. Figure 8.8 shows the same plot
overlaid with the exact solution found by solving the differential equation
(see Eq. (8.20) for the exact step response). The agreement between the
exact and the numerical solutions is excellent until time $T_0/2$. After $T_0/2$,
the response to the pulse train drops down, that is there is a discontinuity in
the output of the system, while the exact solution settles to the steady state
value of 1.

```
steprespdemo.m
% steprespdemo.m
% Pulse-train response
% Step 1 - create input
```

```
fs=1000;
To=10;
t=0:1/fs:To-1/fs;
x=1/2+square(2*pi*/To)/2;
% Step 2 - compute spectrum of input
[X,f]=fdomain(x,fs);
% Step 3 - multiply input spectrum by frequency
% response
omegan=10;
zeta=0.1;
K=1;
omega=2*pi*f;
H=1./(1-(omega/omegan).2+j*2zeta*omega/omegan);
Y=X.*H;
% Step 4 - compute and plot output time waveform
y=tdomain(Y,fs);
plot(t,y,'k');
grid on;
xlabel('Time (sec)');
ylabel('y(t)');

fdomain.m
function [X,f]=fxomain(x,Fs)
% FDOMAIN Function to compute the Fourier
% coefficients from vector x
% and the corresponding frequencies (two-sided)
% usage:
% [X,f]=fdomain(x,fs)
%    x=vector of time domain samples
%    fs=sampling rate (in Hz)
%    X=vector of complex Fourier coefficients
%    f=vector of corresponding frequencies
%     (two-sided)
N=length(x);
if mod(N,2)==0
  k=-N/2:N/2-1; % N even
else
  K=-(N-1)/2:(N-1)/2; % N odd
end
```

```
T=N/Fs;
f=k/T;
X=fft(x)/N; % make up for the lack of 1/N in
% Matlab FFT
X=fftshift(X);

tdomain.m
function [x,t]=tdomain(X,Fs)
% TDOMAIN computes the real time waveform vector
% x corresponding to the
%   Fourier coefficients (two-sided)
% usage: [x,t]=tdomain(X,Fs)
N=length(X);
n=0:N-1;
t=n/Fs;
X=ifftshift(X);
x=real(N*ifft(X)); % make up for the 1/N in
% Matlab IFFT
```

Incidentally, since we know the exact solution, the question one may ask is why are we going to all the trouble of finding the step response with the Fourier transform? Whenever we use a numerical approach to solve a problem, it is wise to test the approach on a simple case where we can solve the problem analytically. If the numerical approach is sound, we should be able to get the same answer numerically. If we do not recover the analytical result, then we know that there is an error in our application of the numerical approach. Finally, the numerical approach of solving the problem is very versatile because it allows us to tackle problems that are not amenable to analytical analyses.

Not surprisingly, the period T_0 of the pulse train has a significant effect on the quality of the numerical results. Previously we considered the case of $T_0 = 10$ s. If we increase T_0 to 20 s, the agreement between the numerical (solid line) and the analytical (dotted line) solution becomes even better, as shown in Figure 8.9(a). But if we decrease T_0 to 3 s, the output of the pulse train does not have time to settle before the next edge is encountered, and the Fourier approach yields a solution that is substantially different from the exact result as shown in Figure 8.9(b). This simple example illustrates the importance of making the period long enough.

We have shown that the frequency response approach can be used to compute the step response of an underdamped second-order system. In

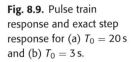

Fig. 8.9. Pulse train response and exact step response for (a) $T_0 = 20\,\text{s}$ and (b) $T_0 = 3\,\text{s}$.

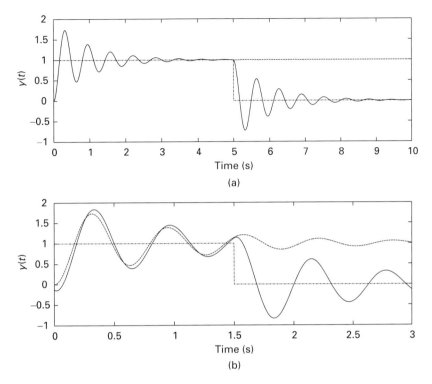

fact, the same approach can be extended to compute the transient response for a system of any order, provided we choose the period to be sufficiently large.

The numerical approach of determining the response of a system is extremely powerful. If we wish to find the response of the system to a different input, we only need to change the input. For example, we could find the response to a sawtooth waveform. Or we could even record data from an earthquake with a seismometer and find the response of the system to that set of data. The Fourier approach works for both simple signals that we could approach analytically and for complicated signals that we can only process numerically. With the advent of modern computers, the numerical approach of solving for the response to complicated inputs becomes more practical.

In another example, suppose we recorded a piece of music and discovered that it contains an annoying tone due to faulty equipment or operator error. Instead of recording it again, we can simply filter out the unwanted tone and avoid the expense of getting all the musicians together again. We would like to filter out the tone to restore a clean signal. The first step is to find the frequency of the tone. We can do this by plotting the spectrum, which

Fig. 8.10. Spectrum of
music with annoying
1500 Hz tone.

```
>> load handeltone;
>> soundsc (x, Fs);
>> [X, f] =fdomain (x, Fs);
>> plot (f,abs(X));
>> xlabel ('Frequency (Hz)');
>> ylabel ('Magnitude');
```

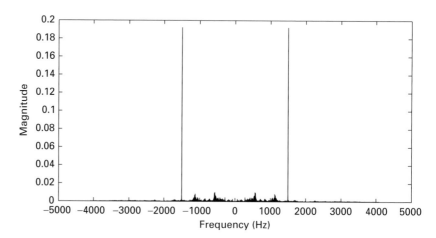

shows the annoying tone is at 1500 Hz, as shown in Figure 8.10. We could
now run the signal through a system or filter that eliminates the annoying
tone. The system should have a frequency response that is one everywhere
except at the tone frequency and zero at the tone frequency. We can create
such a *notch filter* using the command

```
H=(abs(f)<1450)+(abs(f)>1550)
```

This creates a notch that is 100 Hz wide centered at 1500 Hz. Now we can
find the system output using

```
Y = X.*H;
[y,t] = tdomain(Y,Fs);
```

The new signal y no longer has the annoying tone. Figure 8.11(a), Figure 8.11(b), and Figure 8.11(c) show the original spectrum with the tones,
the notch filter, and the filtered spectrum, respectively. Notice that the filtered spectrum has a small gap where the notch eliminated part of the signal
as well as the tone. If this is a problem, we could use a narrower notch filter
(< 100 Hz wide).

Fig. 8.11. Filtered audio signal.

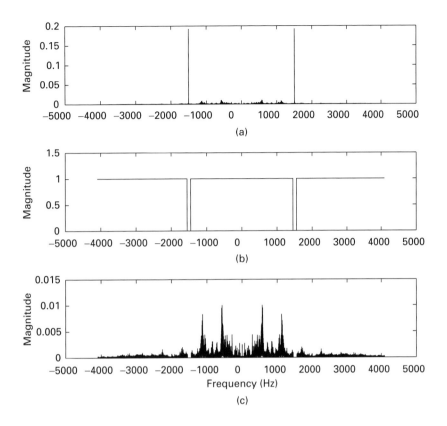

(a)

(b)

Frequency (Hz)

(c)

8.5 Frequency response of discrete-time systems

The FFT approach is very powerful. Its drawback is that we must have the entire dataset before we begin. This is clearly not practical for something like processing cell phone signals because we would like to hear what the person on the other end has to say before the transmission is completed. This motivates us to investigate digital signal processing techniques that can work in real time. To do that, we need to develop an understanding of the frequency response of discrete-time systems.

Recall that a continuous-time system can be described by a governing equation in the form of a differential equation. The continuous-time frequency response function $H(j\omega)$ is defined as the ratio of the complex output amplitude to the complex input amplitude when the input is a complex exponential. We can construct an analogous approach for discrete-time systems. The governing equation for a discrete-time system appears in the form of a difference equation. For instance, a simple example of a difference

282 **Digital signal processing**

Fig. 8.12. Sampling a
complex exponential.

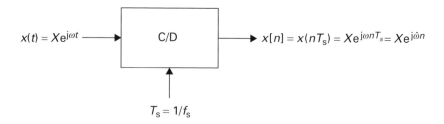

$x(t) = Xe^{j\omega t}$ → C/D → $x[n] = x(nT_s) = Xe^{j\omega n T_s} = Xe^{j\hat{\omega} n}$

$T_s = 1/f_s$

equation is given by

$$y[n] = x[n] + 2x[n-1] + x[n-2]. \tag{8.23}$$

Equation (8.23) is called *a nonrecursive difference equation* because the
output $y[n]$ depends only on the present and past values of the input x and
not on the previous values of y.

We would like to find something analogous to the frequency response
function for difference equations. Let us suppose our input is a complex
exponential sampled at a period T_s to form a discrete-time signal. This is
illustrated in Figure 8.12 and the input can be written as

$$x[n] = x(nT_s) = Xe^{j\omega n T_s} = Xe^{j\hat{\omega} n}, \tag{8.24}$$

where $\hat{\omega} = \omega T_s = 2\pi f/f_s = 2\pi \hat{f}$. If the input to a discrete-time system
is a complex exponential of Eq. (8.24), then the output is also a complex
exponential that may be written as

$$y[n] = Ye^{j\hat{\omega} n}. \tag{8.25}$$

Substituting Eqs. (8.24) and (8.25) into our difference equation of Eq. (8.23)
gives

$$Ye^{j\hat{\omega} n} = Xe^{j\hat{\omega} n} + 2Xe^{j\hat{\omega}(n-1)} + Xe^{j\hat{\omega}(n-2)}. \tag{8.26}$$

We can factor this into the following form:

$$Ye^{j\hat{\omega} n} = (1 + 2e^{-j\hat{\omega}} + e^{-2j\hat{\omega}})Xe^{j\hat{\omega} n}. \tag{8.27}$$

Because the complex exponential never vanishes, we can divide it out of
both sides. Like before, we can take the ratio of the output amplitude to the
input amplitude to find the FRF. This discrete-time frequency response is
usually denoted by either $H(\hat{\omega})$ or $H(e^{j\hat{\omega}})$. From Eq. (8.27), we find

$$H(\hat{\omega}) = H(e^{j\hat{\omega}}) = \frac{Y}{X} = 1 + 2e^{-j\hat{\omega}} + e^{-2j\hat{\omega}}. \tag{8.28}$$

Equation (8.28) constitutes the FRF of the difference equation of Eq. (8.23).

Consider now a *recursive discrete-time system* whose difference equation is given by

$$y[n] - y[n-1] = x[n] + x[n-1] + x[n-2]. \tag{8.29}$$

To find the FRF, we assume the input and the output are complex exponentials of the form given by Eqs. (8.24) and (8.25). Thus, we find

$$Y e^{j\hat{\omega}n} - Y e^{j\hat{\omega}(n-1)} = X e^{j\hat{\omega}n} + X e^{j\hat{\omega}(n-1)} + X e^{j\hat{\omega}(n-2)}. \tag{8.30}$$

We can again factor and divide out the complex exponential to find the desired frequency response

$$H(\hat{\omega}) = \frac{Y}{X} = \frac{1 + e^{-j\hat{\omega}} + e^{-2j\hat{\omega}}}{1 - e^{-j\hat{\omega}}}. \tag{8.31}$$

Observe that the frequency responses are just rational functions of polynomials in $e^{-j\hat{\omega}}$. The gain and phase of the FRF can always be determined by the standard manipulations of complex numbers. Specifically, we can always express the FRF as

$$H(\hat{\omega}) = a + bj, \tag{8.32}$$

from which we can easily compute the gain and phase. However, to find expressions for a and b, the real and imaginary parts of the FRF, requires lots of algebraic manipulations, and can be rather messy if the difference equation is complicated.

Fortunately, there is a simpler approach for determining the magnitude and phase when the sequence of the coefficients is either *symmetric* or *antisymmetric* about a central point. The FRF for a discrete-time system can often be expressed in the following general form:

$$H(\hat{\omega}) = \sum_{k=0}^{M} b_k e^{-j\hat{\omega}k}, \tag{8.33}$$

where M is known as the filter order. The coefficients b_k are said to be symmetric when they are related as follows:

$$b_k = b_{M-k} \tag{8.34}$$

or $b_0 = b_M, b_1 = b_{M-1}$, etc. The coefficients b_k are antisymmetric when

$$b_k = -b_{M-k} \tag{8.35}$$

or $b_0 = -b_M, b_1 = -b_{M-1}$, etc. If the sequence of coefficients is either symmetric or antisymmetric about a central point, we can easily obtain the

Fig. 8.13. (a) Gain and (b) phase of discrete-time frequency response function of Eq. (8.37).

(a)

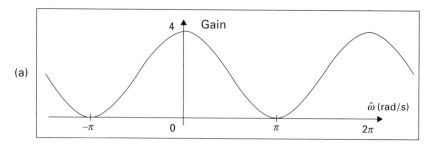

(b)

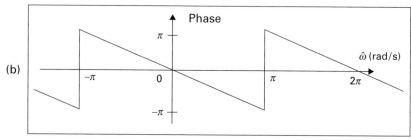

gain and phase of the FRF by factoring out an exponential whose phase is half the filter order times $-\hat{\omega}$.

Consider the frequency response function of Eq. (8.28). Comparing Eq. (8.28) to Eq. (8.33), we note immediately that the filter order is $M = 2$. Factoring out an exponential whose phase is half the filter order times $-\hat{\omega}$, we obtain

$$
\begin{aligned}
H(\hat{\omega}) &= e^{-j\hat{\omega}}(e^{j\hat{\omega}} + 2 + e^{-j\hat{\omega}}) \\
&= 2e^{-j\hat{\omega}} \left(1 + \frac{e^{j\hat{\omega}} + e^{-j\hat{\omega}}}{2} \right) \\
&= 2e^{-j\hat{\omega}} (1 + \cos \hat{\omega}).
\end{aligned}
\tag{8.36}
$$

Because of the symmetry condition, the exponentials inside the parentheses can be combined to yield $\cos \hat{\omega}$. By inspection, the gain and phase of the FRF are simply

$$
\begin{aligned}
|H(\hat{\omega})| &= 2(1 + \cos \hat{\omega}) \\
\angle H(\hat{\omega}) &= -\hat{\omega}.
\end{aligned}
\tag{8.37}
$$

Figure 8.13 plots the gain and phase of this frequency response against $\hat{\omega}$. Unlike continuous-time systems where the gain and phase are usually plotted as a function of ω on a logarithmic scale, for discrete-time systems, the gain and phase are often plotted as a function of $\hat{\omega}$ on a linear scale. Note that the gain of Eq. (8.37) is bounded between 0 and 4 with a period of 2π, and the phase of Eq. (8.37) decreases with increasing frequency. Because the phase is periodic with a period of 2π, the phase plot is bounded between $-\pi$ and π. Thus, phase less than $-\pi$ is equivalent to phase just less than π, and phase greater than π is equivalent to phase just greater than $-\pi$.

Much information is embedded in the gain and phase plots. Consider again the gain plot of Figure 8.13. We note that because the gain at low frequencies (near $\hat{\omega} = 0$) is larger compared to the gains at higher frequencies (near $\hat{\omega} = \pi$), the system functions as a *low-pass filter*. Specifically, the system amplifies the response at low frequencies and attenuates the response at high frequencies.

The FRF for a discrete-time system, given by $H(\hat{\omega})$, is a complex-valued function that depends on $\hat{\omega}$. It has several interesting properties that can be used to simplify the sketching of the gain and phase of $H(\hat{\omega})$. We first note that $H(\hat{\omega})$ is periodic with a period of 2π. This is true because the FRF depends on complex exponentials that also have a period of 2π. All discrete-time systems possess this periodic characteristic,

$$H(\hat{\omega}) = H(\hat{\omega} + 2\pi), \tag{8.38}$$

which implies that it is always sufficient to specify the discrete-time FRF over an interval of 2π only. Typically, we define $H(\hat{\omega})$ over $-\pi < \hat{\omega} \le \pi$. It should be noted that the FRF for a continuous-time system does not enjoy this periodic property. The frequency response $H(\hat{\omega})$ also possesses the characteristic of conjugate symmetry,

$$H(-\hat{\omega}) = H^*(\hat{\omega}), \tag{8.39}$$

which is true whenever the coefficients b_k are real, that is $b_k = b_k^*$. Because of the conjugate symmetry property, we can easily show that the gain of the frequency response is an even function, and that the phase is an odd function:[3]

$$|H(-\hat{\omega})| = |H(\hat{\omega})|$$
$$\angle H(-\hat{\omega}) = -\angle H(\hat{\omega}). \tag{8.40}$$

Thus, when plotting the gain and phase of discrete-time systems, we only need to consider positive frequencies, because the gain and phase

---

[3] A function $f(t)$ is said to be *even* if $f(t) = f(-t)$, and *odd* if $f(t) = -f(-t)$. Thus, $\cos t$ is an even function and $\sin t$ is an odd function.

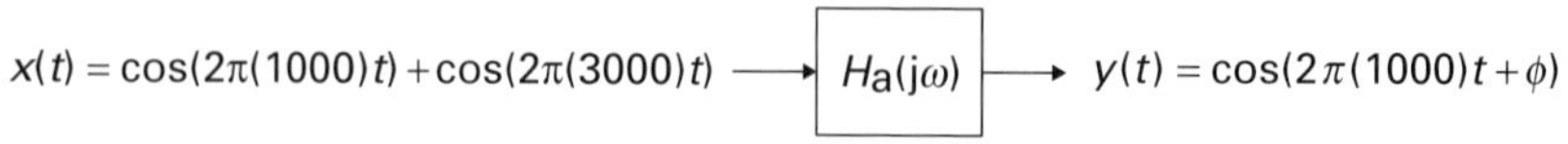

Fig. 8.14. System rejecting
high frequencies.

$x(t) = \cos(2\pi(1000)t) + \cos(2\pi(3000)t) \longrightarrow \boxed{H_a(j\omega)} \longrightarrow y(t) = \cos(2\pi(1000)t + \phi)$

for negative frequencies can be extracted by using the symmetry condition. Figure 8.13 illustrates these symmetry properties. Incidentally, for a continuous-time system whose governing equation has real coefficients, its FRF also has the property of

$$H(-j\omega) = H^*(j\omega). \tag{8.41}$$

Consider a discrete-time system whose difference equation has real coefficients. Its discrete-time FRF is given by $H(\hat{\omega})$. Using Eq. (8.39), we can show that its response to a discrete-time sinusoidal input is another sinusoid whose amplitude is scaled by the gain of the FRF, and whose phase relative to the input is given by the phase of the FRF. Specifically, if the discrete-time input $x[n]$ is given by

$$x[n] = X \cos(\hat{\omega}n + \theta) \tag{8.42}$$

then the response of the discrete-time system is

$$y[n] = X|H(e^{j\hat{\omega}})| \cos(\hat{\omega}n + \theta + \angle H(e^{j\hat{\omega}})). \tag{8.43}$$

Note the similarity to the steady state response of a continuous-time system to a sinusoidal input (see Eq. (7.86)).

8.5.1 Application: discrete-time filter

Suppose we wish to build a system that keeps the lower frequencies in a signal but rejects the higher frequencies. For example, suppose the input signal to the system is given by

$$x(t) = \cos(2\pi(1000)t) + \cos(2\pi(3000)t) \tag{8.44}$$

and for a given application, we wish to design a system that produces an output signal of

$$y(t) = \cos(2\pi(1000)t + \phi). \tag{8.45}$$

Notice that we are willing to accept some change in the phase of the frequency that we pass.

Fig. 8.15. Digital signal
processing system.

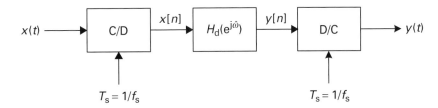

Figure 8.14 shows the block diagram of a continuous-time system with a continuous-time frequency response of $H_a(j\omega)$, where a denotes analog. Because the output amplitude is simply the input amplitude scaled by the gain of the frequency response, in order to meet our design objectives, the gain of the frequency response of the system must satisfy the following requirements:

$$|H_a\,(j2\pi\,(1000))| = 1$$
$$|H_a\,(j2\pi\,(3000))| = 0. \tag{8.46}$$

One approach to accomplishing the task would be to build some network of resistors, capacitors, and inductors with a suitable frequency response function that acts as a low-pass filter. This approach is hardware based. While conceptually simple, selecting the required system parameters may be nontrivial. Moreover, once built, this filter cannot be easily altered. Another approach is to design a digital system to meet the design objectives. This approach is software based, and unlike the hardware-based design, the digital system can be easily modified. Such a digital signal processing (DSP) system is illustrated in Figure 8.15. The system uses a C/D converter to digitize the input signal $x(t)$ into a discrete-time signal $x[n]$ at a suitable sampling rate. A discrete-time system with a discrete-time FRF of $H_d(e^{j\hat{\omega}})$ then processes the data to produce $y[n]$. Finally, the output is converted back to the continuous-time signal $y(t)$ using a D/C converter. Utilizing this approach, we only need to design the appropriate discrete-time system low-pass filter.

Consulting a textbook on digital filter design,[4] we find that we might be able to get an acceptable discrete-time frequency response using a discrete-time system whose governing equation is given by the following difference equation:

$$y[n] = ax[n] + bx[n-1] + ax[n-2], \tag{8.47}$$

---

[4] For example, see *Digital Filters* by R. W. Hamming, Prentice Hall, New Jersey, 1989 or *Filter Design for Signal Processing Using* MATLAB *and Mathematica* by M. D. Lutovac, D. V. Tosic and B. L. Evans, Prentice Hall, New Jersey, 2001.

where a and b are constants that must be selected to meet the design specifications. This system has a discrete-time frequency response given by

$$H_d(e^{j\hat{\omega}}) = H(e^{j\hat{\omega}}) = a + be^{-j\hat{\omega}} + ae^{-j2\hat{\omega}} = e^{-j\hat{\omega}}(b + 2a \cos \hat{\omega}). \quad (8.48)$$

In general, we need a sampling rate at least twice that of the highest frequency in the input signal to avoid aliasing. Thus, we choose a sampling rate of $f_s = 8\,\text{kHz}$, and our sampled input signal becomes

$$x[n] = x(nT_s) = \cos\left(\frac{2\pi\,(1000)\,n}{8000}\right) + \cos\left(\frac{2\pi\,(3000)\,n}{8000}\right)$$

$$= \cos\left(\underbrace{\frac{\pi}{4}n}_{\hat{\omega}_1}\right) + \cos\left(\underbrace{\frac{3\pi}{4}n}_{\hat{\omega}_2}\right). \quad (8.49)$$

According to our design constraints, the gain of the frequency response should be 1 at $\hat{\omega}_1 = \pi/4$ and 0 at $\hat{\omega}_2 = 3\pi/4$. Substituting these constraints into Eq. (8.48), we obtain the following requirements for the gain of the discrete-time frequency response function:

$$\left|H_d\left(e^{j\frac{\pi}{4}}\right)\right| = \left|b + 2a \cos\left(\frac{\pi}{4}\right)\right| = 1$$

$$\left|H_d\left(e^{j\frac{3\pi}{4}}\right)\right| = \left|b + 2a \cos\left(\frac{3\pi}{4}\right)\right| = 0. \quad (8.50)$$

For simplicity, we assume the quantities inside the absolute value to be strictly positive. We now have a pair of linear equations in terms of the two unknowns a and b:

$$b + 2a \cos\left(\frac{\pi}{4}\right) = 1$$

$$b + 2a \cos\left(\frac{3\pi}{4}\right) = 0. \quad (8.51)$$

Solving Eq. (8.51) simultaneously, we obtain $a = 1/(2\sqrt{2})$, $b = 1/2$.

Figure 8.16(a) shows the gain and phase of the frequency response as a function of $\hat{\omega}$ from -3π to 3π. Observe that the response repeats with a period of 2π just as we expected. Moreover, notice that the gain is an even function and the phase is an odd function. The gain has some low-pass behavior in that it is large at frequencies near 0 and small at higher frequencies. Figure 8.16(b) shows the gain plotted against the frequency in Kilohertz (kHz) which is obtained using $f = \hat{\omega}f_s/(2\pi)$ and is periodic with a period of f_s.

Fig. 8.16. (a) Gain and phase of the discrete-time frequency response function of Eq. (8.48) vs. $\hat{\omega}$; (b) gain vs. f in khz.

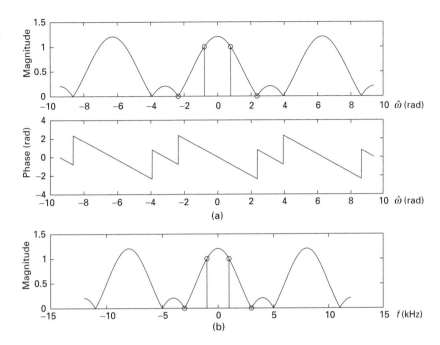

The following code evaluates the digital filter acting on the input. It first computes the input at the sampling frequency. It then calls the built-in MATLAB command filter, which evaluates the difference equation from Eq. (8.47). Figure 8.17(a) shows the two tones in the spectrum of the input signal $x[n]$, while Figure 8.17(b) shows the single tone in the spectrum of the filtered output signal $y[n]$. Note that the digital filter successfully rejects the 3000 Hz component, as specified by the design constraint.

```
% digifilt.m: create two sinusoids and filter out
% the higher frequency
b=1/2;
a=1/(2*sqrt(2));
fs=8000;
t=0:1/fs:2-1/fs;
x=cos(2*pi*1000*t)+cos(2*pi*3000*t);
h=[a,b,a];
y=filter(h,1,x);
```

This simple example is rather contrived because it is strictly tailored to a pair of arbitrary frequencies. Nevertheless, it demonstrates some of the advantages of using discrete-time systems for signal processing. Suppose

Fig. 8.17. Spectrum of the
signal a) before and b)
after filtering.

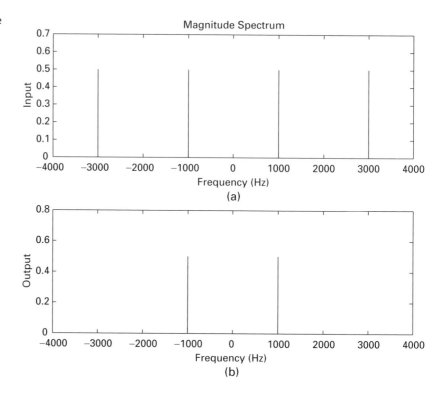

we wish to reverse the filter to pass the 3000 Hz signal and reject the 1000 Hz
signal. In a continuous-time system, we would need to build a new circuit
with different resistors, capacitors, and inductors. In the discrete-time system, we would just need to resolve Eq. (8.51) for the filter coefficients based
on the new desired performance. Specifically, we solve the following set of
algebraic equations:

$$b + 2a \cos(\tfrac{\pi}{4}) = 0$$

$$b + 2a \cos(\tfrac{3\pi}{4}) = 1.$$

(8.52)

After some algebra, we find $a = -1/(2\sqrt{2})$, $b = 1/2$. Figure 8.18 plots the
frequency response of the new filter. Note that the gain of the new frequency
response is now 0 at $f_1 = 1000$ Hz and 1 at $f_2 = 3000$ Hz.

The previous examples demonstrate the power of using discrete-time
systems for signal processing. We note that changing the behavior of the
filter only involves algebraic manipulations (i.e., finding the new coefficients), and not rebuilding the entire physical hardware. Moreover, complete
programmable digital signal processing systems are readily available and
can be easily purchased.

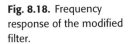

Fig. 8.18. Frequency response of the modified filter.

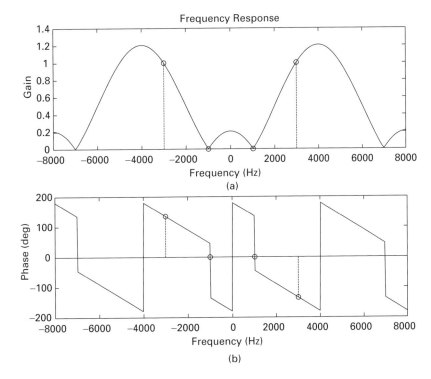

Why did we go to all the trouble of assuming a particular difference equation and solving for the constants a and b? We could have achieved whatever frequency response we wanted by defining the FRF, recording the data, performing the FFT, multiplying by the frequency response, and evaluating the IFFT. However, the FFT and IFFT require that we record the entire dataset before we begin processing. On the other hand, with the difference equation we can start generating output as soon as we have received three inputs (see Eq. (8.47)). The speed of our data processing limits the maximum sample rate. Today's state of the art processing units acquire data at around a gigasample per second (10^9 Hz).

8.6 Relating continuous-time and discrete-time frequency response

Another example is shown in Figure 8.19. Consider processing the input signal $x_a(t)$ with a continuous-time filter and with a discrete-time system. The signal $x_a(t)$ must be periodic. It may be inherently periodic, or we can make it periodic by repeating it after the time interval of interest T_0.

Fig. 8.19. Continuous- and
discrete-time systems for
comparison.

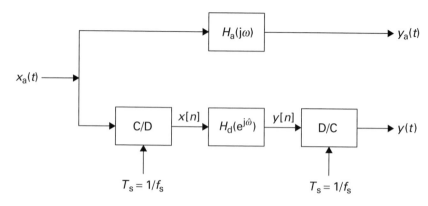

Fig. 8.20.
Continuous-time input
spectrum.

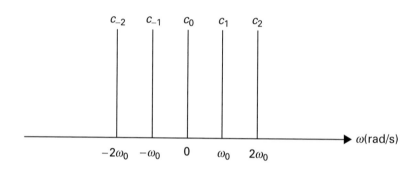

Regardless, the signal $x_a(t)$ must satisfy

$$x_a(t) = x_a(t + T_0). \tag{8.53}$$

Figure 8.20 shows the continuous-time spectrum of the input signal, $x_a(t)$.
For simplicity, let us assume that it contains a constant (DC) component
and two components at higher frequencies. Thus, we could write the signal
as

$$x_a(t) = \sum_{k=-2}^{2} c_k e^{jk\omega_0 t}, \tag{8.54}$$

where

$$\omega_0 = \frac{2\pi}{T_0}. \tag{8.55}$$

Figure 8.21(a) shows the frequency response for a continuous-time system.
For definiteness, the FRF is assumed to be triangular as shown. When the
input spectrum is multiplied by the frequency response, the continuous-time

Fig. 8.21. (a)
Continuous-time
frequency response and
(b) output spectrum.

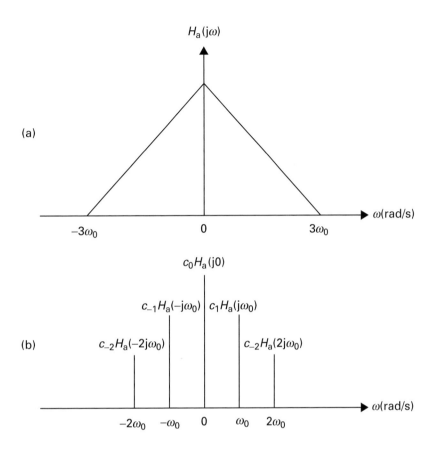

system generates an output spectrum shown in Figure 8.21(b). The output
signal can be expressed as

$$y_a(t) = y_a(t + T_0) = \sum_{k=-2}^{2} c_k H_a(jk\omega_0)e^{jk\omega_0 t}. \tag{8.56}$$

The discrete-time spectrum of $x[n] = x_a(nT_s)$ is like the continuous-time
spectrum, but it repeats after every sampling frequency ω_s, as shown in
Figure 8.22. The discrete-time spectrum can also be plotted against the
discrete-time frequency using the relationship

$$\hat{\omega} = \omega T_s = \frac{\omega}{f_s} = 2\pi \frac{\omega}{\omega_s}. \tag{8.57}$$

Thus, we observe that the discrete-time signal must repeat after an interval
of 2π. This makes sense: both discrete-time signals and discrete-time frequency responses repeat after 2π. For convenience, we often only plot the
signals and responses from $-\pi$ to π.

Fig. 8.22. Discrete-time input spectrum.

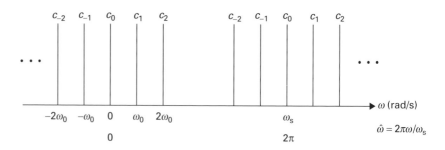

Fig. 8.23. (a) Discrete-time frequency response and (b) output spectrum.

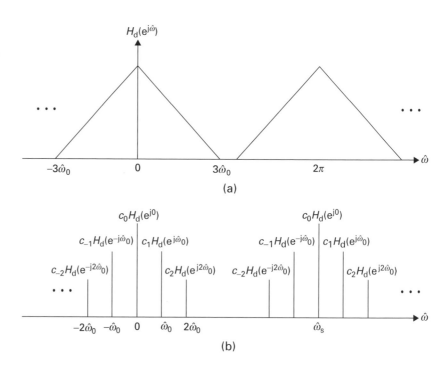

Figure 8.23(a) shows the discrete-time frequency response and Figure 8.23(b) plots the spectrum of $y[n]$. Observe that we obtain the output spectrum by multiplying the discrete-time input spectrum by the discrete-time frequency response.

Finally, the D/C converter regenerates the continuous-time output $y(t)$ using the discrete-time spectrum between $-\pi$ and π. From Figure 8.23, we see the output must be

$$y(t) = \sum_{k=-2}^{2} c_k H_{\mathrm{d}}(e^{jk\hat{\omega}_0})e^{jk\omega_0 t}. \qquad (8.58)$$

Equation (8.58) is nearly identical to the output of the continuous-time system given in Eq. (8.56). They are equal if we sample fast enough, that is if we sample with f_s that is at least twice the highest frequency in the signal to avoid aliasing, and if the discrete-time filter is identical to the continuous-time signal over the range $\hat{\omega} = -\pi$ to π, that is $f = -f_2/2$ to $f_s/2$.

The discrete-time system at first seems more complicated because it has the extra C/D and D/C components. However, in the continuous-time system we must build a physical device with electrical or mechanical components. The device is susceptible to all the real-world effects such as wear and tear, environmental conditions, unavoidable manufacturing tolerances, etc. On the other hand, the discrete-time system is simply an algorithm. The system is exact (or at least the rounding error can be arbitrarily small) and altering the system simply means changing the algorithm.

8.6.1 Application: low-pass filter

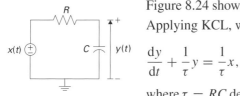

Fig. 8.24. *RC* circuit: a first-order low-pass filter.

Figure 8.24 shows a low-pass *RC* filter circuit, a classic first-order system. Applying KCL, we find that the governing equation is

$$\frac{dy}{dt} + \frac{1}{\tau}y = \frac{1}{\tau}x, \qquad (8.59)$$

where $\tau = RC$ denotes the time constant for this first-order system. Suppose we wish to design a discrete-time system that could simulate the *RC* circuit or give a response approximately equal to that of the circuit. One approach is to use the *backward difference* approximation. If we sample the input and output at an interval T_s, the first derivative with respect to time can be approximated as

$$\frac{dy}{dt} \approx \frac{y(nT_s) - y((n-1)T_s)}{T_s}. \qquad (8.60)$$

This approximation becomes quite good as the sampling time T_s approaches zero. Substituting Eq. (8.60) into Eq. (8.59), we obtain

$$\frac{y(nT_s) - y((n-1)T_s)}{T_s} + \frac{1}{\tau}y(nT_s) = \frac{1}{\tau}x(nT_s). \qquad (8.61)$$

Using the notation $y[n] = y(nT_s)$ and $x[n] = x(nT_s)$, we can rewrite Equation (8.61) as

$$y[n] = \frac{1}{1 + T_s/\tau}y[n-1] + \frac{T_s/\tau}{1 + T_s/\tau}x[n]. \qquad (8.62)$$

There are many numerical techniques for obtaining approximations to the solution of a differential equation. The backward difference approximation is relatively simple to apply and leads to a reasonably accurate solution. Notice that Eq. (8.62) is a recursive difference equation because the current output $y[n]$ depends on the previous values of the output as well as on the current value of the input. The difference equation depends on $y[n-1]$ because the system is of first-order. An nth order system with nth derivatives would depend on the previous n values of y. We can find the discrete-time frequency response by assuming the input and output are complex exponentials. Substituting

$$x[n] = Xe^{j\hat{\omega}n}$$
$$y[n] = Ye^{j\hat{\omega}n}$$
(8.63)

into Eq. (8.62), we find

$$H(e^{j\hat{\omega}}) = \frac{Y}{X} = \frac{b}{1 - ae^{-j\hat{\omega}}},$$
(8.64)

where the constants a and b are given by

$$a = \frac{1}{1 + T_s/\tau}$$
$$b = \frac{T_s/\tau}{1 + T_s/\tau}.$$
(8.65)

Suppose we wish to design the system of Figure 8.24 with a corner frequency of 1000 Hz. Recall that for a first-order system, its corner frequency is related to the time constant as $\omega_c = 1/\tau$. Thus, the corner frequency specification immediately dictates the time constant of the system to be

$$\omega_c = \frac{1}{\tau} = 2\pi f_c \Longrightarrow \tau = RC = \frac{1}{2\pi f_c} = \frac{1}{2\pi(1000)} = 159.2 \ \mu s. \quad (8.66)$$

Let us choose a sampling frequency much higher than the frequency of interest so that we can approximate the system well and avoid aliasing. A common choice for sampling audio signals is $f_s = 44.1$ kHz (used in CD players).

The discrete-time FRF of Eq. (8.62) is given by Eq. (8.64). The continuous-time frequency response for the system of Figure 8.24 is given by

$$H(j\omega) = \frac{1}{1 + j\omega\tau}.$$
(8.67)

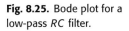

Fig. 8.25. Bode plot for a low-pass *RC* filter.

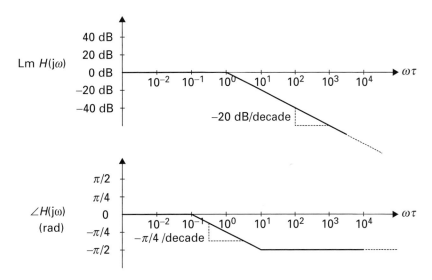

This continuous-time system has a Bode plot shown in Figure 8.25. By inspection, we recognize that it is a low-pass filter. Let us now compare the discrete-time and continuous-time frequency responses. This will help us understand how accurately the discrete-time system approximates the continuous-time system. Figure 8.26 shows the two frequency responses. The plots show the magnitude and phase versus the frequency f on a linear scale. Note that the top plot shows the magnitude or gain of the frequency response function, $|H(e^{j\hat{\omega}})|$ or $|H(j\omega)|$, and not the logarithmic magnitude in decibels. The gains of the discrete-time and continuous-time frequency responses agree very well for frequencies out to half the sampling frequency (22.05 kHz). The phases, on the other hand, agree well only for frequencies near the origin (out to about 1 kHz), but they deviate substantially from one another for large frequencies. This may be critical depending on the application.

As mentioned, the plot of Figure 8.26 is shown on a linear scale. This is common for discrete-time frequency responses, which are periodic in frequency. Figure 8.27 shows the same responses plotted on a logarithmic scale for frequency. Note also that the gain is now plotted using logarithmic magnitude in decibels. Thus, Figure 8.27 constitutes the familiar Bode plot. The logarithmic magnitude plot for the discrete-time system is similar to the classic first-order continuous-time system. At low frequencies, the phase plot for the discrete-time system is also similar to the phase plot of a continuous-time first-order system, but it deviates from the continuous-time

Fig. 8.26. Frequency responses of continuous-time and discrete-time systems (linear scale).

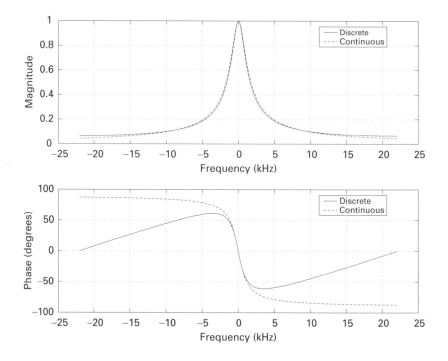

Fig. 8.27. Frequency responses of continuous-time and discrete-time systems (log scale).

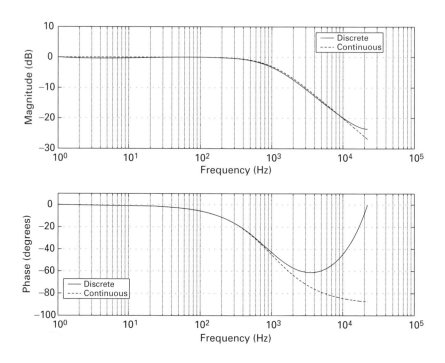

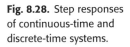

Fig. 8.28. Step responses of continuous-time and discrete-time systems.

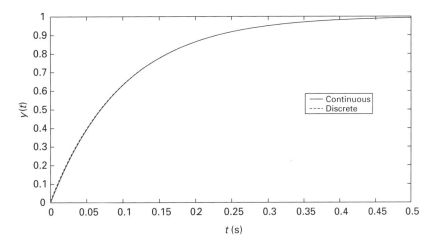

phase plot at high frequencies. Figure 8.28 compares the step responses found with the continuous and discrete-time systems using a time constant of $\tau = 0.1$ s and a sampling interval of $T_s = 0.001$ s. Observe the excellent agreement between the two step responses.

Note that for the difference equation of Eq. (8.62), it is very easy to calculate the output $y[n]$ from the input $x[n]$ and the previous output value $y[n-1]$. It is not necessary to take the FFT and IFFT, so the discrete-time system can be used in real time. In general, such recursive difference equations are very powerful ways of processing signals. However, recursive equations are an example of a feedback system. Under certain conditions, feedback systems can become unstable and the response can oscillate wildly. There is a good deal of theory to help determine when a feedback system is stable, but it is beyond the scope of this text.[5]

8.7 Finite impulse response filters

An alternative is to use *nonrecursive* difference equations in which the output depends only on the input values and not on the previous output values. Such nonrecursive equations are often called *finite impulse response* (*FIR*) filters. The output may be written as a weighted sum of the past and

[5] See *Modeling, Analysis, and Control of Dynamic Systems* by W. J. Palm III, John Wiley & Sons, New York, 1999 or *Control and Dynamic Systems* by Y. Takahashi, M. J. Rabins and D. M. Auslander, Addison-Wesley, Massachusetts, 1972.

future values of the input as shown:

$$y[n] = \sum_{k=-K}^{K} a_k x[n-k]$$
$$= a_{-K} x[n+K] + \cdots + a_0 x[n] + \cdots + a_K x[n-K]. \qquad (8.68)$$

Unfortunately, the output of Eq. (8.68) depends on past and *future* values of the input. The form is called *noncausal*. This clearly becomes problematic if we need to know the output immediately. On the other hand, if we are willing to wait a bit (specifically, wait for K samples), that is if we can tolerate some delay, we can manipulate the output $y[n]$ to a *causal* output $y_1[n]$ that only depends on the previous input values as shown:

$$y_1[n] = y[n-K]$$
$$= \sum_{k=-K}^{K} a_k x[n-K-k] \qquad (8.69)$$
$$= a_{-K} x[n] + \cdots + a_0 x[n-K] + \cdots + a_K x[n-2K].$$

The delay that is introduced may be acceptable for some systems such as television broadcast. However, it is a problem for other systems such as two-way voice communication, where the delay manifests itself as an undesirable pause between sentences.

By assuming the input and output are complex exponentials, we find the discrete-time frequency responses of these two signals as

$$H\left(e^{j\hat\omega}\right) = \frac{Y}{X} = \sum_{k=-K}^{K} a_k e^{-jk\hat\omega} \qquad (8.70)$$

and

$$H_1\left(e^{j\hat\omega}\right) = \frac{Y_1}{X} = \sum_{k=-K}^{K} a_k e^{-j(k+K)\hat\omega} = e^{-jK\hat\omega} \sum_{k=-K}^{K} a_k e^{-jk\hat\omega}. \qquad (8.71)$$

Observe that delaying the signal by K samples is equivalent to multiplying the frequency response by $e^{-jK\hat\omega}$. This quantity has a magnitude of 1 and a phase that decreases linearly with frequency. Thus, we often start by designing a noncausal filter with the desired frequency response characteristics, then delay the output to make the filter causal and to preserve the magnitude (though not phase) of the frequency response.

In general, we would like to find the set of FIR coefficients $\{a_k\}$ that give us a desired frequency response. We begin with our desired discrete-time frequency response function $H(e^{j\hat\omega})$, which is a periodic function with

Fig. 8.29. Desired
frequency response
function.

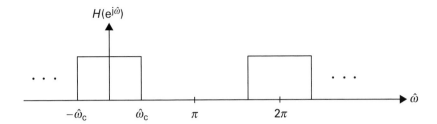

a period of 2π. We would like to determine the coefficients so that the
frequency response can be expressed in the form given by Eq. (8.70). At
first glance it appears that finding these coefficients may be difficult. But
notice that this is just like the Fourier series in which we find the weights
of the exponentials to approximate a periodic signal. Recall that we can
approximate a periodic signal $x(t)$ with $\hat{x}(t)$ using a finite number of terms
as follows:

$$x(t) = x(t + T_0) \approx \hat{x}(t) = \sum_{k=-K}^{K} c_k e^{jk\omega_0 t}, \qquad (8.72)$$

where the coefficients or weights c_k are found by direct integration:

$$c_k = \frac{1}{T_0} \int_{T_0} x(t) e^{-jk\omega_0 t} dt. \qquad (8.73)$$

Comparing Eqs. (8.72) and (8.70), we notice the only difference is that
the exponentials are positive in the Fourier series of Eq. (8.72), while they
are negative in Eq. (8.70). Therefore, we can use Eq. (8.73) to compute
the coefficients as we did for the Fourier coefficients, except that we must
change the sign on the exponentials, that is make the exponentials positive,
as follows:

$$a_k = \frac{1}{2\pi} \int_{2\pi} H\left(e^{j\hat{\omega}}\right) e^{jk\hat{\omega}} d\hat{\omega}. \qquad (8.74)$$

Thus, using Eqs. (8.70) and (8.74), we can design a filter with arbitrary
frequency response characteristics.

For example, suppose we would like to design an ideal low-pass filter
with an abrupt transition. The discrete-time frequency response $H(e^{j\hat{\omega}})$ is
shown in Figure 8.29. As expected, $H(e^{j\hat{\omega}})$ is periodic with period 2π. We
can solve for the filter coefficients using Eq. (8.74), but we note that this is
just the Fourier series of a pulse train, whose coefficients we have computed

Fig. 8.30. Frequency response of discrete-time low-pass filter.

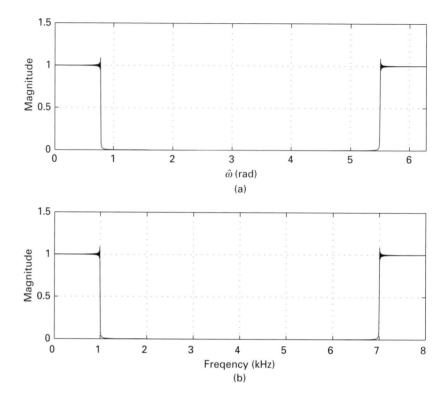

in Chapter 2. Recalling our previous results (see Eq. (2.46)) or integrating Eq. (8.74), we find the coefficients are given by the following sinc functions:

$$a_k = \frac{\hat{\omega}_c}{\pi} \frac{\sin(k\hat{\omega}_c)}{k\hat{\omega}_c} = \frac{\hat{\omega}_c}{\pi} \text{ sinc}\left(\frac{k\hat{\omega}_c}{\pi}\right) = 2\hat{f}_c \text{ sinc}(2k\hat{f}_c), \qquad (8.75)$$

where

$$\hat{\omega}_c = \omega_c T_s = \frac{\omega_c}{f_s}$$

$$\hat{f}_c = \frac{\hat{\omega}_c}{2\pi} = \frac{f_c}{f_s}. \qquad (8.76)$$

Figure 8.30 plots the frequency response. We must select a sampling frequency that is sufficiently high. For definiteness, let us choose 8 kHz and consider a low-pass filter with a cutoff frequency of 1 kHz. We will use $K = 1001$ points. Figure 8.30(a) shows the response plotted against discrete-time angular frequency $\hat{\omega}$, while Figure 8.30(b) is the same response plotted against continuous-time frequency f. Note that we get the desired cutoff at 1 kHz. The filter satisfies the desired frequency response, except for the ringing and overshoot near the cutoff frequency. This is the same Gibbs phenomenon that we saw in the Fourier series.

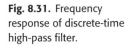

Fig. 8.31. Frequency response of discrete-time high-pass filter.

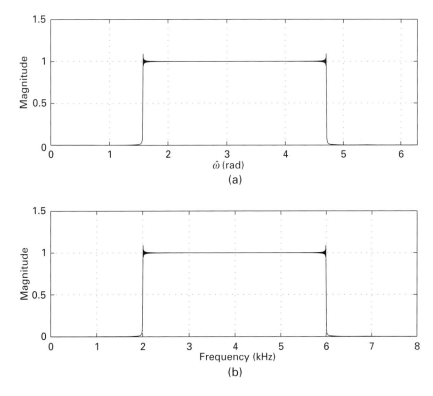

Figure 8.31 shows a high-pass filter response with a cutoff frequency of 2 kHz. Again, we choose a sampling frequency of 8 kHz and we let $K = 1001$. Note that the discrete-time filter satisfies the desired frequency response.

Using the FIR technique, we can construct digital filters with any frequency response characteristics that we want. They have an advantage over the FFT technique because they filter in real time rather than waiting for all of the dataset to arrive. There is, however, an inherent trade-off between the number of coefficients and the required computation time. Using more coefficients creates a more accurate approximation of the desired frequency response but at the expense of more computation. For example, Figure 8.32 shows a low-pass filter with only 25 coefficients. The transition between the passband and stopband is substantially less sharp and the filter is not as flat in either band.

In another example, suppose we want to create the "world-famous" M filter[6] with the frequency response shown in Figure 8.33. It would be

---

[6] Or at least the filter is famous among close friends and immediate relatives of Dr Molinder!

Fig. 8.32. Low-pass filter
with 25 coefficients.

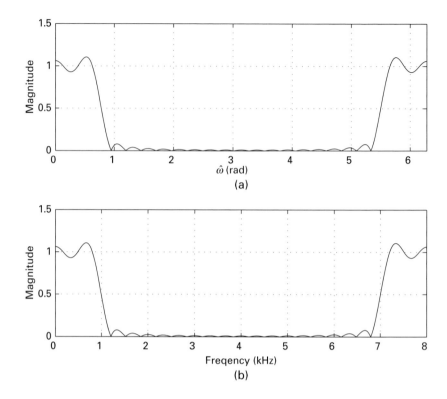

Fig. 8.33. Frequency
response of
"world-famous" M filter.

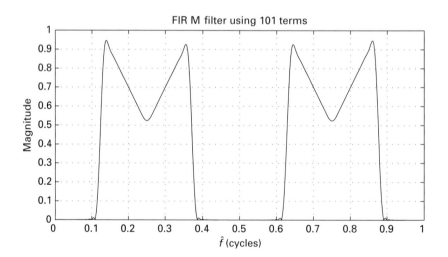

Fig. 8.34. Mixer:
multiplication of signals.

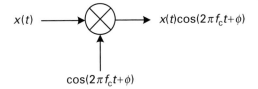

$x(t)$ → ⊗ → $x(t)\cos(2\pi f_c t + \phi)$

$\cos(2\pi f_c t + \phi)$

hideously complicated, if not impossible, to try to connect a series of resistors, capacitors, and inductors to construct a system with such a frequency response. However, using digital signal processing, we can easily find the coefficients for a FIR filter that satisfies this frequency response by evaluating Eq. (8.74) or using the IFFT.

8.8 The mixer

All of the systems we have explored so far are linear and time-invariant. The mixer shown in Figure 8.34 is an example of a system that is linear but not time-invariant. Multiplication by a sinusoid has an interesting effect in the frequency domain. Suppose the information bearing or message signal, $x(t)$, is periodic as follows:

$$x(t) = x(t + T_0) = \sum_{k=-\infty}^{\infty} c_k e^{j2\pi k f_0 t}. \tag{8.77}$$

The spectrum of the output signal is then given by

$$x(t)\cos(2\pi f_c t + \phi) = \left[\sum_{K=-\infty}^{\infty} c_k e^{j2\pi k f_0 t}\right] \frac{e^{j(2\pi f_c t + \phi)} + e^{-j(2\pi f_c t + \phi)}}{2} \tag{8.78}$$

or

$$x(t)\cos(2\pi f_c t + \phi) = \sum_{K=-\infty}^{\infty} \frac{1}{2} e^{j\phi} c_k e^{j2\pi (k f_0 + f_c)t}$$

$$+ \sum_{K=-\infty}^{\infty} \frac{1}{2} e^{-j\phi} c_k e^{j2\pi (k f_0 - f_c)t}. \tag{8.79}$$

From the result of Eq. (8.79), we note that by mixing the input signal $x(t)$ with $\cos(2\pi f_c t + \phi)$, the spectrum of $x(t)$ will be:

• Frequency shifted up and down by the frequency f_c with half of the amplitude; alternatively, two copies of the input spectrum with half of the

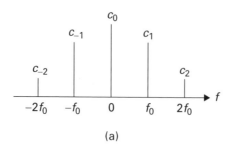

(a)

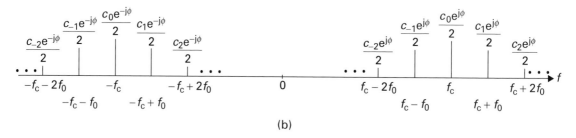

(b)

Fig. 8.35. (a) Input and (b) output spectrum of a mixer.

amplitude are created, one centered at $f = f_c$ and the other centered at $f = -f_c$.

• Phase shifted by ϕ (or multiplied by $e^{j\phi}$) for $f > 0$, and phase shifted by $-\phi$ (or multiplied by $e^{-j\phi}$) for $f < 0$.

For example, if the input spectrum $x(t)$ is shown in Figure 8.35(a), then the spectrum of the output signal $y(t) = x(t)\cos(2\pi f_c t + \phi)$ is shown in Figure 8.35(b). Looking at the spectrum of the output we see that it contains frequencies that were not present in the input signal. This means the system cannot be LTI since the output of an LTI system contains only frequencies present in the input.

In summary, the mixer creates two copies of the spectrum of $x(t)$ centered around $f = f_c$ and $f = -f_c$ with half the amplitude each. It will sometimes be convenient to simply draw an outline, obtained by connecting the tips of the spectral lines, as shown in Figure 8.36. Since it is now impossible to specify both the amplitude and phase of each frequency component, the outline represents the magnitude only. As shown, we can now use shading to map the relationship of input and output frequencies. Why would we want to do this? Sometimes, as we will discuss in the next chapter, we would like to transmit the signal as an electromagnetic wave. From physics we know that in order to transmit a signal, we need an antenna with dimensions of at least about $1/10$ of the wavelength λ of the signal. Recall also from physics

Fig. 8.36. (a) Generic input spectrum for $x(t)$ and (b) the spectrum for its associated output $y(t)$.

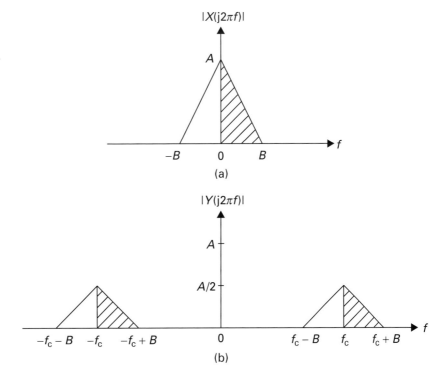

that the product of the signal wavelength and the frequency is the speed of light:

$$\lambda f = c \Longrightarrow \lambda = \frac{c}{f}. \tag{8.80}$$

Thus, mixing with a high frequency sinusoid permits transmitting at higher frequencies and allows the use of a shorter (and cheaper) antenna. Comparing the required minimum antenna dimensions for frequencies of 10 kHz, 1 MHz, and 1 GHz, we see that being able to shift frequencies is very handy. Of course, we will usually want to recover the original signal $x(t)$ from the output of the mixer $y(t)$. As we will see in Chapter 9, another mixer and a filter can be used to accomplish the task of recovering the original signal $x(t)$.

8.9 Problems

8-1 A linear time-invariant system is described by the following difference equation:

$$y[n] = 2x[n] - 3x[n-1] + 2x[n-2].$$

(a) Find the frequency response function $H(\hat{\omega})$ of the system. Determine expressions for the magnitude and phase.

(b) Find the frequencies $\hat{\omega}$ for which the output response to the input $x[n] = e^{j\hat{\omega}n}$ is zero.

(c) Use MATLAB to plot the magnitude and phase of $H(\hat{\omega})$ as a function of $\hat{\omega}$, for $-\pi < \hat{\omega} < 3\pi$.

(d) If the input to the system is $x[n] = \sin(\pi n/13)$, determine the output signal and express it in the form $y[n] = A\cos(\hat{\omega}_0 n + \phi)$.

8-2 A linear time-invariant filter is described by the difference equation

$$y[n] = x[n] + 2x[n-1] + x[n-2].$$

(a) Obtain an expression for the frequency response function $H(\hat{\omega})$ of the filter.

(b) Use MATLAB to plot the magnitude and phase of the frequency response function.

(c) Determine the output when the input is

$$x[n] = 10 + 4\cos(0.5\pi n + \pi/4).$$

(d) Determine and sketch the output when the input is the unit-impulse sequence

$$x[n] = \delta[n] = \begin{cases} 1, & n = 0 \\ 0, & n \neq 0. \end{cases}$$

(e) Determine and sketch the output when the input is the unit-step sequence

$$x[n] = u[n] = \begin{cases} 0, & n < 0 \\ 1, & n \geq 0. \end{cases}$$

8-3 A linear time-invariant filter is described by the difference equation

$$y[n] = \frac{1}{4}(x[n] + x[n-1] + x[n-2] + x[n-3]) = \frac{1}{4}\sum_{k=0}^{3} x[n-k].$$

(a) Obtain an expression for the frequency response function $H(\hat{\omega})$ for this system.

(b) Use MATLAB to plot the magnitude and phase of the frequency response function.

Fig. 8.37. Figure for
Problem 8-4.

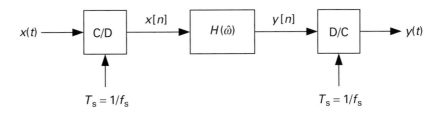

Fig. 8.37. Figure for
Problem 8-4.

(c) Suppose that the input is

$$x[n] = 5 + 4\cos(0.2\pi n) + 3\cos(0.5\pi n + \pi/4).$$

Obtain an expression for the output in the form

$$y[n] = A + B\cos(\hat{\omega}_0 n + \phi_0).$$

8-4 A system for filtering continuous-time signals is shown in Figure 8.37.
It uses a linear time-invariant filter whose difference equation is given
by

$$y[n] = \frac{1}{4}(x[n] + x[n-1] + x[n-2] + x[n-3])$$

$$= \frac{1}{4}\sum_{k=0}^{3} x[n-k].$$

The input to the C/D converter in this system is

$$x(t) = 10 + 8\cos(200\pi t) + 6\cos(500\pi t + \pi/4).$$

If $f_s = 1000$ samples/s, determine an expression for the output $y(t)$.

9 Applications

In this chapter we will consider various applications using the important concepts in signals and systems that we have covered thus far in the text.

9.1 Communication systems

The general process of embedding an information bearing signal into a second signal is typically referred to as *modulation*. Extracting the information bearing signal is known as *demodulation*. Many communication systems rely on the concept of sinusoidal *amplitude modulation*, in which a sinusoidal signal $c(t)$ has its amplitude multiplied or modulated by the information bearing signal $x(t)$. Amplitude modulation is often used to transmit a signal with low frequency content using a high frequency transmission channel. The high frequency of the carrier wave gives it the ability to travel great distances when radiated from an antenna.

9.2 Modulation

Amplitude modulation is performed by multiplying a high frequency signal (called the carrier) by a low frequency message signal $x(t)$. We can do this with the mixer shown in Figure 9.1 Repeating our analysis from Chapter 8, the spectrum of $x(t)$ is shown in Figure 9.2, where again we have used the outline of the spectral lines to represent the spectrum of $x(t)$. After passing through the mixer, the output signal has a spectrum given by Figure 9.3. Knowing the spectrum of the output signal $y(t)$, we now wish to recover the spectrum of $x(t)$ and thus the input signal $x(t)$ itself. Surprisingly, this

Fig. 9.1. Mixing input with carrier wave.

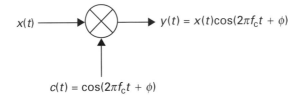

task is conceptually fairly trivial (in practice it is much more challenging). To recover the spectrum of $x(t)$, we simply mix $y(t)$ with the same carrier signal, $c(t)$, and then pass the new output signal $z(t)$ through a low-pass filter to eliminate all of the high frequency components in the spectrum of $z(t)$. Schematically this process is illustrated in Figure 9.4(a), and the resulting spectrum $\hat{x}(t)$ is shown in Figure 9.4(b). In Figure 9.4(b), the dotted triangular waveforms centered at f_c and $-f_c$ correspond to the spectrum of $y(t)$. The triangular wave forms centered at $-2f_c$, 0 and $2f_c$ correspond to the spectrum of $z(t)$. The rectangular dashed line centered at $f = 0$ indicates the frequency response of a low-pass filter (LPF) used to extract the demodulated signal. If the low-pass filter has a gain of two and a cutoff frequency between B and $-B$ we will recover $x(t)$.

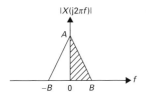

Fig. 9.2. Spectrum of $x(t)$.

9.3 AM radio

In addition to making it possible to use an antenna of reasonable size, AM radio uses mixers to separate the transmitted signals from various stations in terms of frequency. Figure 9.5 shows the AM radio band consisting of stations distributed over the AM band of 540 kHz up to 1600 kHz. The frequency of the station describes the carrier, f_c. Of course the spectrum also contains the corresponding negative frequencies. Stations are distributed on

Fig. 9.3. Spectrum of $y(t)$.

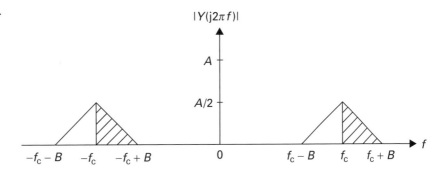

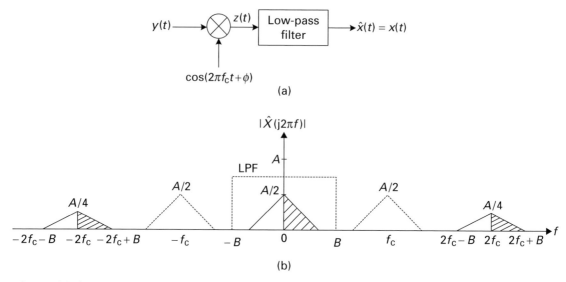

Fig. 9.4. (a) The demodulation process and (b) the resulting spectrum.

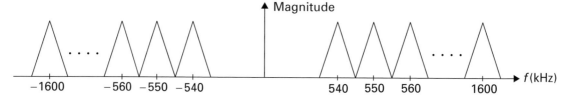

Fig. 9.5. AM radio band.

a 10 kHz spacing and are not allowed to extend outside their band because they would interfere with another station. The Federal Communications Commission (FCC) requires that the center frequency wander by no more than 20 Hz.

Because each station is allocated a 10 kHz bandwidth, it must only contain frequencies in the range ± 5 kHz, or $\pm f_m$. This is why music sounds so poor on AM radio: the higher frequency components of the music must be filtered out before transmission. Tuning an AM radio involves selecting the center frequency that corresponds to the station one wishes to listen to. To receive an AM station, we can mix the output signal from the radio station with another cosine at the same carrier frequency to produce an output $z(t)$, as shown in Figure 9.6(a). This produces the spectrum in Figure 9.6(b). One could show using trigonometric identities that the new spectrum contains the spectrum of the original signal centered at the origin but with half the amplitude of the transmitted signal, and an *image* of the original spectrum

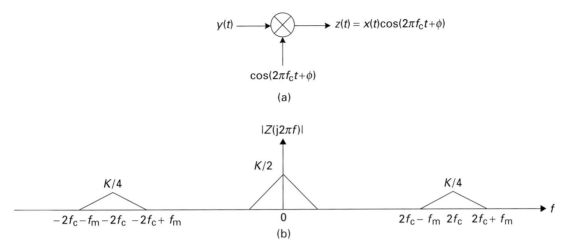

Fig. 9.6. Receiving AM
station.

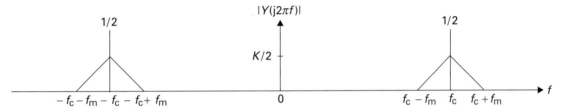

Fig. 9.7. Spectrum of AM
signal.

centered at twice the carrier frequency but with a quarter of the amplitude
of the transmitted signal. We can low-pass filter the output to recover the
spectrum of the original signal (scaled by a factor of 2, which we can always
recover with an amplifier later).

Although this is a legitimate way to recover an AM radio signal, we
must multiply by a cosine with exactly the same frequency and phase as
the transmitter used. When AM radio was first commercialized, this was
not an easy task (it is still challenging but much less expensive these days).
Recovery is much easier if the transmitted signal is offset by a constant as
shown:

$$y(t) = [1 + x(t)] \cos{(2\pi f_c t)}. \tag{9.1}$$

Assume $|x(t)| < 1$ so that $1 + x(t)$ is always positive. The spectrum of the
transmitted signal is shown in Figure 9.7. There is a spectral line in the
middle of each station, at a frequency of f_c, representing the term we have
added, but the spectrum is otherwise unchanged. Figure 9.8 shows $1 + x(t)$
and $y(t)$ in the time domain. The dashed line is called the *envelope* and has

Fig. 9.8. Time domain
plots of a) $x(t)$ and b) $y(t)$.

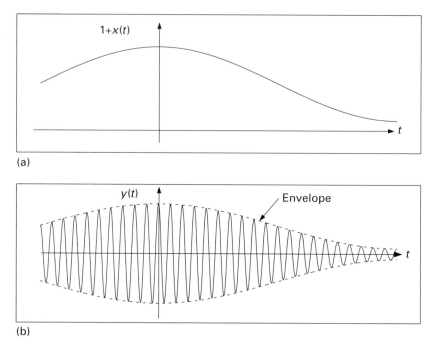

Fig. 9.8. Time domain
plots of a) $x(t)$ and b) $y(t)$.

(a)

(b)

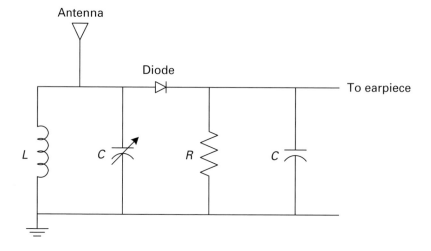

Fig. 9.9. Crystal radio
receiver.

a value of $1 + x(t)$. The signal $y(t)$ oscillates at the carrier frequency f_c between the amplitudes set by the envelope. Note that the maximum frequency contained in $x(t)$ is f_m, which we assume is much less than the carrier frequency f_c.

It is easy and cheap to build a circuit that follows the envelope while filtering out the high-frequency oscillation. Figure 9.9 shows such a circuit

Fig. 9.10. Amplitude modulation with envelope detection.

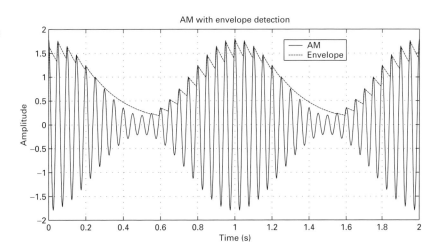

Fig. 9.11. Systems representation of crystal radio receiver.

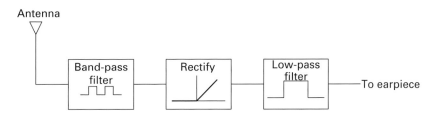

used in a simple AM radio receiver called a *crystal radio*. The antenna on the left is attached to an inductor and a variable capacitor, which form a band-pass filter. By tuning the value of the capacitor, we adjust which station we are receiving. The diode only allows current to flow from left to right. When the input is positive, it charges up the capacitor. When the input drops, the capacitor slowly discharges through the resistor. The resulting signal follows the envelope fairly well, as shown in Figure 9.10. The output goes to an earphone or can be amplified to drive a loudspeaker. Notice that we do not need to mix the signal with a sinusoid at exactly the same frequency and phase as the carrier.

Another detector can be constructed from a band-pass filter, rectifier, and low-pass filter, as shown in Figure 9.11. Again, the band-pass filter selects the station we wish to receive. The *rectifier* passes positive inputs but cuts off negative inputs at zero. Finally the *RC* low-pass filter eliminates the carrier, and higher frequency terms leaving only the envelope.

Figure 9.12 shows the waveforms in the AM radio when this technique is used. Figure 9.12(a) is the amplitude-modulated signal. The original signal

Fig. 9.12. Time-domain
view of AM radio receiver.

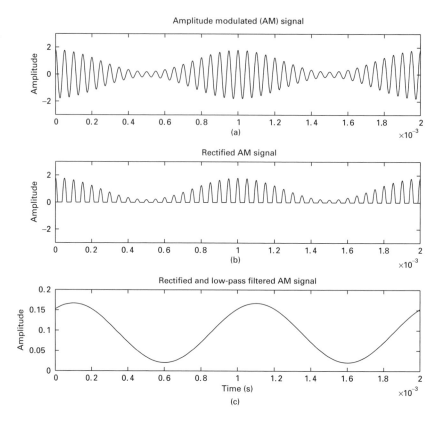

$x(t)$ is a cosine with amplitude of 0.8, so $y(t)$ varies between 0.2 and 1.8 at the carrier frequency (see Eq. (9.1)). Figure 9.12(b) is the signal after rectifying. Figure 9.12(c) is the signal after low-pass filtering. We see that we have recovered the envelope of the signal.

Figure 9.13 shows the effects in the frequency domain. The original signal has a peak at the carrier frequency and two smaller peaks offset by the frequency of $x(t)$ (Figure 9.13(a)). The rectifier is a nonlinear system that produces the spectrum in Figure 9.13(b). Finally, low-pass filtering eliminates everything except the desired output tone, as shown in Figure 9.13(c).

Figure 9.14(a) shows the same input cosine transmitted with an amplitude of 1.2. The recovered envelope is shown in Figure 9.14(b). Observe that it is no longer a cosine after low-pass filtering, as illustrated in Figure 9.14(c). This happened because the input amplitude was too large and distorted the envelope; it is called *overmodulation*. Real AM transmitters must be careful to limit the power of the input signal so that they do not overmodulate and distort their station.

Fig. 9.13.
Frequency-domain view of
AM radio receiver.

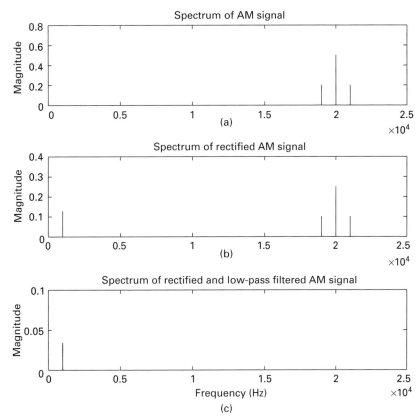

Fig. 9.13.
Frequency-domain view of
AM radio receiver.

9.4 Vibration measuring instruments

An important application of the concept of the frequency response function, $H(j\omega)$, is in the design of devices used to measure vibration. Figure 9.15 shows the essentials of such a vibration measuring device. It consists of a *seismic mass m* connected to a housing through a spring of stiffness k and a viscous damper of damping coefficient c. The housing is rigidly attached to the structure whose motion we wish to measure. A recording instrument (in this case, a pen) on the mass is used to trace out the relative motion between the seismic mass and the housing. This relative motion can be recorded using a rotating drum as illustrated. Suppose we attach the housing to a table that is shaking vertically. The mass will vibrate relative to the drum, so the pen traces out a pattern on the drum that measures the relative motion of the seismic mass. This device is commonly used to record earthquakes. These systems are usually underdamped ($0 < \zeta < 1$).

We start analyzing the system by deriving the equation of motion. Let $x(t)$ denote the vertical displacement of the seismic mass (measured relative to

Fig. 9.14. Time-domain view of AM radio receiver with overmodulation.

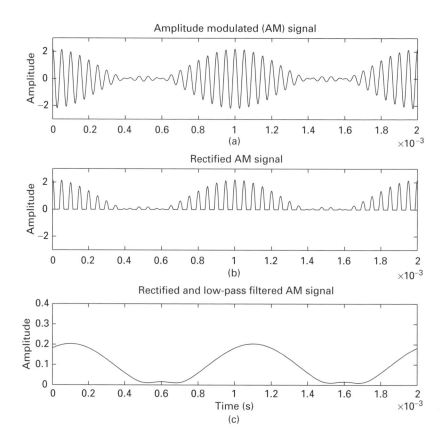

Fig. 9.15. Vibration measuring device.

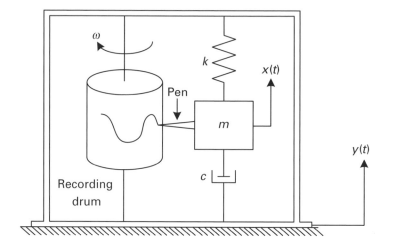

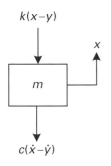

$k(x-y)$

x

m

$c(\dot{x}-\dot{y})$

Fig. 9.16. Free body diagram of seismic mass.

the static equilibrium position), and $y(t)$ represent the vertical displacement of the housing. Figure 9.16 shows the free body diagram for the seismic mass. Applying Newton's Second Law, we obtain the following governing equation:

$$\sum F_y = m\ddot{x} = -k(x - y) - c(\dot{x} - \dot{y}).$$ (9.2)

For such devices, it is generally more convenient to work with the displacement of the seismic mass relative to the housing, or simply the relative displacement, defined as

$$z(t) = x(t) - y(t).$$ (9.3)

In terms of the relative displacement, Eq. (9.2) becomes

$$m\ddot{z} + c\dot{z} + kz = -m\ddot{y}.$$ (9.4)

In standard or canonical form, Eq. (9.4) reduces to

$$\ddot{z} + 2\zeta\omega_n\dot{z} + \omega_n^2 z = -\ddot{y}.$$ (9.5)

To obtain the frequency response function, $H(j\omega)$, we assume that the input and output are complex exponentials:

$$y = Y e^{j\omega t}$$
$$z = Z e^{j\omega t}.$$ (9.6)

Substituting Eq. (9.6) into Eq. (9.5), we find

$$H(j\omega) = \frac{Z}{Y} = \frac{\omega^2}{\omega_n^2 - \omega^2 + 2\zeta\omega_n\omega j}$$ (9.7)

which can also be expressed in terms of the normalized frequency, ω/ω_n, as follows:

$$H(j\omega) = \frac{Z}{Y} = \frac{\left(\dfrac{\omega}{\omega_n}\right)^2}{1 - \left(\dfrac{\omega}{\omega_n}\right)^2 + 2\zeta\left(\dfrac{\omega}{\omega_n}\right)j}.$$ (9.8)

If the structure undergoes simple harmonic motion, then its displacement, which also corresponds to the displacement of the housing, can be written as

$$y(t) = Y \sin(\omega t).$$ (9.9)

Thus, the measured or the relative displacement, that is the output, is given by

$$z(t) = Y |H(\text{j}\omega)| \sin(\omega t + \angle H(\text{j}\omega)), \qquad (9.10)$$

where the gain and the phase of the frequency response function are

$$|H(\text{j}\omega)| = \frac{\left(\dfrac{\omega}{\omega_n}\right)^2}{\sqrt{\left[1 - \left(\dfrac{\omega}{\omega_n}\right)^2\right]^2 + \left(2\zeta\dfrac{\omega}{\omega_n}\right)^2}}$$

$$\angle H(\text{j}\omega) = -\tan^{-1}\left[\frac{2\zeta\dfrac{\omega}{\omega_n}}{1 - \left(\dfrac{\omega}{\omega_n}\right)^2}\right]. \qquad (9.11)$$

The ratio of the output amplitude to the input amplitude is given by the gain, $|H(\text{j}\omega)|$, and is also known as the *magnification factor*. Lots of information is embedded in the magnification factor. While complicated at first glance, $|H(\text{j}\omega)|$ simplifies substantially when we consider various limiting cases of the normalized frequency, ω/ω_n. Depending on the choice of ω/ω_n, we can use this device to measure the displacement, velocity, or acceleration of the input. We now consider the following three limiting cases of ω/ω_n.

9.4.1 Seismometer

If $\omega/\omega_n \gg 1$, then the gain of Eq. (9.11) reduces to

$$|H(\text{j}\omega)| = \frac{Z}{Y} \approx \frac{\left(\dfrac{\omega}{\omega_n}\right)^2}{\left(\dfrac{\omega}{\omega_n}\right)^2} = 1. \qquad (9.12)$$

Thus, the output amplitude Z is approximately equal to the input amplitude Y, and we have a displacement measuring device or a *seismometer*. For this to work, the excitation frequency, ω, of the input must be much greater than the natural frequency, ω_n, of the system. Physically, the mass m remains nearly stationary while the housing moves with the vibrating structure. Since we do not have any control over ω, the frequency of the vibrating body, we must make ω_n, the natural frequency of the device, very small. Because

$$\omega_n = \sqrt{\frac{k}{m}} \qquad (9.13)$$

this implies that a seismometer is either very large in size (large m) or very delicate (small k). Such a device is difficult to build and is not very practical for measuring displacements.

9.4.2 Velocity meter

If $\omega/\omega_n \approx 1$, then the gain of Eq. (9.11) becomes

$$|H(j\omega)| = \frac{Z}{Y} \approx \frac{\left(\frac{\omega}{\omega_n}\right)^2}{2\zeta\frac{\omega}{\omega_n}} = \frac{1}{2\zeta}\frac{\omega}{\omega_n}. \tag{9.14}$$

Thus, if the input is a sinusoid of angular frequency ω, the output is also a sinusoid whose amplitude is given by

$$Z \approx \frac{1}{2\zeta\omega_n}(\omega Y). \tag{9.15}$$

Because the input is $y(t) = Y\sin(\omega t)$, the term (ωY) represents the magnitude of the velocity of the structure. This can be easily checked from examining the units of (ωY) or by simply differentiating the input signal, $y(t)$. Thus, the output amplitude Z is proportional to the velocity of the input, with a proportionality constant of $1/(2\zeta\omega_n)$. This device is known as a *velocity meter*, and it can only be used for $\omega/\omega_n \approx 1$, implying that we need to know the oscillation frequency of the vibrating structure a priori. This severely limits the device's range of application.

9.4.3 Accelerometer

If $\omega/\omega_n \ll 1$, then the gain of Eq. (9.11) simplifies to

$$|H(j\omega)| = \frac{Z}{Y} \approx \left(\frac{\omega}{\omega_n}\right)^2. \tag{9.16}$$

Thus, the output amplitude is

$$Z \approx \frac{1}{\omega_n^2}(\omega^2 Y). \tag{9.17}$$

Because the input is given by Eq. (9.9), $(\omega^2 Y)$ represents the magnitude of the acceleration of the structure, which can be easily checked from the units or by differentiating the input signal twice. Thus, the output amplitude

Fig. 9.17. Gain of
Eq. (9.11) as a function of
ω/ω_n.

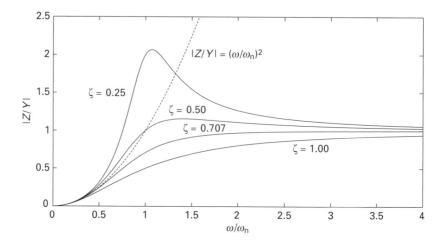

Z is proportional to the acceleration of the input, with a proportionality constant of $1/\omega_n^2$. For this to work, the natural frequency must be much larger in comparison to the excitation frequency of the structure. For a large natural frequency, we need a stiff spring and/or a small mass. This is very feasible in application; we can build tiny accelerometers with very small masses. Indeed, many accelerometers used in airbag sensors are built by micromachining a tiny beam onto a silicon chip.

How much is "much much greater" or "much much less" than 1? Figure 9.17 plots the gain or the magnification factor of the frequency response function (see Eq. (9.11)) against the normalized frequency ω/ω_n in the range of $0 \leq \omega/\omega_n \leq 4$, with the damping factor ζ acting as a parameter. The curve $(\omega/\omega_n)^2$ is also plotted for comparison. For normalized frequencies below about 0.2, the gain grows with the square of the normalized frequency, indicating that we have a good accelerometer. For normalized frequencies above about 3.0, the gain approaches unity, so we have a good seismometer. The normalized frequency must be very close to 1 to have an accurate velocity-measuring device.

Based on the above discussion, the best way to measure motion is with an accelerometer because it is small and reliable. Thus, if the interest lies in finding the velocity or displacement of the structure, it may prove more desirable to use an accelerometer to measure the structure's acceleration, then integrate it once or twice with respect to time, respectively, to obtain the velocity or displacement.

9.5 Undamped vibration absorbers

Another important application of $H(j\omega)$ is in the design of vibration absorbers, which are often used to reduce the vibration of a system. When
rotating machinery operates at a frequency very close to resonance, violent
vibration is induced. Assuming that the system can be represented by a simple spring–mass system subjected to a harmonic excitation, the undesirable
near resonant or resonant condition can be alleviated by changing either the
mass or the spring constant to modify the natural frequency of this spring–
mass system. There are times, however, that such a solution may not be
feasible. In this case, a second spring–mass can be attached to the original
system, thus producing a combined system that has two degrees of freedom.
The added spring–mass system is called the *absorber*. The parameters of
the absorber mass and stiffness are chosen or tuned such that the motion
of the original mass is drastically reduced or completely eliminated. This
is often accompanied by substantial motion of the added absorber system.
Devices that absorb energy in this way are found in a variety of applications.
Examples include sanders, compactors, reciprocating tools, etc. Probably
the most visible vibration absorbers can be seen on transmission lines and
telephone lines. A vibration absorber is often used on such wires to provide vibration isolation against wind blowing, which can cause the wire to
oscillate at its natural frequency, and can lead to large vibrations and high
stresses. The presence of the vibration absorber prevents the wire from vibrating too much. This is accomplished by transferring the energy that was
vibrating the wire to vibrating the absorber.

Figure 9.18(a) shows a vibration absorber attached to an optical table.
The vibration absorber protects the optical table from excess vibration. The
optical table and its legs are modeled as a system with a primary mass, m,
and two springs each with a stiffness of $k/2$ (see Figure 9.18(b)). Because
the system consists of two masses, we need to apply Newton's Second Law
twice to find the governing equation associated with each mass. Sketching
the free body diagrams for both masses, we find that the equations of motion
are given by

$$\sum F_m = m\ddot{y} = -ky - k_a(y - y_a) + f(t) \tag{9.18}$$
$$\sum F_{m_a} = m_a\ddot{y}_a = -k_a(y_a - y) \tag{9.19}$$

Fig. 9.18. Model of a vibration absorber attached to an optical table.

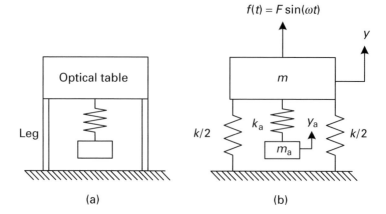

(a)   (b)

or

$$m\ddot{y} + (k + k_a)y - k_a y_a = f(t) \tag{9.20}$$

$$m_a \ddot{y}_a + k_a y_a = k_a y. \tag{9.21}$$

From Eq. (9.20), we can find an expression of y_a in terms of y, $f(t)$ and the relevant system parameters. Substituting the resulting expression into Eq. (9.21), we get, after some lengthy algebra, the following fourth-order governing equation for the primary mass:

$$m_a m \frac{d^4 y}{dt^4} + m_a(k + k_a)\frac{d^2 y}{dt^2} + k_a m \frac{d^2 y}{dt^2} + k_a k y = m_a \frac{d^2 f}{dt^2} + k_a f. \tag{9.22}$$

To find the frequency response, we let

$$f(t) = F e^{j\omega t}$$
$$y(t) = Y e^{j\omega t}. \tag{9.23}$$

Substituting Eq. (9.23) into Eq. (9.22), we get

$$H(j\omega) = \frac{Y}{F} = \frac{k_a - m_a \omega^2}{m_a m \omega^4 - [m_a(k + k_a) + k_a m]\omega^2 + k_a k}. \tag{9.24}$$

For an input of $f(t) = F \sin \omega t$, the displacement of the primary mass is

$$y(t) = F |H(j\omega)| \sin(\omega t + \angle H(j\omega)), \tag{9.25}$$

where $|H(j\omega)|$ and $\angle H(j\omega)$ represent the gain and phase of $H(j\omega)$, respectively. To eliminate the vibration of the primary mass completely, we choose the absorber parameters so that the gain of the frequency response function is exactly zero, that is $|H(j\omega)| = 0$. Physically, we can accomplish this by

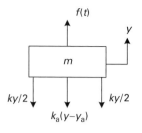

Fig. 9.19. Free body diagram for the system of Figure 9.18.

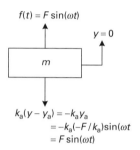

Fig. 9.20. Free body diagram for the system of Figure 9.18 when $x = 0$.

setting the numerator of Eq. (9.24) equal to zero, from which we find that the required absorber parameters must satisfy

$$\frac{k_a}{m_a} = \omega^2. \tag{9.26}$$

It should be noted that the selection of the absorber parameters is not unique. The actual choice is generally dictated by limitations placed on the vibration amplitude of the absorber mass. Moreover, when $k_a/m_a = \omega^2$, the motion of the absorber mass is

$$y_a(t) = -\frac{F}{k_a} \sin \omega t. \tag{9.27}$$

Thus, the absorber mass oscillates at the driving frequency with an amplitude of $Y_a = F/k_a$. The force exerted by the absorber mass (through k_a) on the primary mass can be obtained by considering the free body diagram of m, as shown in Figure 9.19. Because $y = 0$, the restoring force exerted by k_a reduces to $-k_a y_a$, as illustrated in Figure 9.20. Thus, the restoring force exerted by k_a is equal in magnitude and opposite in direction to the input force $f(t)$, thereby keeping the primary mass stationary. Finally, we note that although the vibration absorber is designed for a given operating frequency, ω, the absorber can perform satisfactorily for operating frequencies close to ω. In this case, the motion of m is not completely zero, but its vibration amplitude is drastically reduced.

9.6 JPEG compression

So far we have studied one-dimensional continuous and discrete-time signals. Images are an important example of two-dimensional signals, and image processing and compression are important examples of two-dimensional systems. We will explore the JPEG image compression. JPEG is an international image compression standard developed by the Joint Photographic Experts Group. It can commonly compress a 1 MB image down to only 50 kB, a 20 times compression factor. JPEG has been very important for the success of graphic-rich web pages because they can be transmitted over low-speed dial-up connections in a shorter amount of time with compression. It also allows storage of more photos on a digital camera. The JPEG algorithm is based on the two-dimensional *discrete cosine transform*, which is closely related to the discrete Fourier transform.

9.6.1 JPEG overview

Compression reduces the amount of data required to describe a signal. There are two classes of compression algorithms: lossless and lossy. Lossless compression allows exact reconstruction of the original signal. For example, a signal consisting of a run of ten 1s might be represented by the number 1 followed by a code indicating that it should repeat ten times. For signals with many repeating values, this compression is effective. Lossy compression throws out information. Fortunately, the information discarded is often imperceptible. For example, computer graphics usually have a space of 2^{24} different colors. One could compress an image by only allowing 2^8 different colors and rounding to the nearest allowable color. If the subset of allowable colors is chosen judiciously, the differences may be too small for the eye to detect. Such lossy compression reduces the data by a factor of 3 (24 bits down to 8 for each pixel).

JPEG uses both lossless and lossy compression. It begins with a stage of lossy compression. It transforms the image to find the different frequencies and discards the high frequencies if they have small amplitudes (and hence are unimportant to the image). The choice of "small amplitude" affects the degree of compression but also the amount of loss. JPEG then uses lossless compression to replace runs of a constant value (for example, 0) with a code indicating the run.

JPEG operates on images broken into eight by eight pixel blocks. It finds the frequency components in each block using the two-dimensional discrete cosine transform. We can view the discrete cosine transform as an extension of what we have already learned about the discrete Fourier transform.

9.6.2 Discrete Cosine transform

Recall that using the Fourier series, we can approximate a periodic signal as a weighted sum of complex exponentials. These complex exponentials are called the basis functions and are given by

$$\phi_k(t) = e^{jk\omega_0 t} \tag{9.28}$$

and the periodic signal can be expressed as an infinite sum of basis functions with weights c_k:

$$x(t) \approx \hat{x}(t) = \sum_{k=-K}^{K} c_k \phi_k = \sum_{k=-K}^{K} c_k e^{jk\omega_0 t}. \tag{9.29}$$

To find the weights that approximate the signal with the least integrated squared error, we use the orthogonality principle (OP), which states that the approximation should satisfy (see Chapter 2 for detailed discussion)

$$\int_{T_0} [x(t) - \hat{x}(t)]\phi_k^*(t)\mathrm{d}t = 0, \quad \text{for all } k. \tag{9.30}$$

The OP of Eq. (9.30) remains valid for any set of basis functions, but it simplifies substantially when the basis functions are orthogonal, thereby satisfying

$$\int_{T_0} \phi_l(t)\phi_k^*(t)\mathrm{d}t = \begin{cases} K_k & k = l \\ 0 & k \neq l. \end{cases} \tag{9.31}$$

In this case, Eq. (9.30) reduces to

$$c_k = \frac{1}{K_k} \int_{T_0} x(t)\phi_k^*(t)\mathrm{d}t \tag{9.32}$$

and each coefficient can be determined separately instead of solving a system of simultaneous equations. Complex exponentials form a set of orthogonal basis functions. Using complex exponentials as basis functions, the weights can be found individually as follows:

$$c_k = \frac{1}{2\pi} \int_{T_0} x(t)e^{-jk\omega_0 t}\mathrm{d}t. \tag{9.33}$$

Equations (9.33) and (9.29) are called the Fourier coefficients and Fourier series, respectively.

    This cosine series is similar to the Fourier series but uses cosines instead of complex exponentials as basis functions. Specifically, it uses cosines of frequencies $0, \omega_0/2, \omega_0, 3\omega_0/2, 2\omega_0, 5\omega_0/2$, etc., where ω_0 is the fundamental frequency. Thus, the basis functions for the cosine transform are given by

$$\phi_k(t) = \cos\left(\frac{k}{2}\omega_0 t\right). \tag{9.34}$$

Because cosine is an even function, that is

$$\cos\left(-\frac{k}{2}\omega_0 t\right) = \cos\left(\frac{k}{2}\omega_0 t\right) \tag{9.35}$$

we only need to consider positive values of k. Thus, we now approximate our periodic signal as a finite sum of cosine functions with weights c_k:

$$x(t) \approx \hat{x}(t) = \sum_{k=0}^{K-1} c_k \phi_k = \sum_{k=0}^{K-1} c_k \cos\left(\frac{k}{2}\omega_0 t\right). \tag{9.36}$$

Again we need to apply the orthogonality principle to solve for the coefficients that best approximate the signal. Fortunately, the set of cosine functions is also orthogonal; a bit of trigonometry shows us that the K_k of Eq. (9.31) are

$$K_k = \int_0^{T_0} \cos^2\left(\frac{k}{2}\omega_0 t\right) dt = \begin{cases} 2\pi & k = 0 \\ \pi & k > 0. \end{cases} \tag{9.37}$$

For convenience, let us introduce a term w_k defined as

$$w_k = \begin{cases} 1 & k = 0 \\ 2 & k > 0. \end{cases} \tag{9.38}$$

Thus, the formula for the coefficients of this cosine series can be expressed as

$$c_k = \frac{w_k}{2\pi} \int_{T_0} x(t) \cos\left(\frac{k}{2}\omega_0 t\right) dt. \tag{9.39}$$

An intuitive explanation on w_k is that because we do not use negative values of k, the coefficients for the positive values should have twice the magnitude, giving the factor of 2. We call Eq. (9.39) the cosine coefficients and Eq. (9.36) the cosine series.

9.6.3 Discrete cosine transform

In image processing, we are dealing with pixels, which are discrete (sampled) data. Therefore, we need transforms that operate on discrete-time signals. We will first review the discrete Fourier transform (DFT) and then extend it to the discrete cosine transform (DCT).

In discrete-time we would like to approximate a signal $x[n]$. The discrete Fourier transform uses basis sequences of the following form:

$$\phi_k[n] = e^{jk(\frac{2\pi}{N})n}. \tag{9.40}$$

The inverse DFT $(x[n])$ represents our signal as

$$x[n] = \sum_{k=0}^{N-1} X[k] e^{jk(\frac{2\pi}{N})n}. \tag{9.41}$$

The DFT $(X[k])$ finds the best weights $X[k]$,

$$X[k] = \frac{1}{N} \sum_{n=0}^{N-1} x[n] e^{-jk(\frac{2\pi}{N})n}. \tag{9.42}$$

Fig. 9.21. Basis functions
for the discrete cosine
transform (DCT).

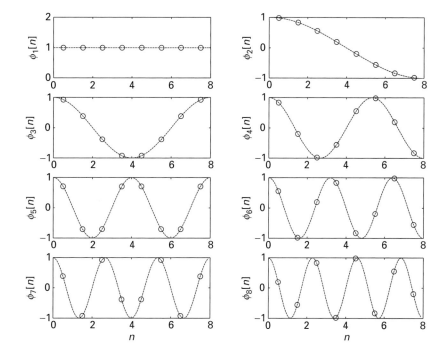

Similarly, the DCT uses basis functions:

$$\phi_k[n] = \cos\left(\frac{k}{2}\frac{2\pi}{N}\left(n+\frac{1}{2}\right)\right).$$

(9.43)

One might wonder why we use $n+1/2$ rather than simply n. To answer this
question, consider Figure 9.21, which plots the various basis functions. We
see that adding the $1/2$ term makes the points more symmetric, which allows
us to approximate a symmetric image accurately with fewer coefficients.

Now we can formulate expressions for the inverse DCT ($x[n]$) and DCT
($X[k]$) to go between the coefficients and the samples, using the same w_k
that was defined earlier:

$$x[n] = \sum_{k=0}^{N-1} X[k]\cos\left(\frac{k}{2}\frac{2\pi}{N}\left(n+\frac{1}{2}\right)\right)$$

(9.44)

$$X[k] = \frac{w_k}{N}\sum_{n=0}^{N-1} x[n]\cos\left(\frac{k}{2}\frac{2\pi}{N}\left(n+\frac{1}{2}\right)\right).$$

(9.45)

The DCT and DFT both transform N samples into N coefficients. Why
would one be preferable to the other? One difference is that for a real
signal, the DFT coefficients will be complex but the DCT coefficients are
real. This might suggest that the DCT takes half the storage space. But

Fig. 9.22. Approximating a slowly changing signal; $x[n]$ vs n.

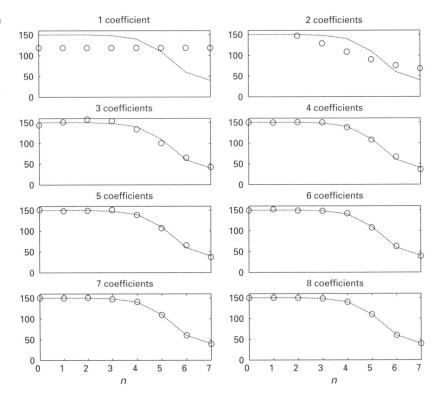

on careful thought, recall that for a real signal, the upper half of the DFT coefficients are just a mirror image of the conjugates of the lower half; they need not be stored. A more important difference is that the DCT involves computations with real numbers while the DFT uses complex numbers. Therefore, the DCT is generally faster.

Both the DFT and DCT coefficients contain information about the frequency components in the signal. The DFT also contains information about phase; this is one reason it is very useful for many signal processing applications. The DCT does not directly convey phase information, but does give frequency information on a finer scale because it covers half-integer multiples as well as integer multiples of the fundamental.

How many DCT coefficients are required to reconstruct a signal reasonably well? This depends on the highest frequency content in the signal; a slowly changing signal requires fewer coefficients than a rapidly changing signal. For example, Figure 9.22 shows approximations of a relatively slowly changing signal using only the first M coefficients:

$$x[n] \approx \sum_{k=0}^{M-1} X[k] \cos \left(\frac{k}{2} \frac{2\pi}{N} \left(n + \frac{1}{2} \right) \right). \tag{9.46}$$

Table 9.1. Coefficients for slowly changing signal

k	$X[k]$
0	118.5000
1	51.6255
2	−27.4186
3	7.5569
4	1.4142
5	−3.1756
6	2.1728
7	−0.7822

Table 9.2. Coefficients for abruptly changing signal

k	$X[k]$
0	45.0000
1	−16.6671
2	−66.9133
3	29.4236
4	21.2132
5	−5.8527
6	4.7554
7	−24.9441

The desired signal is shown with the approximate samples given with the open circles. The full set of coefficients is given in Table 9.1. Notice that the first three coefficients are much larger than the others. In other words, the signal has little energy at high frequencies. We observe that an approximation using the first three terms fits the signal quite well.

In contrast, Figure 9.23 shows an abruptly changing signal. The coefficients are given in Table 9.2. There is significant energy in even the highest frequencies, so all eight coefficients are needed to accurately reconstruct the signal.

9.6.4 Two-dimensional discrete cosine transform

An image is a two-dimensional signal. A continuous-time two-dimensional signal may be represented as a function of two variables: $x(s, t)$. Similarly, a discrete-time two-dimensional signal uses two indices: $x[m, n]$. Transforms

Fig. 9.23. Approximating an abruptly changing signal; $x[n]$ vs n.

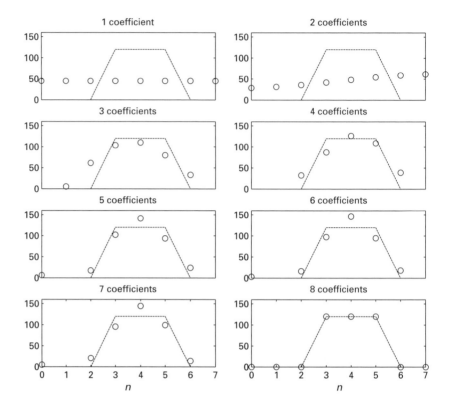

can be generalized to multiple dimensions as well. For example, a two-dimensional transform uses basis functions $\phi_{k,l}(s, t)$ or $\phi_{k,l}[m, n]$. The two-dimensional DCT uses a set of N^2 basis functions as follows:

$$\phi_{k,l}[m, n] = \cos\left(\frac{k}{2}\frac{2\pi}{N}\left(m + \frac{1}{2}\right)\right)\cos\left(\frac{l}{2}\frac{2\pi}{N}\left(n + \frac{1}{2}\right)\right). \qquad (9.47)$$

Figure 9.24 plots some of these basis functions (for ease of visualization, m and n are treated as continuous variables). The basis function $\phi_{0,0}$ is a constant. The basis functions $\phi_{k,0}$ only vary in one dimension, while $\phi_{0,l}$ only vary in the other dimension, and $\phi_{k,l}$ vary in both dimensions. Observe how the frequency is 0, 1/2, or 1 in each dimension.

The two-dimensional inverse DCT ($x[m, n]$) describes the signal as the weighted sum of the basis functions. The two-dimensional DCT ($X[k, l]$) describes how to compute these weights. They are given by the following expressions:

$$x[m, n] = \sum_{k=0}^{N-1}\sum_{l=0}^{N-1} X[k, l] \cos\left(\frac{k}{2}\frac{2\pi}{N}\left(m + \frac{1}{2}\right)\right)\cos\left(\frac{l}{2}\frac{2\pi}{N}\left(n + \frac{1}{2}\right)\right)$$

$$(9.48)$$

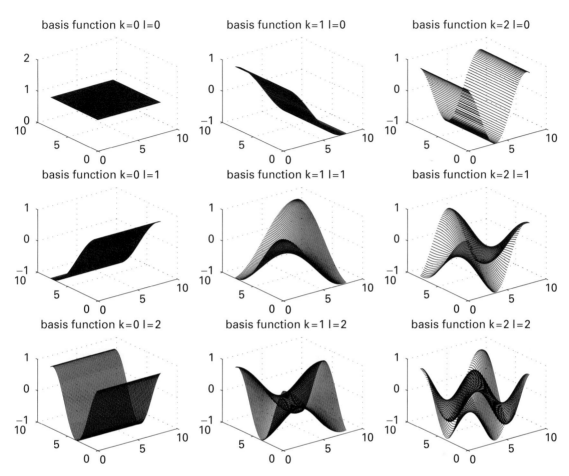

Fig. 9.24. Basis functions
for two-dimensional DCT
as a function or m, n.

and

$$X[k, l] = \frac{w_k w_l}{N^2} \sum_{m=0}^{N-1} \sum_{n=0}^{N-1} x[m, n] \cos\left(\frac{k}{2}\frac{2\pi}{N}\left(m + \frac{1}{2}\right)\right)$$

$$\times \cos\left(\frac{l}{2}\frac{2\pi}{N}\left(n + \frac{1}{2}\right)\right).$$

(9.49)

JPEG begins by breaking an image into eight by eight blocks and performing the two-dimensional DCT on each ($N = 8$). The signals being transformed are the color coordinates. Most photographs have color that slowly varies across the picture. Therefore, it is often sufficient to keep only a few of the 64 coefficients computed. On the other hand, line art has sharply defined edges so JPEG compression does not work as well. The degree of compression

Fig. 9.25. Compressed
photographs.

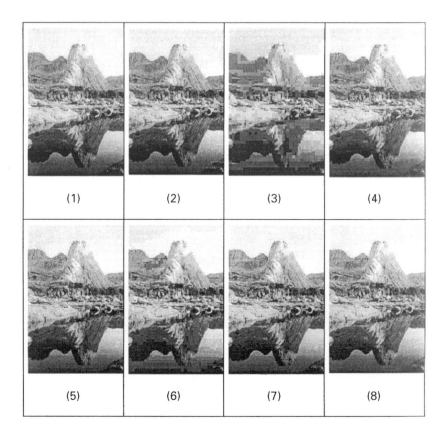

depends on how ruthlessly the components are discarded, so there is a
trade-off between the quality of the image and the amount of data. Once
the coefficients are computed, lossless compression is used to compact the
long runs of 0s in the higher frequencies.

Most computers represent images using three color components: red,
green, and blue (RGB). JPEG does better compression by first converting
the image to the luminosity and chrominance color system (YCrCb). The
chrominance components (Cr and Cb) describe the redness and blueness; if
they are both zero, the image is grayscale. The luminosity (Y) component
describes the brightness. Because the eye is more sensitive to brightness than
exact shade, the Cr and Cb components can be compressed more heavily
without introducing visible artifacts.

As an example of JPEG compression, Figure 9.25 shows a photograph
compressed to various degrees from an original 106 kB bitmap. Try to visually rank the photographs from highest compression to lowest. What
photographs are indistinguishable to the eye? Table 9.3 shows the size of

Table 9.3. File sizes of photographs shown in Figure 9.25

(1) 91 kB	(2) 12 kB	(3) 2 kB	(4) 5 kB
(5) 8 kB	(6) 3 kB	(7) 17 kB	(8) 26 kB

Fig. 9.26. Figure for Problem 9-1.

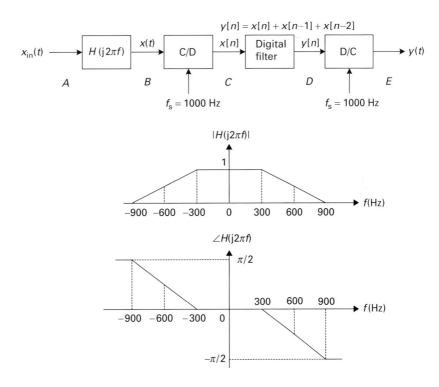

each file. The file can be compressed to about 8 kB before it starts showing visible banding in the sky.

9.7 Problems

9-1 The data acquisition system shown in Figure 9.26 is designed to record the output $x_{in}(t)$ of a vibration sensor during a rocket engine test firing. To verify that the system is operating correctly, the following test signal is used as the input:

$$x_{in}(t) = 2\cos(2\pi(300)t) + 4\cos\left(2\pi(600)t - \frac{\pi}{4}\right) + 6\cos(2\pi(900)t)).$$

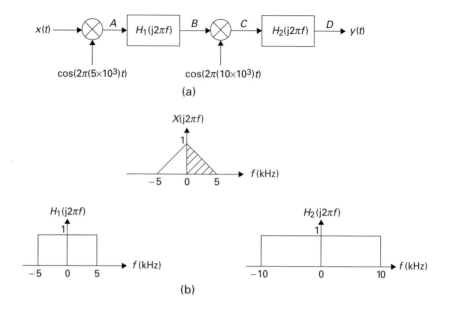

Fig. 9.27. Figure for Problem 9-2.

(a) Show that the frequency response function of the digital filter is given by

$$H(e^{j2\pi\hat{f}}) = e^{-j2\pi\hat{f}}(1 + 2\cos(2\pi\hat{f})).$$

(b) Sketch and dimension the spectrum at points A, B, C, D and E.
(c) Determine an expression for $y(t)$.

9-2 A system that generates a single-sideband (SSB) AM signal is shown in Figure 9.27(a). The spectrum of the input signal $x(t)$ is represented by $X(j2\pi f)$ and is shown below, along with the frequency response functions $H_1(j2\pi f)$ and $H_2(j2\pi f)$ (see Figure 9.27(b)). Sketch and dimension the spectra of the signals at points A, B, C and D.

9-3 A new movie theater with fancy digital audio capabilities was found to excite an annoying room resonance at low frequencies. An acoustical engineer was called to improve the situation. The engineer decides to address the problem by adjusting the digital filter to provide lower gain at the low frequencies than at the high frequencies to compensate for the resonance. Suppose the desired frequency response function is given in Figure 9.28. The digital audio is provided at $f_s = 44$ kHz. The frequencies of interest, f_1 and f_2, are both less than $f_s/2$. Compute the

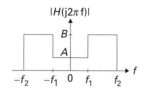

Fig. 9.28. Figure for Problem 9-3.

Fig. 9.29. Figure for
Problem 9-4.

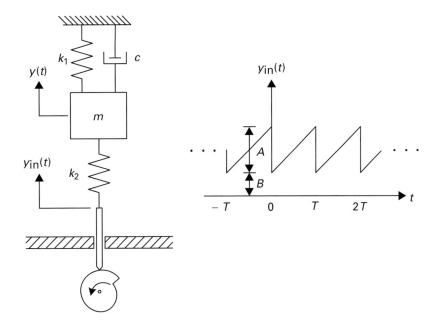

coefficients of the FIR filter to best approximate the desired frequency
response. Leave your answer in terms of A, B, f_s, f_1 and f_2.

9-4 The cam shown in Figure 9.29 imparts a displacement $y_{in}(t)$ in the
form of a periodic sawtooth function to the lower end of the system.
The output of the system corresponds to the vertical displacement of m.
(a) Show that the frequency response function of the system is given by

$$H(j\omega) = \frac{Y}{Y_{in}} = \frac{k_2/m}{\omega_n^2 - \omega^2 + 2\zeta\omega\omega_n j}.$$

(b) Find the expressions for ω_n and ζ.
(c) Sketch the spectrum for the continuous-time input $y_{in}(t)$.
(d) Determine the steady state response of the system to the given
sawtooth input $y_{in}(t)$.

9-5 The brake pedal of an automobile can be modeled by the system shown
in Figure 9.30(a). Consider the lever as a massless rod and the pedal as
a lumped mass at the end of the rod. During braking, the driver exerts
an input horizontal force on the mass, m, as shown. Assume small
motions and assume the spring to be unstretched at $\theta = 0$.
(a) Determine the governing equation for the angular displacement,
$\theta(t)$, of the system.

Fig. 9.30. Figure for
Problem 9-5.

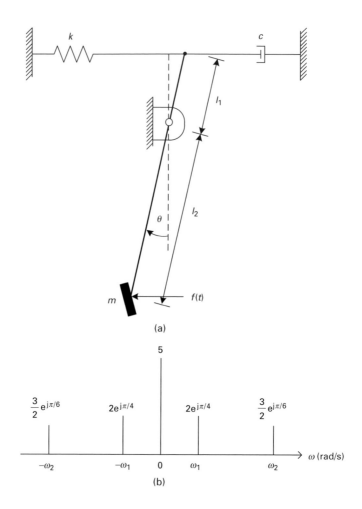

(a)

(b)

(b) Find the undamped natural frequency, ω_n, and the damping factor, ζ, in terms of the given system parameters.

(c) Determine the frequency response function of the system,

(d) During braking on an icy road, the driver pumps the brake pedal with a force $f(t)$, whose continuous-time spectrum is shown in Figure 9.30(b). Determine the steady state response of the system.

9-6 The system shown in Figure 9.31(a) is used to acquire vibration data from a burning building. To test the system, an input of

$$x_{in}(t) = 4\cos(2\pi(1000)t)$$

is applied to the system. The anti-aliasing filter consists of the *RLC* circuit shown in Figure 9.31(b).

Fig. 9.31. Figure for
Problem 9-6.

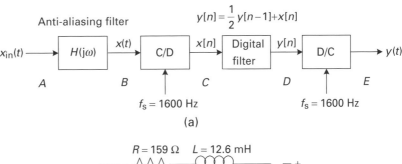

Anti-aliasing filter $y[n] = \dfrac{1}{2}y[n-1] + x[n]$

(a)

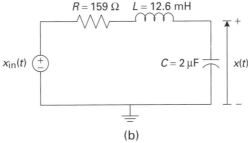

(b)

(a) Show that the frequency response function of the anti-aliasing
filter is given by

$$H(j\omega) = \cfrac{1}{1 - \left(\dfrac{\omega}{\omega_n}\right)^2 + 2\zeta \dfrac{\omega}{\omega_n}j}.$$

Find the numerical values for ω_n and ζ.

(b) Give an expression for $x(t)$.

(c) Show that the frequency response function of the digital filter is
given by

$$H(e^{j\hat\omega}) = \cfrac{1}{1 - \dfrac{1}{2}e^{j\hat\omega}}.$$

(d) Sketch the spectrum of the signals at points A, B, C, D and E.
Also indicate the values of the complex amplitudes.

(e) Give an expression for the output $y(t)$.

9-7 The system shown in Figure 9.32(a) is used to collect vibration data
from a turbine engine. To test the system, the following input

$$x_{in}(t) = 50 + 4\sqrt{2}\,\cos(2\pi(1500)t)$$

is applied to the system. The anti-aliasing filter for the system consists
of the RC circuit shown in Figure 9.32(b).

Fig. 9.32. Figure for
Problem 9-7.

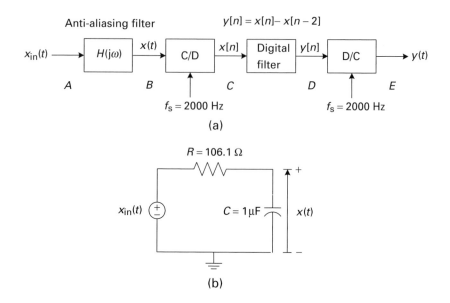

(a) Determine the frequency response function of the anti-aliasing filter.
(b) Give an expression for $x(t)$.
(c) Show that the frequency response function of the digital filter is given by

$$H(e^{j\hat{\omega}}) = 2je^{-j\hat{\omega}} \sin(\hat{\omega}).$$

(d) Sketch the spectra of the signals at points A, B, C, D and E. Also indicate the values of the complex amplitudes.
(e) Give an expression for the output $y(t)$.

10 Summary

This text has explored signals and systems. We will conclude with reviewing the material and constructing the big picture.

We began with continuous-time signals, which are just functions of a single variable (usually time, but perhaps position, etc.). It is often easier to understand a complicated signal by breaking it down into a weighted sum of simpler signals. When the signal is periodic (or if the signal is only of interest over a finite amount of time), it is particularly useful to represent the signal as a weighted sum of complex exponentials. The Fourier coefficients tell us how to find the weights and the Fourier series reconstructs the signal from the weights.

The expression for the Fourier coefficients involves an integral, so it only has closed-form solutions for fairly simple signals. For more complicated signals, we would like to find the weights numerically. A computer can process only a finite number of data points, so we must sample the continuous-time signal at some sampling frequency to produce a discrete-time signal. We can then find the weights numerically using the discrete Fourier transform (DFT) and reconstruct the original signal from the weights using the inverse discrete Fourier transform (IDFT). There is a clever computer algorithm for computing the DFT called the fast Fourier transform (FFT). If we do not sample the continuous-time signal fast enough, we may distort it. This is called aliasing or folding. The sampling theorem tells us that we must sample more than twice as fast as the highest frequency in the signal.

We then moved on to systems that take an input signal and produce an output signal. We are particularly interested in lumped-parameter linear time-invariant (LTI) systems because they are easy to analyze and are sufficient to approximate the real world in many situations. We focused on mechanical and electrical LTI systems. Mechanical systems consist of springs, dampers and masses driven by force or displacement inputs. Electrical systems consist of inductors, resistors and capacitors driven by current or voltage source inputs.

The same mathematics describes both mechanical and electrical systems (and also thermal, hydraulic, and other LTI systems). We begin by writing a governing equation, which is a differential equation relating the input to the output. We are often interested in computing either the transient step response or the steady-state sinusoidal response. The step response is found by solving the differential equation for a step input, with all the initial conditions set equal to zero. For a first-order system, if we write the differential equation in canonical form, we can extract the time constant of the system. Similarly, for a second-order system, if we write the differential equation in canonical form, we can read the natural frequency and damping factor straight from the equation and immediately determine quantities such as rise time and percentage overshoot. The sinusoidal steady-state (or frequency) response is found more easily by assuming the input and output are complex exponentials and computing the frequency response function. The Bode plot shows both magnitude and phase versus frequency on a logarithmic scale.

When the input is more complicated than a step or sinusoid, the easiest way to compute the output is to break the input into a weighted sum of sinusoids (or complex exponentials), find the response to each, and sum the responses (applying the principle of superposition). We can do this numerically by sampling the signal, taking the FFT, multiplying by the frequency response function, taking the inverse fast Fourier transform (IFFT), and reconstructing the continuous-time signal. Here we are exploiting the full power of the Fourier techniques we have developed. Moreover, we can obtain any frequency response we want (for example, an ideal low-pass filter), not just those easily found from connections of mechanical or electrical components. This works well if we already know the complete signal and can process it at our leisure. On the other hand, it would also be nice to be able to process signals as they arrive in real time. Finite impulse response (FIR) filters are discrete-time systems that produce an output that depends on a weighted sum of current and previous values of the input. We can compute a discrete-time frequency response for an FIR filter and may use the FFT to compute the best weights to obtain a particular frequency response. Both the FFT and FIR filters are examples of digital signal processing (DSP). Finally, we explored a number of applications of signals and systems including AM radio, accelerometers, vibration absorbers, and JPEG image compression.

We will now devote the remainder of this chapter to reviewing these ideas in more depth and working some examples.

10.1 Continuous-time signals

We would like to approximate a signal $x(t)$ as a weighted sum of other building blocks $\phi_k(t)$ (sometimes called basis functions),

$$x(t) \approx \hat{x}(t) = \sum_k c_k \phi_k(t). \tag{10.1}$$

We would also like to find the weights c_k that best approximate the function. One reasonable definition for "best" is to minimize the integral of the squared difference between the signal and its approximation. The integrated squared error (ISE) is defined as

$$\text{ISE} = \int_T |x(t) - \hat{x}(t)|^2 \, dt = \int_T \left| x(t) - \sum_k c_k \phi_k(t) \right|^2 dt. \tag{10.2}$$

The orthogonality principle (OP) says that the weights that minimize the ISE must satisfy

$$\int_T \left[x(t) - \sum_n c_n \phi_n(t) \right] \phi_k^*(t) dt = 0, \qquad \text{for all } k, \tag{10.3}$$

where the asterisk denotes the complex conjugate. Evaluating the integrals gives us a set of simultaneous equations that we could solve for the weights c_k, but this involves lots of complicated algebra. However, the mathematics becomes easier for certain sets of basis functions called orthogonal basis functions. A set of functions is orthogonal if it satisfies the condition

$$\int_T \phi_l(t)\phi_k^*(t) dt = \begin{cases} K_k & k = l \\ 0 & k \neq l. \end{cases} \tag{10.4}$$

In such a case, the weights can be found individually by evaluating the following integral:

$$c_k = \frac{1}{K_k} \int_T x(t)\phi_k^*(t) dt. \tag{10.5}$$

When the signal is periodic, one particularly useful set of orthogonal basis functions are the complex exponentials. Suppose the signal is periodic with period $T_0 = 1/f_0 = 2\pi/\omega_0$. Even if the signal is not periodic, we can make it periodic if we are only interested in its value for some specified time

344

Summary

interval by assuming the signal repeats itself outside the duration of interest. We choose basis functions that are complex exponentials at multiples of this frequency and approximate our signal as

$$x(t) = x(t + T_0) \approx \hat{x}(t) = \sum_{k=-K}^{K} c_k e^{jk\omega_0 t}. \tag{10.6}$$

Equation (10.6) is known as the Fourier series. The complex exponentials are orthogonal with $K_k = T_0$. The expression given by Eq. (10.5), allows us to compute the Fourier coefficients using the following integral:

$$c_k = \frac{1}{T_0} \int_{T_0} x(t) e^{-jk\omega_0 t} \, dt. \tag{10.7}$$

With an infinite number of terms, we can drive the ISE to zero.

Sometimes the weights are trivial to find by inspection. For example, if

$$x(t) = 4 \cos\left[2\pi(3)t\right] + 6 \cos\left[2\pi(7)t + \frac{\pi}{3}\right] \tag{10.8}$$

the signal is periodic with period $T_0 = 1$ s, frequency $f_0 = 1$ Hz, and angular frequency $\omega_0 = 2\pi$ rad/s. By writing the cosines as complex exponentials, we can immediately see that the coefficients are

Fig. 10.1. Spectrum of sum of sinusoids.

$$c_3 = 2$$
$$c_{-3} = 2$$
$$c_7 = 3 e^{j\frac{\pi}{3}} \tag{10.9}$$
$$c_{-7} = 3 e^{-j\frac{\pi}{3}}.$$

The spectrum of the signal of Eq. (10.8) is shown in Figure 10.1.

On the other hand, suppose our input is a unit step function, $x(t) = u(t)$, as shown in Figure 10.2(a) and that we are interested in the time interval from -10 to 10 seconds. Let us assume the signal is periodic outside this interval, as shown in Figure 10.2(b). The period is $T_0 = 20$ s, frequency $f_0 = 1/20$ Hz and angular frequency $\omega_0 = 2\pi/20$ rad/s. Now we integrate Eq. (10.7) to find the weights, which are found to be

$$c_k = \frac{1}{20} \int_{-10}^{10} x(t) e^{-jk\frac{2\pi}{20}t} dt = \frac{1}{20} \int_0^{10} e^{-jk\frac{\pi}{10}t} dt$$

$$= \begin{cases} \dfrac{1}{2} & k = 0 \\ \dfrac{j}{\pi k} & k \text{ odd} \\ 0 & k \text{ even.} \end{cases} \tag{10.10}$$

Fig. 10.2. Spectrum of pulse train.

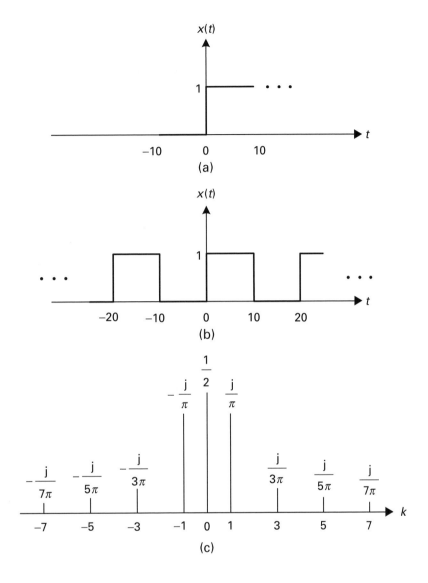

We can also plot the spectrum, as shown in Figure 10.2(c). Note that c_0 is the average value. For real-valued signals, the positive and negative coefficients are complex conjugates, that is $c_k = c_{-k}^*$.

10.2 Discrete-time signals

Evaluating the integral of Eq. (10.7) was a hassle. For many real-world signals (an earthquake recording for example) evaluating the integral by hand would be impossible. In general, it would be more convenient to let a computer process data numerically. If we sample the continuous-time signal $x(t)$ with a continuous-to-discrete (C/D) converter at a sampling frequency f_s (or sampling period $T_s = 1/f_s$), we obtain a discrete-time signal $x[n]$, defined as

$$x[n] = x(nT_s). \tag{10.11}$$

Consider sampling a periodic signal with N samples evenly spaced over the period T_0. Thus, the sampling period is defined as $T_s = T_0/N$. We can still write the signal as a weighted sum of complex exponentials. In this case, the exponentials are also sampled and the weights are called $X[k]$. This is called the inverse discrete Fourier transform (IDFT or $x[n]$). The DFT ($X[k]$) tells us how to compute the best weights. Expressions for IDFT and DFT are as follows:

$$x[n] = \sum_{k=0}^{N-1} X[k] e^{jk\frac{2\pi}{N}n} \tag{10.12}$$

and

$$X[k] = \frac{1}{N} \sum_{n=0}^{N-1} x[n] e^{-j\frac{2\pi}{N}n}. \tag{10.13}$$

Both $x[n]$ and $X[k]$ are periodic with period N. Therefore the discrete-time spectrum also repeats every f_s.

Figure 10.3(a) shows the spectrum of $x(t) = 6 \cos[7(2\pi)t + \pi/3]$ sampled at 16 Hz. The original signal $x(t)$ can be reconstructed from the sampled data. Figure 10.3(b) shows the spectrum if the input is sampled at 12 Hz. This illustrates folding and we would reconstruct the signal as $6 \cos[5(2\pi)t - \pi/3]$. Figure 10.3(c) shows the spectrum if $x(t)$ is sampled at 6 Hz. This illustrates aliasing, and we would reconstruct the signal as $6 \cos[(2\pi)t + \pi/3]$. In summary, we must sample at a frequency more than twice the maximum frequency contained in the signal in order to reconstruct the signal correctly. If we sample between the frequency and twice the frequency, we fold, thus reconstructing the signal with the wrong frequency and phase. If we sample below the signal frequency, we alias, thus reconstructing the signal with the wrong frequency but correct phase.

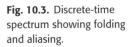

Fig. 10.3. Discrete-time spectrum showing folding and aliasing.

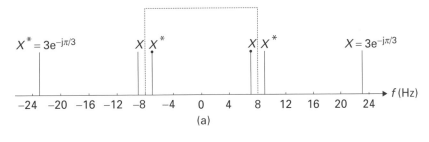

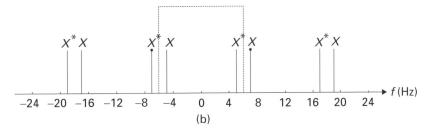

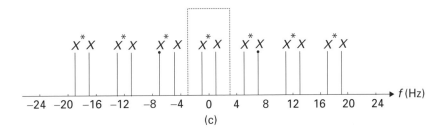

10.3 Lumped element modeling of mechanical and electrical systems

A system is linear if the response to a sum of two inputs is the sum of the responses to each input. It is time-invariant if the response to a delayed input is the delayed response to the input. If a system is linear and time-invariant (LTI), we can define a frequency response function and use it to analyze the response of the system to any input. Mechanical and electrical systems are two examples of common LTI systems. Although real systems are distributed (that is, a spring has some mass distributed along its length and an inductor has some resistance distributed along its length), they are much easier to model if we treat them as lumped into separate components.

In lumped element modeling, we model a mechanical system as being composed of ideal springs, dampers and masses. The inputs are forces or displacements as a function of time. We model an electrical system as

Table 10.1. Analogy between electrical and mechanical elements

Electrical	Mechanical	Relationship
$i_L = \frac{1}{L} \int_{-\infty}^{t} (v_a - v_b) \, d\tau$	$F_s = k(x_a - x_b) = k \int_{-\infty}^{t} (u_a - u_b) \, d\tau$	$\frac{1}{L} \Leftrightarrow k$
$i_R = \frac{1}{R}(v_a - v_b)$	$F_d = c\frac{d}{dt}(x_a - x_b) = c(u_a - u_b)$	$\frac{1}{R} \Leftrightarrow c$
$i_C = C\frac{d}{dt}(v_a - v_b)$	$F_m = m\frac{d^2}{dt^2}(x_a - x_b) = m\frac{d}{dt}(u_a - u_b)$	$C \Leftrightarrow m$

Fig. 10.4. (a) An electrical and (b) a mechanical system.

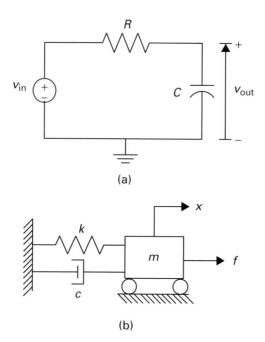

(a)

(b)

being composed of ideal inductors, resistors and capacitors. The inputs are currents and voltages as a function of time. The constitutive relationships for each element are listed in Table 10.1. A direct analogy exists between current (i) and force (F) and between voltage (v) and velocity ($u = \dot{x}$). From these relationships, we can derive the governing equations relating the inputs and outputs. To derive the equation for a mechanical system, we must apply Newton's Second Law by summing either the forces or the torques, depending on if the system is under translational or rotational motion, respectively. For an electrical system, we apply Kirchhoff's current

Fig. 10.5. Transient
response characteristics.

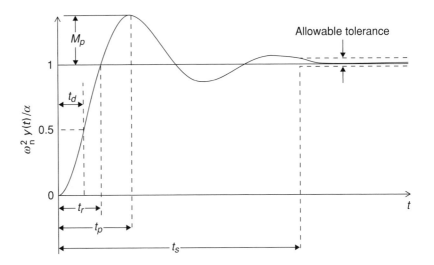

and voltage laws, which state that the sum of currents into a node must be zero and that the sum of voltage drops around a closed loop must also be zero, respectively.

Figure 10.4 shows a first-order electrical system and a second-order mechanical system. The governing equations are

$$\dot{v}_{out}(t) + \frac{1}{RC} v_{out}(t) = \frac{1}{RC} v_{in} \tag{10.14}$$

and

$$m\ddot{x} + c\dot{x} + kx = f(t). \tag{10.15}$$

We are interested in finding the response of first and second-order systems to step inputs and to sinusoidal inputs. If we put the differential equations in canonical form, we need only solve them once. Let $y(t)$ and $x(t)$ represent the output and input, respectively. The canonical form for a first-order system is

$$\dot{y} + \frac{1}{\tau} y = \alpha x(t). \tag{10.16}$$

For the electrical system of Eq. (10.14), $y(t) = v_{out}(t)$, $\tau = RC$, $x(t) = v_{in}(t)$ and $\alpha = 1/\tau$. The canonical form for a second-order system is

$$\ddot{y} + 2\zeta\omega_{n}\dot{y} + \omega_{n}^{2}y = \alpha x(t). \tag{10.17}$$

Fig. 10.6. Bode plot of a
first-order system in the
denominator using
straight-line
approximation.

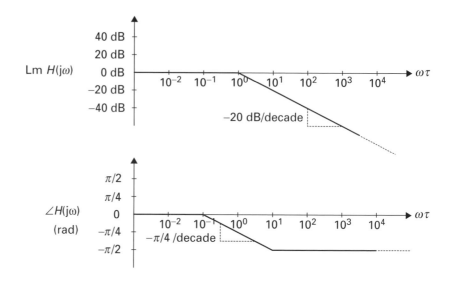

For the mechanical system of Eq. (10.15), $\omega_n = \sqrt{k/m}$, $\zeta = c/(2m\omega_n)$, $x(t) = f(t)$ and $\alpha = 1/m$.

10.4 Transient response

We are often interested in computing the response to a changing input, such as a step input. We can compute the response by solving the differential equation. The step response of a first-order system is an exponential with time constant τ. The step response of a second-order system depends on the damping factor ζ. The most interesting response is for underdamped systems $(0 < \zeta < 1)$ shown in Figure 10.5. These systems overshoot the final value and oscillate around it at a damped natural frequency of

$$\omega_d = \omega_n\sqrt{1 - \zeta^2}. \tag{10.18}$$

The performance of any order system can be characterized by the rise time t_r, the maximum percentage overshoot M_p, the time at which this peak occurs t_p, and the settling time t_s to within ϵ of the final value (for example, $\epsilon = 0.02$ is the 2 percent settling time). For an underdamped, second-order system, the transient specifications in terms of the canonical parameters ζ

and ω_n are given by

$$t_r = \frac{1}{\omega_d} \left[\pi - \tan^{-1} \left(\frac{\sqrt{1 - \zeta^2}}{\zeta} \right) \right] \tag{10.19}$$

$$M_p = 100 e^{-\frac{\pi \zeta}{\sqrt{1 - \zeta^2}}} \% \tag{10.20}$$

$$t_p = \frac{\pi}{\omega_n \sqrt{1 - \zeta^2}} \tag{10.21}$$

$$t_s = \frac{- \ln \left(\epsilon \sqrt{1 - \zeta^2} \right)}{\zeta \omega_n}. \tag{10.22}$$

10.5 Frequency response

We are also interested in the steady state response to a sinusoidal input. If the system is stable, the steady state response also corresponds to the particular solution. Because the system is linear and time-invariant, we know that the output must also be a sinusoid of the same frequency differing only in amplitude and phase. If the input is a complex exponential of the form

$$x(t) = X e^{j \omega t} \tag{10.23}$$

then the output of the system will also be a complex exponential

$$y(t) = Y e^{j \omega t}. \tag{10.24}$$

The ratio of the output amplitude to the input amplitude is called the frequency response function (FRF), and is denoted by $H(j\omega)$. Thus, if the input and output of a system are given by Eqs. (10.23) and (10.24), then the system's FRF is given by

$$H(j\omega) = \frac{Y}{X}. \tag{10.25}$$

The FRFs for the first and second-order systems of Eqs. (10.14) and (10.15) are

$$H(j\omega) = \frac{V_{out}}{V_{in}} = \frac{1}{1 + j\omega\tau} \tag{10.26}$$

and

$$H(j\omega) = \frac{X}{F} = \frac{\frac{1}{k}}{1 - \left(\frac{\omega}{\omega_n}\right)^2 + 2\zeta\frac{\omega}{\omega_n}j}. \tag{10.27}$$

If the input is a sinusoid,

$$x(t) = X\cos(\omega t + \phi), \tag{10.28}$$

then the output will also be a sinusoid scaled in amplitude by the gain of the FRF, $|H(j\omega)|$, and phase shifted by the phase of the FRF, $\angle H(j\omega)$:

$$y(t) = X|H(j\omega)|\cos(\omega t + \phi + \angle H(j\omega)). \tag{10.29}$$

Bode plots show the magnitude and phase of the FRF across many orders of magnitude of frequency. They are easy to construct by considering the limiting cases where ω is very small or very large compared to other terms (for example, the time constant for a first-order system or the natural frequency for a second-order system). The magnitude plot shows the logarithmic magnitude Lm (in decibels, dB), defined as

$$\text{Lm } H(j\omega) = 20\log_{10}|H(j\omega)| \tag{10.30}$$

and the phase just shows the phase in degrees or radians. For example, Figure 10.6 shows a Bode plot for the first-order system of the form

$$H(j\omega) = \frac{1}{j\omega\tau + 1}. \tag{10.31}$$

The magnitude is one below the corner frequency ($\omega_c = 1/\tau$) and rolls off at -20 dB per decade above ω_c. The phase is zero for frequencies well below ω_c, exactly $-\pi/4$ at ω_c, and $-\pi/2$ for frequencies well above ω_c. Figure 10.7 shows a Bode plot for an underdamped second-order system of the form

$$H(j\omega) = \frac{1}{1 - \left(\frac{\omega}{\omega_n}\right)^2 + 2\zeta\frac{\omega}{\omega_n}j}. \tag{10.32}$$

The magnitude is again one below the corner frequency at $\omega_c = \omega_n$, but drops off at -40 dB/decade above ω_c. The phase is zero for frequencies well below ω_c, exactly $-\pi/2$ at ω_c, and $-\pi$ for frequencies well above ω_c. The exact FRF is a bit more rounded at the corners but fits the Bode plot quite well. Bode plots are very useful because when the input is a

Fig. 10.7. Bode plot of an underdamped second-order system in the denominator using straight-line approximation.

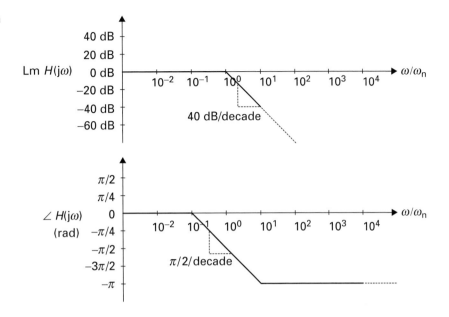

Fig. 10.8. Series and parallel combinations, and voltage dividers.

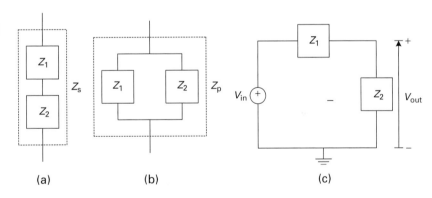

(a)   (b)   (c)

sinusoid of a particular frequency, we know the output will be too and we can immediately read off the magnitude and phase.

10.6 Impedance

Analyzing circuits with resistors is easier than analyzing those with inductors and capacitors, because circuits with resistors only involve algebraic equations, not differential equations. We can find the FRF more easily if we define an impedance for each element that depends only on the frequency. Then each element can be treated like a resistor and we can use rules of

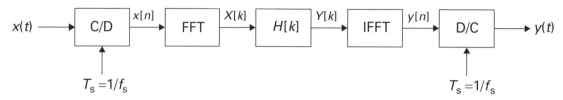

Fig. 10.9. DSP with FFT.

parallel and series combinations, and voltage dividers. The impedance is defined as the ratio of the voltage drop across an element to the current through the element, assuming the voltage and the current are complex exponentials of frequency ω. Thus, the impedances for a resistor, inductor and capacitor are given by

$$
\begin{aligned}
Z_R &= R \\
Z_L &= j\omega L \\
Z_C &= \frac{1}{j\omega C}.
\end{aligned}
\tag{10.33}
$$

Figure 10.8 shows elements in parallel and series. The equivalent impedance for two elements in series (see Figure 10.8(a)) is simply the sum of their impedances

$$
Z_s = Z_1 + Z_2.
\tag{10.34}
$$

The reciprocal of the equivalent impedance for two elements in parallel (see Figure 10.8(b)) is the sum of the reciprocals of their impedances. This simplifies to the following parallel combination rule

$$
\frac{1}{Z_p} = \frac{1}{Z_1} + \frac{1}{Z_2}.
\tag{10.35}
$$

We can often use the parallel and series rules to reduce a circuit to the form in Figure 10.8(c) called a voltage divider. Then the ratio of the output to the input voltage is easily found as

$$
H(j\omega) = \frac{V_{\text{out}}}{V_{\text{in}}} = \frac{Z_2}{Z_1 + Z_2}.
\tag{10.36}
$$

10.7 Digital signal processing

We are often interested in the transient response of a system to an input that is more complicated than just a step. Because the integration is often

Fig. 10.10. DSP with FIR.

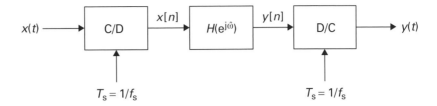

hard to evaluate, we would like to solve the problem numerically using a computer. This is called digital signal processing (DSP) and combines our knowledge of signals and systems.

One straightforward way to find the output is to break the input into a weighted sum of complex exponentials, find the response to each (given the FRF of the system) complex exponential, and then find the output as the sum of the responses. Because we are solving the problem numerically, we must first sample the input, then convert the output back to continuous-time. This complete process is shown in Figure 10.9.

In other cases, we cannot wait to have the entire signal before we begin processing it. Such real-time systems include cell phones and GPS receivers. In these cases, we can use finite impulse response (FIR) filters to process the data. A FIR filter computes the output as a weighted sum of the inputs:

$$y[n] = \sum_{k=-K}^{K} a_k x[n-k]. \tag{10.37}$$

We can find the discrete-time frequency response by assuming that the input and output are discrete-time exponentials. Thus, the frequency response of the discrete-time system of Eq. (10.37) is given by

$$H(e^{j\hat{\omega}}) = \sum_{k=-K}^{K} a_k e^{-jk\hat{\omega}}. \tag{10.38}$$

Suppose we wish the system to possess a specified FRF. The natural question is how to choose the coefficients a_k to get the frequency response we want. Note that Eq. (10.38) looks very much like the Fourier series (but with a sign reversal). It is not surprising that the best values for the coefficients to get a particular response is thus very much like the Fourier series.

$$a_k = \frac{1}{2\pi} \int_{2\pi} H(e^{jk\hat{\omega}}) e^{jk\hat{\omega}} \, d\hat{\omega}. \tag{10.39}$$

Now we can process the signal directly with the FIR filter shown in Figure 10.10 without having to compute the integral. Of course there is a trade-off in the value of K; bigger values take longer to process but better approximate a desired frequency response.

10.8 Transition to more advanced texts (or, what's next?)

Now that you understand the fundamental principles of signals and systems you may want to consult other and possibly more advanced texts. If you do, you will encounter a number of terms and concepts that may seem unrelated to, or different from, those covered thus far in this text. You will see terms including Fourier transform, Laplace transform, and z-transform. Some terms like the discrete-time Fourier transform can be easily confused with the discrete Fourier transform (DFT) discussed in this text (although, as you will see, they are closely related). Systems with feedback, stability, transfer functions, impulse response, convolution, pole-zero diagrams, and root locus are among the concepts we have not discussed. The purpose of this final chapter is to ease the transition and, more importantly, show that these terms and concepts can be understood and related to the fundamental principles we have studied thus far. While we will not be able to go into detail, we hope the following material will allow you to get a feel for the key relationships and enable you to use results and techniques from other texts with confidence.

10.8.1 Spectra of non-periodic signals

For the most part we have limited our study of the frequency spectrum to periodic signals. This is not much of a restriction since, as we have discussed, any signal defined over a finite time interval can be repeated to produce a periodic signal. The DFT is always periodic, so any analysis using it (or the FFT algorithm for computing it) results in a periodic sequence. So you really have all the background you need to compute the spectrum of most real-world signals. But what happens if we want to consider a non-periodic signal? One approach is to start with the Fourier series of a periodic signal and then let the period become arbitrarily large. The result for continuous-time signals, assuming the integrals converge, is the Fourier transform pair given below:[1]

$$x(t) = \frac{1}{2\pi} \int_{-\infty}^{\infty} X(j\omega)e^{j\omega t}\, d\omega = \int_{-\infty}^{\infty} X(j2\pi f)e^{j2\pi ft}\, df \qquad (10.40)$$

and

$$X(j\omega) = \int_{-\infty}^{\infty} x(t)e^{-j\omega t}\, dt = \int_{-\infty}^{\infty} x(t)e^{-j2\pi ft}\, dt, \qquad (10.41)$$

where $\omega = 2\pi f$. Thus, a non-periodic signal has a continuous spectrum $X(j\omega)$ instead of the lines at discrete frequencies that are the case for the spectrum of a periodic signal. You can easily show, however, that the following relationship holds for the Fourier coefficients a_k of a periodic signal $x(t) = x(t + T_0)$:

$$a_k = \frac{1}{T_0} X(jk\omega_0), \qquad (10.42)$$

where $\omega_0 = 2\pi/T_0$, and

$$X(j\omega) = \int_{T_0} x(t)e^{-j\omega t}\, dt. \qquad (10.43)$$

This means that the Fourier coefficients of a periodic signal are samples scaled by $1/T_0$ of the spectrum of a non-periodic signal consisting of one period of the periodic signal!

For discrete-time signals, letting the period become arbitrarily large yields the following:

$$x[n] = \frac{1}{2\pi} \int_{2\pi} X(e^{j\hat\omega})d\hat\omega \qquad (10.44)$$

and

$$X(e^{j\hat\omega}) = X(e^{j(\hat\omega+2\pi)}) = \sum_{n=-\infty}^{\infty} x[n]e^{-j\hat\omega n}. \qquad (10.45)$$

Thus, the spectrum $X(e^{j\hat\omega})$ (or the discrete-time Fourier transform) of a non-periodic discrete-time signal is also continuous (and periodic). For periodic sequences $x[n] = x[n + N]$, the following relationships hold:

$$X[k] = \frac{1}{N} X(e^{j(2\pi/N)k}), \qquad (10.46)$$

where

$$X(e^{j\hat\omega}) = \sum_{N} x[n]e^{-j\hat\omega n}. \qquad (10.47)$$

---

[1] See *Signals and Systems* by A. V. Oppenheim, A. S. Willsky and H. Nawab, Prentice Hall, New Jersey, 1997.

Once again, we see that the Fourier coefficients (or DFT) of a periodic discrete-time signal are samples of the spectrum (or discrete-time Fourier transform) of a non-periodic signal consisting of one period of the periodic signal!

10.8.2 Laplace and z-transforms

An important issue arises once we allow non-periodic signals of possibly infinite length. It is possible that the signals become unbounded. For example, the continuous-time signal $x(t) = e^t$ becomes arbitrarily large as t approaches infinity. This means the Fourier transform of Eq. (10.41) does not exist (the integral blows up). A similar problem occurs for discrete-time signals such as $x[n] = 2^n$. In this case the discrete-time Fourier transform of Eq. (10.45) does not exist (the infinite sum diverges). To handle such cases the Laplace transform for continuous-time signals is given by

$$X(s) = \int_{-\infty}^{\infty} x(t)e^{-st}\, dt, \tag{10.48}$$

where $s = \sigma + j\omega$ (σ and ω are both real), and the inverse Laplace transform is defined as

$$x(t) = \frac{1}{2\pi j} \int_{\sigma-j\infty}^{\sigma+j\infty} X(s)e^{st}\, ds. \tag{10.49}$$

Similarly, the z-transform for discrete-time signals is as follows:

$$X(z) = \sum_{n=-\infty}^{\infty} x[n]z^{-n}, \tag{10.50}$$

where $z = re^{j\hat{\omega}}$ (r and ω are both real), and the inverse z-transform is

$$x[n] = \frac{1}{2\pi j} \int X(z)z^{n-1}\, dz. \tag{10.51}$$

Adjusting the parameters σ and r makes it possible to handle a large class of non-periodic signals, including some that become arbitrarily large. These strange looking integrals are evaluated along special contours in the complex plane, a topic beyond the scope of this text (in fact the need to evaluate the integrals is usually avoided by using a table of Laplace or z-transform pairs).[2] Suffice to say, if $\sigma = 0$ and $r = 1$, the Laplace transform and the

[2] See *Signals and Systems* by A. V. Oppenheim, A. S. Willsky and H. Nawab, Prentice Hall, New Jersey, 1997.

z-transform reduce to the Fourier transform and the discrete-time Fourier transform, respectively, assuming the Fourier transforms do, in fact, exist (see the discussion on stability in Section 10.8.4).

10.8.3 Transfer functions

In addition to FRFs other texts will talk about transfer functions. The transfer function of a system is defined as the Laplace or z-transform of the output divided by the Laplace or z-transform of the input with zero initial conditions for continuous-time and discrete-time systems, respectively. Usually the transfer function for a continuous-time system is the ratio of two polynomials in s while the transfer function for a discrete-time system will be the ratio of two polynomials in z. As long as the system is stable (see Section 10.8.4), the relationships between the transfer function and the frequency response are as follows:

$$H(j\omega) = H(s)|_{s\, =\, j\omega} \qquad\qquad (10.52)$$

$$H(e^{j\hat{\omega}}) = H(z)|_{z\, =\, e^{j\hat{\omega}}}. \qquad\qquad (10.53)$$

10.8.4 Stability, pole-zero diagrams, and root locus diagrams

The relationships between the transfer function and the frequency response function given in Eqs. (10.52) and (10.53) hold only if the system is stable. By this we mean that a bounded input will produce a bounded output (referred to as BIBO stable). This may not be the case for a system with feedback.[3] Of course, in the real world the output will not actually become arbitrarily large (even though the mathematical model predicts it). Either the system will become nonlinear (a mechanical actuator hits the limits of its travel for example) and thus the linear model is no longer valid, or something will break (again invalidating the linear model). So how can you tell when the relationships given above can be used? Because a good deal of the theory is beyond the scope of this text we will just give you the answer and encourage you to study further if you are interested (it is cool stuff!).

A continuous-time system with a transfer function consisting of a ratio of polynomials in s is stable as long as the roots of the denominator are located in the left half of the complex plane, that is $\sigma < 0$. On the other

---

[3] See *Modern Control Engineering* by K. Ogata, Prentice Hall, New Jersey, 2002.

hand, a discrete-time system with a transfer function consisting of a ratio of polynomials in z is stable as long as the roots of the denominator are located inside the unit circle of the complex plane, that is $r < 1$. The locations of the roots of the denominator polynomial are called *poles* (since the transfer function blows up at these points), while the locations of the roots of the numerator polynomial are called *zeros* (because the transfer function is zero at these points). A diagram showing the locations of the poles and zeros of a transfer function is naturally called a pole-zero diagram and contains a good deal more information than just whether or not the system is stable. In fact, a system is often designed by specifying the desired locations of the poles and zeros of its transfer function. A plot of their locations as a function of various system parameters is referred to as a root-locus plot and is also a very important design tool.

10.8.5 Impulse response and convolution

The transfer function of a system is defined as the ratio of the Laplace transform of the output to the Laplace transform of the input. This is equivalent to saying that the Laplace transform of the output is the product of the Laplace transform of the input multiplied by the transfer function of the system. For continuous-time systems, this means that the output and input are related as follows in the s-domain:

$$Y(s) = H(s)X(s), \tag{10.54}$$

where $Y(s)$ is the Laplace transform of the output, $H(s)$ is the transfer function of the system, and $X(s)$ is the Laplace transform of the input. Similarly, for discrete-time systems, the output and input are related as follows in the z-domain:

$$Y(z) = H(z)X(z), \tag{10.55}$$

where $Y(z)$ is the z-transform of the output, $H(z)$ is the transfer function of the system, and $X(z)$ is the z-transform of the input. It can be shown that the equivalent relationship between the output and the input for continuous-time systems in the t-domain is simply

$$y(t) = \int_{-\infty}^{\infty} h(\tau)x(t-\tau)d\tau = \int_{-\infty}^{\infty} h(t-\tau)x(\tau)d\tau, \tag{10.56}$$

where $y(t)$ is the output, $h(t)$ is the inverse Laplace transform of the transfer function $H(s)$ ($h(t)$ is also known as the impulse response since it is the output when the input is a continuous-time impulse), and $x(t)$ is the input. The integrals are referred to as *convolution integrals*. For discrete-time systems, the output and input relationship is

$$y[n] = \sum_{k=-\infty}^{\infty} h[k]x[n-k] = \sum_{k=-\infty}^{\infty} h[n-k]x[k], \qquad (10.57)$$

where $y[n]$ is the output, $h[n]$ is the inverse z-transform of the transfer function $H(z)$ ($h[n]$ is also known as the impulse response since it is the output when the input is a discrete-time impulse), and $x[n]$ is the input. The sums are referred to as *convolution sums*.

Although we have just skimmed the surface, hopefully you can see the parallels to the fundamental concepts covered in this text. As an example, suppose you want to design a digital filter. A more advanced text has a design algorithm (or maybe a table of transfer functions) for various types of filters. Even if you do not understand the details of the design process you can easily see if the result meets your needs. First, check if the transfer function represents a stable system by determining the roots of the denominator (pole locations). Assuming it is stable (and any valid design technique will make sure it is, but check anyway), substitute $e^{j\hat\omega}$ for z and plot the FRF to see if it meets your specifications. Similarly, you can plot the FRF for a continuous-time system by first checking for stability and then substituting $j\omega$ for s in the transfer function.

After studying this text we hope you are excited about the many applications of signals and systems and are motivated to continue your study of this increasingly important area.

LABORATORY EXERCISES

A goal of this text is to integrate the traditional topics covered in signals and systems with the use of MATLAB, a modern computational software package. The recent trend in engineering is to include computational software as part of the learning experience at every level. This text utilizes MATLAB extensively. Because of this integration of MATLAB into the development of the standard materials, more design topics can be covered and more sophisticated problems can be analyzed. The authors believe that such an integration allows students to learn the materials in a more interactive way and gives the students the opportunity to visualize the results on their computer screens. In addition, we hope that the use of computational software will enhance the learning experience by encouraging students to try numerous cases, answer "what if?" questions, and most importantly, stimulate the students' interest in signals and systems. Finally, the authors hope that the students will also reap the rewards of applying MATLAB to solve problems he or she may encounter in other courses in science or engineering.

A total of seven MATLAB laboratory exercises are presented, and each exercise is followed by a series of problems. We encourage the students to do all of the exercises in order to gain a deeper understanding and mastery of the topics covered in the text. Each exercise may require materials covered in multiple chapters, and the following table summarizes the content of each exercise and the required chapters for each.

Table L. Laboratory exercises and relevant chapters

Laboratory exercise	Title	Required chapters
1	Introduction to MATLAB	None
2	Synthesize music	1, 2
3	DFT and IDFT	1, 2, 3
4	FFT and IFFT	1, 2, 3, 5
5	Frequency response	1, 2, 3, 4, 6, 7
6	DTMF	1, 2, 3, 7, 8
7	AM radio	1, 2, 3, 7, 8, 9

Laboratory exercise 1

L1.1 Objective

MATLAB is a computing language devoted to the analysis of matrices of numbers. The primary objective of this first exercise[1] is to guide you in developing a working knowledge of MATLAB. Specifically, in this first exercise you will learn to use MATLAB to create and manipulate signals, and to output these signals as text, graphics or audio. You will experiment with all the programming constructs that you will need to know for future exercises. With a good working knowledge of MATLAB, we can analyze very sophisticated problems that are not amenable to hand calculations.

L1.2 Guided introduction to MATLAB

Starting MATLAB brings up the Command Window in which you can enter simple commands. Other windows, which may appear, are the Launch Pad, the Workspace, the Command History and the Current Directory windows. You can clear all these extra windows for the time being, but you should browse through the Current Directory until you have set it to your own personal directory, which contains (or will soon contain) all your own personal MATLAB files. To gain a basic understanding on the working knowledge of MATLAB, please read through this guided introduction and try all the examples.

---

[1] The tutorial is provided by Prof. Tony Bright at Harvey Mudd College. The sumcos exercise is borrowed from *DSP First* by J. H. McClellan, R. W. Schafer and M. A. Yoder, © 1998. Reprinted by permission of Pearson Education Inc., Upper Saddle River, NJ.

L1.3 Vector and matrix manipulation

MATLAB works on matrices of numbers. For example, to enter a two by four matrix, we type

```
>> x = [ 0 .1 .2 .3 ; .3 .4 .5 .6 ]
```

Note that we have placed brackets around the data and separated the rows with semicolons. When we press enter, MATLAB responds with the following:

```
x =
      0 0.1000 0.2000 0.3000
0.3000 0.4000 0.5000 0.6000
```

The colon command ":" enables the easy generation of matrices. For example, the following command:

```
>> y = 0: pi/3: pi
```

results in a one by four row vector starting at 0 and ending at π in increments of $\pi/3$, as shown below:

```
y =
0 1.0472 2.0944 3.1416
```

The following command shows the length, or the number of elements, in the vector **y**:

```
>> length(y)
ans =
4
```

The colon by itself refers to the entire range of that index, for example x (: ,2) yields the second column of the matrix x, as illustrated below:

```
>> x1 = x(:,2)
x1 =
0.1000
0.4000
```

Matrix addition and multiplication are particularly simple in MATLAB. Of course, the matrices must be dimensionally compatible in order to perform these manipulations. Consider the following examples:

```
>> x2=x+[y;y]
x2 =
     0 1.1472 2.2944 3.4416
0.3000 1.4472 2.5944 3.7416
>> x3=x*y'
x3 =
1.4661
3.3510
```

Note the use of the apostrophe ' to transpose vector **y** before multiplying. Matrices may be concatenated or joined together with scalars or other matrices as shown below:

```
>> [2 y -1]
ans =
2.0000 0 1.0472 2.0944 3.1416 -1.0000
```

Element-by-element operations are prefixed with a dot (e.g., .*, ./, .^) and only work on matrices of the same size. For instance,

```
>> ysquared = y.*y
ysquared =
0 1.0966 4.3865 9.8696
```

Do not confuse this element-by-element multiplication with the dot product, which is performed by multiplying a row and a column vector. The dot product results in a scalar quantity. The column is often obtained from a row vector with the transpose operation:

```
>> ydoty=y*y'
ydoty =
15.3527
```

Most functions operate on all elements of a vector or matrix. For example,

```
>> sin(y)
ans =
0 0.8660 0.8660 0.0000
```

Complex numbers in MATLAB are expressed in Cartesian form using either i or j. The commands sqrt(x) and exp(x) correspond to the mathematical operations of $\sqrt{x}$ and e^x, respectively. The Cartesian coordinates or the polar coordinates may be extracted from the complex number. Consider the following examples:

```
>> C=3+2*sqrt(-1)
C =
3.000 + 2.000i
>> [real(C); imag(C); abs(C); angle(C)]
ans =
3.0000
2.0000
3.6056
0.5880
>> C=j*exp(j*11*pi/4)
C =
-0.7071 - 0.7071i
```

A collection of special matrices can be looked up in MATLAB's online help, for example,

```
>> help ones
ONES Ones array.
ONES(N) is an N-by-N matrix of ones.
ONES(M,N) or ONES([M,N]) is an M-by-N matrix of
  ones.
ONES(M,N,P,...) or ONES([M N P ...]) is an
  M-by-N-by-P-by-...
array of ones.
ONES(SIZE(A)) is the same size as A and all ones.
See also ZEROS.
>> A = zeros(3,3)
A =
0 0 0
0 0 0
0 0 0
```

More interesting matrices and vectors may be generated, such as the following alternating sequence:

```
>> D=(-2*ones(1,4)).^(1:4)
D =
-2 4 -8 16
```

Relational operators return a matrix of elements that satisfy the relation. The relational operator returns a value of 1 when the relation is true, and 0 when the relation is false. They include > (greater than), >= (greater than or equal to), < (less than), and <= (less than or equal to), along with == (equal) and ~= (not equal).

```
>> E=(D>0)
E =
0 1 0 1
>> F = D.*(D>0)
F =
0 4 0 16
```

The transpose (.') and conjugate transpose(') operators are useful for linear algebra. Be careful which one you want to use when manipulating complex numbers. The following example illustrates the difference between the two transpositions:

```
>> G = exp((0:6)*j*pi/3)
G =
1.0000 0.5000 + 0.8660i - 0.5000 + 0.8660i
  -1.0000 + 0.0000i -0.5000 - 0.8660i 0.5000 -
  0.8660i 1.0000 - 0.0000i
>> G'
ans =
 1.0000
 0.5000 - 0.8660i
-0.5000 - 0.8660i
-1.0000 - 0.0000i
-0.5000 + 0.8660i
 0.5000 + 0.8660i
 1.0000 + 0.0000i
>> G.'
ans =
 1.0000
 0.5000 + 0.8660i
-0.5000 + 0.8660i
-1.0000 + 0.0000i
-0.5000 - 0.8660i
 0.5000 - 0.8660i
```

```
 1.0000 - 0.0000i
>> G*G'
ans =
7.0000 - 0.0000i
```

L1.4 Variables

All variables in the current MATLAB session are saved in the
Workspace. Type whos to list the current variables in the MATLAB
Workspace.

```
>> whos
Name Size Bytes Class
A 3x3 72 double array
C 1x1 16 double array (complex)
x 2x4 64 double array
x1 2x1 16 double array
x2 2x4 64 double array
x3 2x1 16 double array
y 1x4 32 double array
```

The grand total is 34 elements using 280 bytes.

The default numerical data type in MATLAB is double. This refers
to a double precision, 64-bit (8-byte), floating point number. The floating
point relative accuracy is given by

```
>> eps
ans =
2.2204e-016
```

Listed below are some useful tips for command window sessions:

- The format command controls the numeric format of the values displayed in the command window. Consider the following example:
  ```
  >> x=4/3
  x =
  1.3333
  >> format long; x
  x =
  ```

```
1.33333333333333
>> format short e; x
x =
1.3333e+000
```

The `format` command only affects how the numbers are displayed, not how they are saved.

- To enter long command lines, use three periods, ..., followed by `enter`, and continue on the next line, as shown below:

```
>> some_useless_number=exp(0.12345)
+sin(2*pi*0.12345)+...
cos(2*pi*0.12345)
some_useless_number =
2.5455e+000
```

- You can reuse or edit previous command lines using the up arrow ↑ key.
- A semicolon at the end of a command suppresses the output. If/when you get into the unfortunate position of forgetting the semicolon and having reams of numbers endlessly filling your screen, hit `Ctrl-C` for relief.
- *Last, but not least, when all else fails, get help from* `View` > `Help`.

L1.5 Plotting

A simple plot to produce a 5 Hz sinusoidal signal is given by the following commands:

```
>> t = 0:0.001:1;
>> sin5 = sin(2*pi*5*t);
>> plot(t,sin5)
```

The plot is shown in Figure L1.1. By clicking on the edit plot arrow ↖ in the toolbar, you can insert labels, titles, legends, and text. By double clicking on the plot itself, you open the property editor to change axis limits, specify line styles, markers and so on. When you save your figure, it is saved as a FIG-file, which you can load subsequently using the `open` command.

Alternatively, you can set the figure properties from the command line:

Fig. L1.1. Example
MATLAB plot of 5 Hz
sinusoidal waveform.

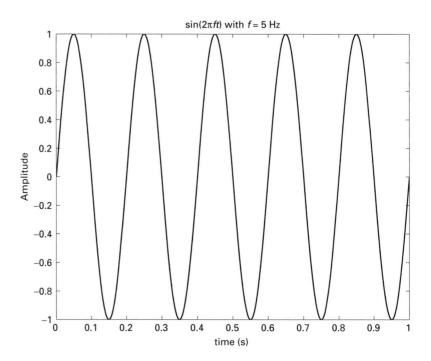

```
>> plot(t,sin5,'LineWidth',1.5)
>> title('sin(2\pi ft) with f = 5 Hz',
   'FontSize',[15])
>> xlabel('time (sec)','FontWeight','bold')
```

You can also print Greek letters on the figures using the TeX convention
\alpha, \beta, \gamma, etc.

You can plot multiple data sets on a single set of axes or can make multiple
separate plots in one window, as illustrated in Figure L1.2. Notice the use
of the matrix operation to produce four cosines with one statement; be sure
you understand how this works.

```
>> t = 0:0.001:1;
>> F = cos(2*pi*(1:4)'*t);
>> subplot(3,1,1)
>> plot(t,F(1,:))
>> hold on
>> plot(t,t)
>> hold off
>> subplot(3,1,2)
```

Fig. L1.2. Examples of MATLAB multiple data-sets and multiple separate plots in one window.

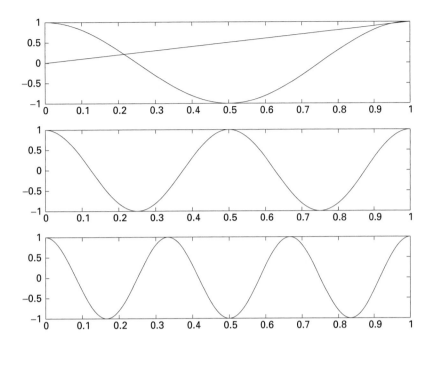

```
>> plot(t,F(2,:))
>> subplot(3,1,3)
>> plot(t,F(3,:))
```

L1.6 M-files

If you do not want to spend time and effort typing and retyping in the command window, you would be well advised to take advantage of M-files. These are sequences of normal MATLAB commands that you create with a text editor. There are two types of M-files: `scripts`, which are just sequences of MATLAB commands; and `functions`, which allow you to create your own user-specific functions. Both `scripts` and `functions` are ordinary ASCII text files.

As an example of a `script` file, create the following text file `sinc5.m`. You can open a text editor in MATLAB by clicking File > New > M-file.

```
% An M-file to calculate and plot
  sin(2*pi*5*t)/(2*pi*5*t), the sinc function
t= -1:0.01:1;
```

Fig. L1.3. Example
MATLAB plot for the
sinc 5.m script file.

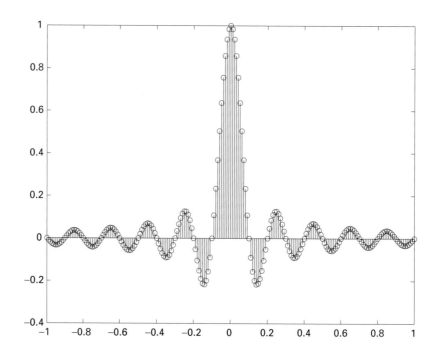

```
i=find(t); j=find(t==0);
sinc(i)=sin(2*pi*5*t(i))./(2*pi*5*t(i));
sinc(j)=1;
stem(t,sinc)
```

You may want to use the MATLAB online help to see what `find` and
`stem` do. Also, note that `./` is used for the element-by-element division of
vectors.

Tell MATLAB where your scripts are by using the Set Path item under
the file menu to point it to your personal directory. Typing the statement
`sinc5` causes MATLAB to execute the commands:

```
>> sinc5
```

and produces Figure L1.3.

An M-file that contains the word `function` at the start is called a
function file. Arguments may be passed to and from function files. They
act like procedures or subroutines to the main MATLAB session. As an
example of a `function` file, create the text file `echos.m`, which adds an
"echo" to a signal.

```
function y = echos(x, T, delay, atten)
% adds an echo to signal x
% y(t)=x(t)+atten*x(t-delay)
%
% inputs: T = sample period
% delay = delay time
% atten = attenuation
%
xz=zeros(1, round(delay/T));
xe=[xz atten*x];
y=x+xe(1:length(x));
```

Use the MATLAB online help to see what `round` does. The comments in the beginning of your function serve as built-in help for the function. Try typing

```
>> help echos
```

Now you can include `echos(x, T, delay, atten)` as part of a command sequence:

```
>> T=0.001; t=0:T:1; x=exp(-5*t).
   *sin(2*pi*10*t);
```

Notice that .* is used for the element-by-element product of vectors.

```
>> delay=0.5; atten=0.5; xecho=echos(x, T,
   delay, atten);
>> plot(t,xecho)
```

The above command sequence produces the plot shown in Figure L1.4.

MATLAB has all the usual programming constructs which you can use in your M-files: `if`, `while`, `for`, `switch`, `continue`, `break`. Get online help if you need to know how to use them. However, to take advantage of MATLAB's superior matrix handling capabilities, you should look for ways to vectorize the algorithms in your M-files to do away with less efficient "loops" in your programs. For example, instead of using the `for` loop

```
i=0;
for t=0:0.01:1
i=i+1
```

Fig. L1.4. Example Matlab
plot for an echoed
waveform.

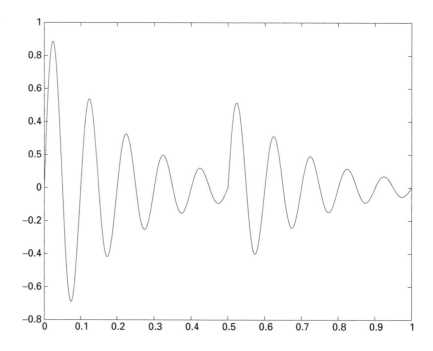

```
x(i)=sin(2*pi*5*t);
end
```

a vectorized version of the same code

```
t=0:0.01:1;x=sin(2*pi*5*t);
```

executes much faster since memory does not have to be reallocated to the
variable x each time through the loop.

The best way to learn to program M-files is to look at and adapt
someone else's programs. For example, here is an M-file function
x=sinegen(f, fs, tdur) for plotting and listening to a sine wave,
which gives an "audio-visual" experience.[2]

```
function x=sinegen(f, fs, tdur)
%
% generates a sine wave, plots ten cycles, and
  plays the tone
% for tdur seconds
%
```

---

[2] Most academic computing machines lack speakers, so you may need to bring a headset or work
at home to hear sounds on later exercises.

```
% inputs: frequency f Hz
% sample frequency fs Hz
% duration tdur s
%
t=0:1/fs:tdur;
x=sin(2*pi*f*t);
tplot=0:1/fs:10/f;
xplot=x(1:length(tplot));
plot(tplot,xplot);
xlabel('time (s)')
str=sprintf(' Ten cycles of %g Hz sine
  wave',f);
title(str)
sound(x)
```

Use the MATLAB online help to see what sprintf and sound do. You may try out the function x=sinegen(f, fs, tdur) with

```
>> x=sinegen(440, 8000, 2);
```

The above plays note A (440 Hz) sampled at 8000 samples per second for a duration of 2 seconds (assuming you have a sound card in your computer) and plots the first ten cycles of the note.

L1.7 Housekeeping

After a few MATLAB sessions, your directory will be flooded with M-files. Use the desktop tools to help you stay organized. The Current Directory window lets you create new folders and paste related M-files into them. The Command History window keeps track of commands you have entered in the current MATLAB session and allows you to select commands to include in M-files. The Launch Pad window provides easy access to help, demos and the MathWorks web page. The Workspace Browser does the same thing as the whos command with pretty graphics.

You can use any text editor to create M-files, but MATLAB's text editor comes complete with an Editor/Debugger, which you can use to set breakpoints, show functions and examine current values of variables to help you test your M-files. When the command window gets cluttered, type `clc`. When the workspace gets cluttered, type `clear`. When your mind gets cluttered, type `quit`. Before quitting, you can `save` variables to a file and `load` them back at the next MATLAB session. For example,

```
>> save firstecho t xecho
```

will save the vectors **t** and **xecho** in a binary file `firstecho.mat`, which will appear in your directory. The next time you start a MATLAB session, you can retrieve the variables

```
>> load firstecho
```

Just to be certain

```
>> whos
Name   Size    Bytes Class
t      1x1001  8008  double array
xecho 1x1001  8008   double array Size Bytes
```

Yes, they are there!

L1.8 Summary of MATLAB commands

When using MATLAB (or any tool), look for the simplest way to accomplish your tasks. Many tasks can be done in one line with suitable application of matrix operations. Write a function if you find yourself frequently repeating a task. The following commands will be used extensively. Use help if you need more information on the commands. MATLAB has many more commands that you are free to use, but if you find yourself searching for exotic functions, step back and look for an easier approach.

- Arithmetic: `+`, `-`, `*`, `/`, `^`
- Element-by-element arithmetic: `.*`, `./`, `.^`
- Relational operations: `<`, `<=`, `>`, `>=`, `==`, `~=`
- Logical operations: `and`, `or`, `not`
- Transpose, conjugate transpose: `.'`, `'`
- Elementary functions: `sqrt`, `exp`, `sin`, `cos`, `tan`
- More functions: `mod`, `floor`, `max`, `min`
- Complex functions: `abs`, `angle`, `real`, `imag`

- Basic matrices: `zeros, ones`
- Matrix dimensions: `length, size`
- Plotting: `plot, subplot, hold, xlabel, ylabel, title`
- Sound: `sound, soundsc`
- Loops and conditionals: `for, end`
- Miscellaneous: `error`

In addition, the following specialized functions will be introduced in later labs:

`specgram, lsqcurvefit, fft, ifft, conv, fir1`

L1.9 Exercises

1. Compute i^i with MATLAB. Compare the MATLAB results with the exact solutions.

2. According to phaser notation,

   $$\text{Re}\{e^{j\phi}\} = \cos\phi.$$

   To verify this, evaluate each side of this expression in MATLAB for $\phi = \pi/3$

3. Write a function `function xx=odd(n)` that returns a row vector of odd numbers

   $$[1 \quad 3 \quad 5 \quad \ldots \quad 2*n+1].$$

   Your function should only have one line in the body. **Hint**: the line should have two colons.

4. Write another function `function xx=funny(n)` that returns a vector of $n+1$ numbers of the form

   $$\frac{j4}{k\pi},$$

   where the values k are those returned by `odd(n)`. Your function again should have only one line in the body. **Hint**: use the `./` command for element-by-element division.

5. We will often represent signals as the sum of multiple sinusoids of different frequencies, amplitudes, and phases. For example, if $\mathbf{f} = [f_n, f_{n-1}, \ldots, f_1]$ is a vector of frequencies and $\mathbf{X} = [X_n, X_{n-1}, \ldots, X_1]$

is a corresponding vector of complex amplitudes (magnitude and phase), we can write the sum of cosines of these frequency and amplitude as

$$x(t) = \sum_{i=1}^{n} \text{Re}\{X_i e^{j2\pi f_i t}\}.$$

This may be expressed as a matrix-vector product

$$\mathbf{x} = \text{Re}\{\mathbf{X} e^{j2\pi \mathbf{f}^T \mathbf{t}}\}.$$

For example, let t be the row vector of times

$$\mathbf{t} = [0.00 \quad 0.25 \quad 0.50 \quad 0.75 \quad 1.00].$$

Suppose we wish to compute

$$x(t) = 7\cos(2\,\text{Hz}) + 5\cos(3\,\text{Hz}),$$

then $\mathbf{f} = [2 \quad 3]$, $\mathbf{X} = [7 \quad 5]$ and we evaluate

$$\mathbf{x} = [x_1 \quad x_2 \quad x_3 \quad x_4 \quad x_5]$$
$$= \text{Re}\left\{[7 \quad 5] e^{j2\pi[\frac{2}{3}][0.00 \quad 0.25 \quad 0.50 \quad 0.75 \quad 1.00]}\right\}$$
$$= \text{Re}\left\{[7 \quad 5] e^{j2\pi\left[\begin{smallmatrix} 0.00 & 0.50 & 1.00 & 1.50 & 2.00 \\ 0.00 & 0.75 & 1.50 & 2.25 & 3.00 \end{smallmatrix}\right]}\right\}$$
$$= \text{Re}\left\{[7 \quad 5]\begin{bmatrix} 1 & -1 & 1 & -1 & 1 \\ 1 & -j & -1 & j & 1 \end{bmatrix}\right\}$$
$$= [12 \quad -7 \quad 2 \quad -7 \quad 12]$$

In general, suppose the time intervals are spaced by 1/fs, the reciprocal of the sampling frequency, from time 0 to dur. We can take advantage of the matrix representation to express the computation very concisely. Complete the last line of the code below to do the computation.

```
function [tt, xx] = sumcos(f, X, fs, dur)
% SUMCOS1 Function to synthesize a sum of
   cosine waves
% usage:
% xx = sumcos(f, X, fs, dur)
% f = vector of frequencies
% X = vector of complex exponentials:
   amplitude*e^(j*phase)
% fs = the sampling rate (in Hz)
% dur = totoal time duration of signal
%
```

```
% Note: f and X must be the same length
if (length(f) ~= length(X))
error ('SUMCOS: f and X must be the same
   length');
end
tt = 0:1/fs:dur;
xx = % complete this line of code
```

Check your work by plotting a 1 Hz cosine using

```
>> [tt,xx] = sumcos(1, 1, 100, 1);
>> plot(tt,xx)
```

and the $7\cos(2\,\text{Hz}) + 5\cos(3\,\text{Hz})$ signal using

```
>> [tt,xx] = sumcos([2 3], [7 5], 100, 1)
>> plot(tt,xx)
```

6. One interesting signal is expressed as the sum of n odd-frequency sinusoids where the amplitude of each sinusoid is given by the funny function you wrote in exercise 4 above. For example, for $n = 2$, it can be generated over three periods with:

```
>> [tt, xx] = sumcos(odd(2), funny(2), 100, 3)
```

Make a three-panel plot showing versions of the signal for $n = 2, 4$ and 12. What does it appear the signal converges to for large n?

Laboratory exercise 2: Synthesize music

L2.1 Objective

In this exercise[1] you will synthesize music from sinusoids. You will learn about the relationships between musical notes on a scale and the frequency and duration of sinusoids. You may need headphones or you may use computers with speakers.

L2.2 Playing sinusoids

Throughout this exercise, use a sampling rate of $f_s = 11\,025$ Hz. You may use your `sumcos` function from Lab 1. Before we start, please consider and answer the following:

1. How many samples do you expect to have in a digital signal with a duration of d and a sampling rate of f_s?
2. If the digital signal is periodic with a frequency of f and is sampled at f_s, how many samples do you expect per period?
3. Use the `sumcos` function to compute a vector **x1** of samples of a sinusoidal signal with $A = 100$, $\omega_0 = 2\pi(1100)$ and $\phi = 0$ for a duration of 2 seconds.
4. How many samples are in **x1** (use the MATLAB command `length(x1)`)? Does this match your expectation? Explain any discrepancies.

[1] This exercise is based on Project 3 from *DSP First* by J. H. McClellan, R. W. Schafer and M. A. Yoder, © 1998. Reprinted by permission of Pearson Education Inc., Upper Saddle River, NJ.

Fig. L2.1. Layout of a piano keyboard. Key numbers are shaded. The notation C_4 means the C key in the fourth octave.

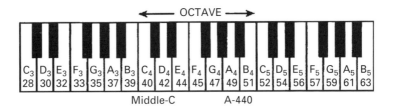

5. Plot **x1**. Zoom in on the plot and determine how many samples occur per period. Does this match your expectation?
6. Play your sinusoid using the command `soundsc(x1,fs)`.
7. Generate a second vector **x2** of a sinusoid with $A = 50$, $\omega_0 = 2\pi(1650)$, and $\phi = \pi/3$ at the same sampling rate and with a duration of 1 second.
8. Generate a signal `xx = [x1 zeros(1, 2000) x2]`.
9. Play your **xx** signal. Does it sound like what you would expect?

L2.3 Generating musical notes

The frequency layout of the piano keyboard follows an interesting mathematical pattern. The keyboard, as shown in Figure L2.1, is divided into octaves; the notes in each octave are at twice the frequency of the same name notes in the next lower octave. For example, the reference note is the A above middle C, which is usually called A-440 (or A_4) because its frequency is 440 Hz. Each octave contains 12 notes (five black keys and seven white keys) and the ratio between the frequencies of the notes is constant between successive notes. Thus this ratio must be $2^{1/12}$. Since middle C is 9 keys below A-440, its frequency is approximately $440 \times 2^{-9/12} \text{ Hz} = 261.6 \text{ Hz}$.

Musical notation, as illustrated in Figure L2.2, shows which notes are to be played and their relative timing. Half notes last twice as long as quarter notes, which in turn last twice as long as eighth notes. The figure shows how the keys on the piano correspond to notes drawn in musical notation. A chord contains notes in both the treble and bass clefs. They may be produced independently and summed to provide an overall sound.

Listed below are steps to guide you through the exercise:

1. Generate a sinusoid of 2 seconds' duration to represent the note E_5 above A-440 (key number 56). Play this sound.

Fig. L2.2. Musical notation is a time–frequency diagram where vertical position indicates the frequency of the note to be played.

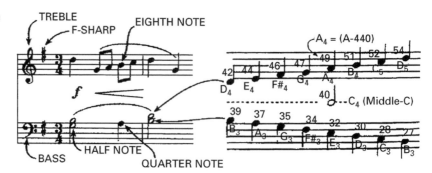

2. Write an M-file to produce a desired note for a given duration. Your M-file should be in the form of a function called `note.m` given below. Use the `sumcos` function within the file.

```
function tone = note(keynum, dur)
% NOTE: produces a sinusoidal waveform
% corresponding
% to a given piano key number
% USAGE: tone = note(keynum, dur)
% tone = the output sinusoidal signal
% keynum = the piano keyboard number of the
% desired note
% dur = the duration (in seconds) of the
% output note
%
fs = 11025;
freq =
tone =
```

For the `freq` = line, use the formulae based on the twelfth root of 2 to determine the frequency for a sinusoid in terms of its key number. For the `tone` = line, generate the actual sinusoid at the proper frequency and duration using `sumcos`. If your `sumcos` function returns two arguments, you may instead wish to begin this line with `[tt,tone]` = ...

After you have completed your M-file, test your program by typing `soundsc(note(56, 2), 11025)`.

3. The following is an incomplete M-file that will play scales. For the `tone` = line, generate the actual sinusoid for `keynum` by making a call to the

Fig. L2.3. The first few bars of Beethoven's *Für Elise*.

function note(keynum,dur) written in the previous exercise. Note that the code in play_scale.m allocates a vector of zeros large enough to hold the entire scale, then adds each note into its proper place in the vector **xx**.

```
% play-scale.m
% Create keys to play and corresponding
% durations
% Keys: C D E F G A B C
% durations: quarter second each
keys = [40 42 44 45 47 49 51 52];
durs = 0.25*ones(1, length(keys));
fs = 11025;
xx = zeros(1, sum(durs)*fs+1);
n1 = 1;
for kk = 1:length(keys)
keynum = keys(kk);
duration = durs(kk);
tone = %  ⇐ FILL IN THIS LINE
n2 = n1 + length(tone) - 1;
xx(n1:n2) = xx(n1:n2) + tone;
n1 = n2;
end
soundsc(xx,fs)
```

L2.4 *Für Elise* project

Copy the fenotes.m[2] file into your working directory with the rest of your lab2 files. Develop a MATLAB program to play

[2] Source: McClellan, J. H., Schafer, R. W. and Yoder, M. A. *DSP First* © 1998. Reprinted by permission of Pearson Education Inc., Upper Saddle River, NJ.

Beethoven's *Für Elise*. The first few bars of the music appear in Figure L2.3 but do not fear, you do not need to read the score! You may load the score by typing `fenotes`, which creates four vectors: **t**, **tdur**, **b**, and **bdur** corresponding to the keys and durations for the treble and base clefs, respectively. Use the `type fenotes.m` command to see the contents of the file. Your MATLAB program should generate two vectors for the two clefs, add them together to produce the overall signal, and play the results. You may model it after the `play_scale` program.

Hint: You may wish to create an `xx = play_scale2(keys, durs)` that generates a sinusoidal signal for the given vectors of keys and durations but does not play the sounds. The function can be called twice to produce the two clefs.

L2.5 Extra credit

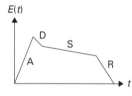

Fig. L2.4. Example ADSR envelope to produce fading.

The musical passage is likely to sound artificial because it is created from pure sinusoids. Therefore, you may want to try to improve the quality of the sound by incorporating some modifications. For example, you could multiply each pure tone signal by the following envelope $E(t)$ (see Figure L2.4):

$$x(t) = E(t)\cos(2\pi f_0 t + \phi)$$

so that each pure tone signal will fade in and out. If an envelope is used, each tone should "fade in" quickly and fade out more slowly. An envelope such as a half-cycle of a sine wave $\sin(\pi t/dur)$ is *not* good because it does not turn on quickly enough, so simultaneous notes of different durations no longer appear to begin at the same time. A standard way to define the envelope function is to divide $E(t)$ into four sections: attack (A), delay (D), sustain (S), and release (R). Together these are called ADSR. The attack is a quickly rising front edge, the delay is a short small-duration drop, the sustain is more or less constant, and the release drops quickly back to zero.

Some other issues that affect the quality of your synthesis include the relative timing of the notes, the correct duration for tempo, rests (pauses) in the appropriate places, the relative amplitudes to emphasize certain notes and make others soft, and harmonics. True piano sounds contain several

frequency components such as the second and third harmonics, but these
are at a lower amplitude than the fundamental.

For extra credit, enhance your *Für Elise* program to make it sound better
(such as adding an envelope or harmonics). Write a short description of
your changes.

L2.6 Exercises

1. Describe your commands and results for parts (1) to (9) from Section
   L2.2.
2. Produce a printout of your note function.
3. Produce a printout of your `play_scale` function. Describe how it
   sounds.
4. Produce a printout of the command used to play *Für Elise*.
5. Write a short paragraph describing your results.
6. Write the extra credit code and description (if applicable).

Place all your MATLAB files in the online folder specified by your instructor. Be sure to test that all the files are there; if they are not, your
instructor will be unable to play your music and you will get no credit for
your program. Include instructions in your report of how to play your music
(e.g., what to type and any special directions). If your instructions are not
clear, the graders will be unable to play your music.

Laboratory exercise 3: DFT and IDFT

L3.1 Objective

In this lab exercise, you will develop MATLAB functions to perform the discrete Fourier transform (DFT) and inverse DFT. You will find that the runtime for the DFT grows quickly with the number of samples you are analyzing. The FFT will perform the same task much more efficiently.

L3.2 The discrete Fourier transform

Recall that the discrete Fourier transform (DFT) acts on a vector of N samples $x[0], x[1], \ldots, x[N-1]$ and produces a vector of N frequencies $X[0], X[1], \ldots, X[N-1]$:

$$X[k] = \frac{1}{N} \sum_{n=0}^{N-1} x[n] e^{-jk\frac{2\pi}{N}n}.$$

MATLAB numbers vectors starting at element 1 rather than element 0. Thus, we can rewrite the DFT as acting on samples 1 through N:

$$X[k] = \frac{1}{N} \sum_{n=1}^{N} x[n] e^{-j(k-1)\frac{2\pi}{N}(n-1)}.$$

One can view the DFT as a matrix multiplication:

$$
\begin{bmatrix} X[1] \\ X[2] \\ \vdots \\ X[N] \end{bmatrix} = \frac{1}{N} \begin{bmatrix} D_{1,1} & D_{1,2} & \cdots & D_{1,N} \\ D_{2,1} & D_{2,2} & \cdots & D_{2,N} \\ \vdots & \vdots & \vdots & \vdots \\ D_{N,1} & D_{N,2} & \cdots & D_{N,N} \end{bmatrix} \begin{bmatrix} x[1] \\ x[2] \\ \vdots \\ x[N] \end{bmatrix},
$$

where

$$D_{k,n} = \mathrm{e}^{-\mathrm{j}(k-1)\frac{2\pi}{N}(n-1)}.$$

In this exercise, you will write a MATLAB function to compute the DFT using this matrix relationship. Note that MATLAB is case-sensitive so we can use lowercase x to indicate the sampled signal in time and uppercase X to indicate the frequency samples.

Your goal is to complete the following MATLAB function:

```
function X = mydft(x)
% MYDFT computes the discrete Fourier transform
% of vector x.
% Written <date> by <name>
N = length(x);
<write code here to compute matrix D>
X = D*x/N;
```

Create a MATLAB script named mydft.m. Enter the function above. You may compute D in whatever fashion you want. As a hint, recall that MATLAB functions acting on a matrix return a matrix of values. Also recall that it is easy to create a matrix of the form

$$\begin{bmatrix} 0 & 0 & 0 & 0 \\ 0 & 1 & 2 & 3 \\ 0 & 2 & 4 & 6 \\ 0 & 3 & 6 & 9 \end{bmatrix}$$

by multiplying a column vector by a row vector. Your code should only require 1–4 lines if you take advantage of MATLAB's matrix operations.

To understand the expected output of the DFT, consider a discrete time cosine wave with a frequency $f_0 = 1\,\mathrm{Hz}$ and a sampling frequency $f_s = 4\,\mathrm{Hz}$. One period of the cosine is shown in Figure L3.1, with the four samples being $[1 \quad 0 \quad -1 \quad 0]$. The continuous-time spectrum of

$$x(t) = \cos(2\pi f_0 t) = \frac{1}{2}\mathrm{e}^{\mathrm{j}2\pi f_0 t} + \frac{1}{2}\mathrm{e}^{-\mathrm{j}2\pi f_0 t}$$

is illustrated in Figure L3.2. The discrete-time Fourier transform spectrum is similar but is periodic with a period of f_s. The dots in Figure L3.3 indicate the original points from the spectrum. The DFT spectrum of Figure L3.4

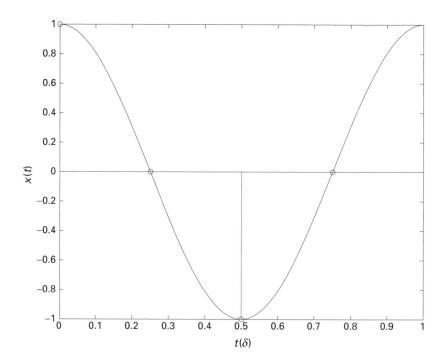

Fig. L3.1. One period of cosine waveform with four samples.

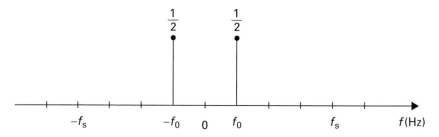

Fig. L3.2. Continuous-time spectrum for Figure L3.1.

consists of the N values from the discrete-time Fourier transform spectrum at times in the range $[0, f_s]$. Thus, the DFT of the cosine at this sampling rate consists of the values $[0 \quad 0.5 \quad 0 \quad 0.5]$.

Remember that the input x to mydft must be a column vector (why?). Test your mydft program on the column vector $x = [3 \quad 0 \quad -3 \quad 0]'$. Remember that ' indicates the conjugate transpose instead of simple transpose in MATLAB; this is important if the values in the vector are complex. Also .' indicates the ordinary transpose. Does your answer make sense? Please explain intuitively.

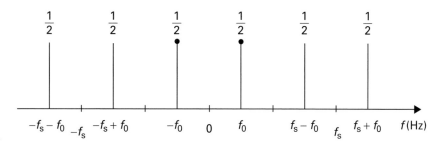

Fig. L3.3. Discrete-time Fourier transform spectrum for Figure L3.1.

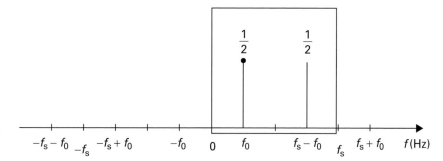

Fig. L3.4. DFT spectrum for Figure L3.1.

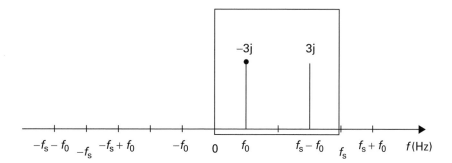

Fig. L3.5. Example spectrum for exercise.

L3.3 The inverse discrete Fourier transform

Recall that the inverse discrete Fourier transform (IDFT) acts on a vector of N frequencies $X[0]$, $X[1]$, ..., $X[N-1]$ and produces a vector of N samples $x[0]$, $x[1]$, ..., $x[N-1]$:

$$x[n] = \sum_{k=0}^{N-1} X[k] e^{jk\frac{2\pi}{N}n}.$$

MATLAB numbers vectors starting at element 1 rather than element 0. Thus, we can rewrite the IDFT as acting on samples 1 through N:

$$x[n] = \sum_{k=1}^{N} X[k]e^{j(k-1)\frac{2\pi}{N}(n-1)}.$$

One can view the IDFT as a matrix multiplication:

$$\begin{bmatrix} x[1] \\ x[2] \\ \vdots \\ x[N] \end{bmatrix} = \begin{bmatrix} I_{1,1} & I_{1,2} & \cdots & I_{1,N} \\ I_{2,1} & I_{2,2} & \cdots & I_{2,N} \\ \vdots & \vdots & \vdots & \vdots \\ I_{N,1} & I_{N,2} & \cdots & I_{N,N} \end{bmatrix} \begin{bmatrix} X[1] \\ X[2] \\ \vdots \\ X[N] \end{bmatrix},$$

where

$$I_{n,k} = e^{j(k-1)\frac{2\pi}{N}(n-1)}.$$

Write another MATLAB function called `myidft`. It should be very similar to your `mydft` from the previous section.

Suppose you are given the spectrum shown in Figure L3.5, along with the information that the sampling frequency is $f_s = 4\,\text{Hz}$. What function has this spectrum? What are the sampled data values? Test `myidft` on the column vector $[0 \quad -3j \quad 0 \quad 3j].'$ corresponding to this spectrum. Explain why your results make sense.

L3.4 The fast Fourier transform

The DFT runtime grows as N^2 for a signal with N samples. Even for a signal of relatively short time duration and a low sampling rate, this can require a significant amount of time. DFTs of signals with more samples may take longer than you are willing to wait. Fortunately the fast Fourier transform (FFT) can be orders of magnitude faster, with a runtime of $N \log N$ for a signal with N samples.

For example, use your `sumcos` function to create the following signal:

$$x(t) = 9\cos(2\pi(4)t) + 5\sin(2\pi(7)t).$$

Choose a sampling rate of 1 kHz and a duration of 3 seconds. Use your `mydft` to find the coefficients. You can time how long this takes[1] with the command

```
t = clock; X=mydft(x'); etime(clock,t)
```

What are the amplitudes of each nonzero coefficient in X? Ignore coefficients that are close enough to zero because they are merely noise. Write $x(t)$ as a sum of complex exponentials and explain how the terms in your sum relate to the coefficients.

Repeat the process, using the built-in FFT command in place of `mydft`. How long does MATLAB take? Are the coefficients the same? Explain any discrepancies. Finally, you can always use the help facility or try new commands, but you should not need to use any MATLAB commands that are not introduced in Lab 1 or described in this exercise.

L3.5 Exercises

1. Provide a copy of `mydft.m` and your results of using it on the column vector. Explain why your results make sense.
2. Provide a copy of `myidft.m` and your results. What function has the spectrum given? What are its four discrete-time sampled values?
3. How long did your `mydft` take for a 3 second signal at a 1 kHz sampling rate? How long did the FFT take?
4. What are the amplitudes of the nonzero coefficients in X? Show how these relate to $x(t)$ expressed as a sum of complex exponentials.
5. Do the coefficients produced by the FFT match those of `mydft`? Explain any discrepancies.

---

[1] If you are using a slow computer or one with too little RAM or if your `mydft` is inefficient and you find this taking more than a minute, you can try a shorter signal of 1 or 2 seconds. Mention the duration you actually use.

Laboratory exercise 4: FFT and IFFT

L4.1 Objective

There are two general ways to analyze a linear time-invariant (LTI) system described by a *governing equation* in the form of a differential equation. One is to directly solve the differential equation. Finding the particular solution involves nasty integrals if the input is a complicated signal. These integrals could be evaluated numerically with a computer, but they do not give much insight on how the system might behave to a different input.

Another powerful way to understand a linear time-invariant system is to find its *frequency response function*, that is the ratio of the output amplitude to the input amplitude when the input is a complex exponential of a particular frequency. Once we know the frequency response function, we can find the response of the system to an arbitrary complicated signal by decomposing the signal into a sum of complex exponentials with the FFT, finding the response to each complex exponential, and summing the responses.

In this exercise, you will explore the frequency response of an *RLC* circuit shown in figure L.4.1. You will first determine an expression for its frequency response function. You will write a MATLAB M-file for this frequency response function and graph its magnitude and phase in a *Bode plot*. You will then explore the response of the circuit to a sweep input. You will use the FFT to decompose the sweep into its frequency components, multiply by the frequency response function, and use the inverse FFT to obtain the output of the circuit. When you listen to the output signal you will hear how the *RLC* acts as a *band-pass filter*.

Fig. L4.1. Example *RLC* circuit.

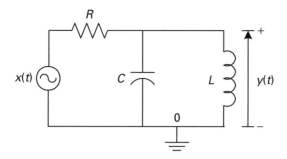

L4.2 Frequency response of a *RLC* circuit

We would like to understand how the output voltage $y(t)$ depends on the input voltage $x(t)$. The governing equation relates the input and output through a differential equation. Recall that resistors, capacitors, and inductors are all linear time-invariant circuit elements. Thus, like any other LTI system, the response of the system to a complex exponential input

$$x(t) = X e^{j\omega t}$$

is another complex exponential of the same frequency

$$y(t) = Y e^{j\omega t}.$$

The ratio of the output amplitude to the input amplitude gives

$$H(j\omega) = \frac{Y}{X}$$

and is called the *frequency response function*.

Show that the frequency response function of the parallel *RLC* circuit is given by

$$H(j\omega) = \frac{2\zeta \dfrac{\omega}{\omega_n} j}{1 - \left(\dfrac{\omega}{\omega_n}\right)^2 + 2\zeta \dfrac{\omega}{\omega_n} j},$$

where ω_n and ζ are constants depending on R, L and C. What is the natural frequency ω_n and the damping factor ζ? Evaluate them numerically for $R = 2.2\,\mathrm{k\Omega}$, $L = 47\,\mathrm{mH}$, $C = 15\,\mathrm{\mu F}$.

Write a function $\mathtt{h} = \mathtt{rlc(omega)}$ that returns $H(j\omega)$ using the component values listed above. Plot the magnitude and phase of your frequency response function for frequencies from $100\,\mathrm{Hz}$ to $1\,\mathrm{kHz}$. Remember to convert to radian frequency when plotting. For the magnitude plot, use

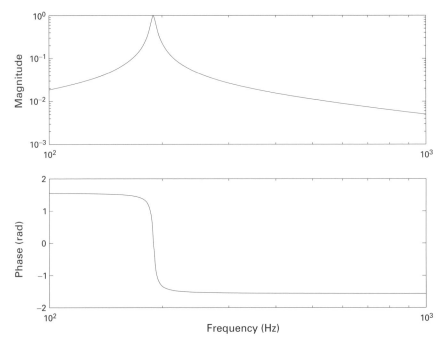

Fig. L4.2. Magnitude and phase plots for the circuit of Figure L4.1.

logarithmic scales on both axes. For the phase plot, use a logarithmic scale on the frequency axis. The MATLAB loglog and semilogx commands will be helpful here; they operate just like plot, but use logarithmic scales on one or both axes. You will likely need a for-loop to compute the frequency response function at each frequency, or better yet, a version of rlc that accepts a vector of frequencies. Your plot should look like Figure L4.2. If it does not, please check your equation and plotting program and try again.

For the *RLC* circuit, the plot shows a resonance just below 200 Hz, with the magnitude of the response dropping off sharply on both sides. Below the resonance, the phase is $\pi/2$. Above the resonance, the phase is $-\pi/2$. The *RLC* circuit is often called a *band-pass filter* because it passes frequencies near the resonance and attenuates the rest.

L4.3 Time response of a parallel *RLC* circuit to a sweep input

Now we will examine the time response of the parallel *RLC* circuit to a sweep input. Copy the sweep.m function to the directory with the rest of your lab files.

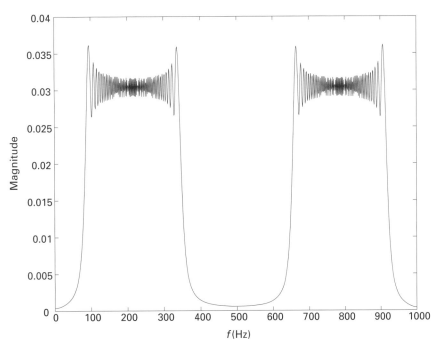

Fig. L4.3. Example sweep
spectrum.

You will need to repeat much of the work you do in this section in the later
sections, so you may wish to create an M-file that contains the commands
you enter.

Produce a 1 second sweep from 80 Hz to 350 Hz at a sample rate of 1 kHz
using the sweep.m function. How many elements are in x? Listen to the
sweep on the speaker.

```
>> fs = 1000;
>> x = sweep(80, 350, fs, 1-1/fs);
```

To find the time response of the circuit, we will take the FFT to find the
frequencies in the sweep, multiply by the frequency response of the circuit,
then take the inverse FFT to go back to the time domain. Use MATLAB's
fft function to produce a vector X that is the FFT of x. Do not forget
the factors of $1/N$. Graph the spectrum of x by plotting the magnitude
of X. It should look like Figure L4.3. The horizontal axis is labeled with
coefficient number rather than frequency. From the spectrum, approximately
what positive and negative frequencies are in the sweep?

An easier way to visualize the spectrum is to use the fdomain function (also on the accompanying CD) that associates the coefficients with

the appropriate frequencies. Generate the spectrum using the following commands:

```
>> [X1,f] = fdomain(x, fs);
>> plot(f, abs(X1));
```

Be sure you understand the relationship between plotting the spectrum labeled with coefficient number and the spectrum labeled with frequency, as you have seen in the last two plots. Given one form of the spectrum plot, you should be able to visualize the other.

Now you will generate the frequency response function $H(j\omega)$ of the *RLC* circuit. It should have the same number of points as X. You will need to carefully understand what elements produced by a FFT correspond to what frequencies so you can calculate the frequency response elements for the same frequencies. Recall that in X, element 1 is the component at frequency 0. Element 500 is at frequency 499 Hz. Element 501 is at frequency -500 Hz. Finally element 1000 is at frequency -1 Hz. Write a loop to generate h=rlc(omega) by computing the frequency response function at the appropriate frequency for each element. You will need the code you developed in the previous part to evaluate the frequency response function at a given excitation frequency. Make a plot of your frequency response function; it should look like Figure L4.4.

The spectrum of the output Y of the *RLC* circuit is simply the product of the input X and the frequency response function $H(j\omega)$. Pay attention to whether your vectors are rows or columns when multiplying. Make a plot of the magnitude of Y. Finally, take the inverse FFT to recreate the signal $y(t)$ at the output of the *RLC* circuit. As before, be careful about the factor of N. The limited numerical accuracy of the computer leads to $y(t)$ being complex with very small imaginary components produced from round-off error. Take the real part of $y(t)$ to discard the imaginary error. Plot your result.

Play your output sound. You should hear the component around 200 Hz at its original volume, but the lower and higher frequencies will be attenuated. If you just hear static, you may have forgotten to take the real part. Make a plot of the real part of $y(t)$. Explain why it has the form you see.

Make a spectrogram of the input and output using commands like that below. The spectrogram indicates time on the x-axis, frequency on the y-axis, and intensity of the frequency at a given time with color. Why do the spectrograms have the frequency components shown?

```
>> specgram(x, 64, fs)
```

Fig. L4.4. Frequency
response function of the
circuit of Figure L4.1.

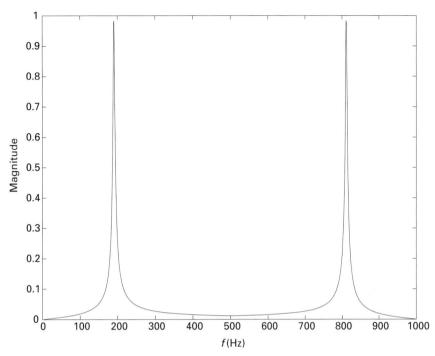

L4.4 Exercises

1. Determine the frequency response function $H(j\omega)$ for the *RLC* circuit.
   What is the natural frequency and damping factor?
2. Using your `h=rlc(omega)` function, plot the magnitude and phase of
   the FRR. It should look like the plot in Figure L4.2.
3. Make a five-panel plot with your original sweep, the magnitude of the
   FFT, the magnitude of the frequency response function, the magnitude
   of the output in the frequency domain, and the real output back in the
   time domain.
4. Make a two-panel plot with the spectrogram of the input and output
   signals. Explain what the spectrograms indicate.

Laboratory exercise 5: Frequency response

L5.1 Objective

In Lab 4 you learned to compute the frequency response function of an electrical circuit, then use the FFT and IFFT to find the time response of the circuit to a complicated input signal. In this exercise you will use the concept of frequency response to explore the behavior of a mechanical system. At this point, you should know how to use MATLAB as a tool and this exercise will focus on the objectives rather than techniques. Some of the questions will be most easily answered with hand calculations while others will be easier using MATLAB. Refer to examples in the previous exercises to refresh your memory on technique. You may find it helpful to write small functions to help your calculations, such as a function that computes the frequency response.

L5.2 Automobile suspension

The base excitation system shown in Figure L5.1 represents a simple model of an automobile suspension. The vehicle is represented by the mass m and the suspension consists of the spring and damper elements in parallel. The vehicle drives over an undulating track so the wheels impose an input displacement of $x(t)$. The vertical displacement of the driver (and mass of the vehicle) constitutes the output of the system and is given by $y(t)$.

L5.3 Frequency response

Derive the equation of motion for the vertical displacement of the vehicle. Express your governing equation in canonical form. What are ζ and ω_n

Fig. L5.1. A simple model of an automobile suspension system.

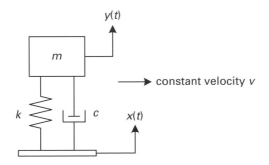

of the system? Find the frequency response function $H(j\omega)$ that relates the output amplitude Y to the input amplitude X. Express the frequency response function in terms of the natural frequency and damping factor.

Suppose the mass of the vehicle is $m = 1460$ kg. What are ζ and ω_n if the system parameters of the suspension are $k = 36$ kN/m and $c = 10$ kN s/m? Sketch a Bode plot of $H(j\omega)$ for frequencies between 0.1 Hz and 100 Hz. You may find the logspace command helpful to create evenly distributed frequencies on a logarithmic scale. For example,

```
omega = 2*pi*logspace(-1, 2, 50)
```

returns 50 points distributed between 10^{-1} Hz and 10^2 Hz.

What is the damped natural frequency, ω_d, of the system? What is the magnitude of the frequency response function at this frequency?

L5.4 Time response to sinusoidal input

Suppose you race the vehicle down a "washboard" dirt road with a regular pattern of bumps and dips. If you have ever driven on such a road, you probably know that the ride is tolerable at very slow speeds. At moderate speeds, the car shakes horribly, but if you drive fast enough, the ride becomes smoother! The road profile at a distance z from the starting line can be expressed as

$$x(z) = 0.005 \cos(2\pi z)\,\text{m}.$$

If you drive at a constant velocity v, find an expression for $x(t)$. Using the concept of frequency response function, determine the vehicle speed at which vibrations will be worst. Alternatively, determine the vehicle speed at which the vertical displacement of the car will be at its largest.

Because the system is linear and time-invariant, the vertical displacement of the driver's seat will be of the form

$$y(t) = |Y| \cos(2\pi vt + \angle Y),$$

where $|Y|$ represents the vibration amplitude that the driver feels. Find an expression for $|Y|$ that depends on the vehicle speed, v. Plot $|Y|$ versus v for velocities from $0\,\text{m/s}$ to $10\,\text{m/s}$. Does the speed where the vibration is worst match your prediction? How does your plot relate to the Bode plot you made in the previous section? What value does $|Y|$ approach for very low frequencies? For very high frequencies?

L5.5 Numerical solution with the Fourier transform

When the input is a sinusoid or complex exponential, it is easy to find the output of an LTI system using the frequency response function. Let the input be described by the phasor X and have angular frequency ω. Then by the definition of frequency response function, the output is described by the phasor $Y = H(j\omega)X = |Y|e^{j\angle Y}$.

When the input is more complicated, solving the differential equation may be infeasible. However, we know from Lab 3 that we can always decompose the input into a weighted sum of complex exponentials using the FFT. Because the system is linear, the output is the sum of the responses to each exponential. This may be computed by multiplying the weight on each exponential by the frequency response at that frequency, then taking the IFFT. If the input is real, the output should also be real. However, rounding error sometimes introduces small imaginary components that should be discarded.

To verify this procedure, predict the amplitude of $y(t)$ if $x(t)$ is a 3 Hz sinusoid. Now let us check your prediction numerically with the following code. As we cannot numerically process continuous signals, we work with discrete-time representations sampled at f_s for a duration T (giving a total of $N = Tf_s$ samples). The only tricky part of the code is determining the appropriate frequencies to use for the frequency response function. Recall that the first half of the frequencies must correspond to 0 to $f_s/2$ while the second half span $-f_s/2$ to 0. Therefore

$$
\omega[n] = \begin{cases} \dfrac{n-1}{N}\, 2\pi f_s & 1 \le n \le \dfrac{N}{2} \\[2ex] \left[\dfrac{n-1}{N} - 1\right] 2\pi f_s & \dfrac{N}{2} + 1 \le n \le N \end{cases}
$$

```
fs = 1000;
T = 5;
N = fs*T;
t = 0:1/fs:T-1/fs;
x = cos(2*pi*t*3);
fdigital = [0:N/2-1 -N/2:-1]/N;
omega = 2*pi*fs*fdigital;
H=secondorder(omega);
X=fft(x);
Y=H.*X;
y=real(ifft(Y));
plot(t,[y;x])
```

Write a `secondorder` function that returns the frequency response function of your second-order system for a vector of frequencies **omega**. Test your function and the MATLAB snippet by verifying that the amplitude of $y(t)$ matches your prediction for the 3 Hz sinusoid.

L5.6 Time response to step input

Suppose the vehicle is driven down a track and encounters a sudden step from height 0 to height $h = 5$ cm at time $t = 5$ seconds (and no other bumps before or afterward). Use the FFT method to compute and graph $x(t)$ and $y(t)$ on the same set of axes. Note that the FFT assumes a periodic signal but that the step is not periodic. Instead use a square wave where the period T is long compared to the response time of the system. From your graph, determine the 0 to 100 percent rise time, t_r? What is the maximum percentage overshoot, M_p? At what time does this overshoot occur? What is the 5 percent settling time, t_s?

L5.7 Optimizing the suspension

Suppose you are redesigning the suspension for a better ride. One way to quantify the quality of the ride is in the way the vehicle vibrates when striking a step in the road. For a gentle ride, the rise time should be slow so the rider feels no sudden acceleration. However, the overshoot should also be limited and the oscillations should rapidly settle; these requirements involve trade-offs. Specifically, you would like the rise time to be more than 1 second, the maximum percentage overshoot less than 40 percent, and the 5 percent settling time less than 6.5 seconds.

You may modify the suspension by changing the spring and damper but must not change the mass of the vehicle. What values do you recommend and what natural frequency and damping factor do they provide? Design for an underdamped system so the rise time is defined. Graph the step response and determine the actual rise time, maximum percent overshoot, and settling time for your components.

L5.8 Exercises

1. Derive the equation of motion and write it in canonical form. What are ζ and ω_n (in general, and for the specific component values)? Give an expression for the frequency response function. What are the damped natural frequency and the magnitude of the frequency response function at this frequency? Make a Bode plot of the frequency response function.
2. Make a plot of $|Y|$ versus v. At what speed is $|Y|$ the biggest? Explain how this plot relates to your Bode plot. What are the limiting values of $|Y|$ at low and at high speeds?
3. Print your secondorder.m function. What amplitude do you predict for a 3 Hz sinusoidal input? Give a plot showing the output.
4. Graph the response to a step input. What is the rise time? What is the maximum percent overshoot and when does it occur? What is the 5 percent settling time?
5. What values of k and c do you recommend to improve the suspension? What are ζ and ω_n? Plot the step response and determine the rise time, maximum percentage overshoot, and settling time.

Laboratory exercise 6: DTMF

L6.1 Objective

In this exercise, you will use MATLAB to decode dual tone multifrequency (DTMF) touchtone telephone signals. You will first construct bandpass filters to identify each of the tones. Then you will use the filters to identify a phone number despite background noise.

L6.2 DTMF dialing

A long time ago you may have heard an example of DTMF tone generation being used to dial a telephone. These tones are the *superposition*, that is the sum, of a pair of sine waves, as shown in Table 1.1. You could build your own dialer by producing these tones with another system such as your computer. Similarly, you could determine which phone number was dialed by looking at the spectrum of the tones and observing what frequencies are present.

Copy all of the files from the Lab 6 directory. Look at `dtmfdial.m` and see how it works. Write a one or two-line MATLAB program that calls `dtmfdial` to play the tones to dial the phone number (123)456-7890.

As you see, there is nothing special about pressing the keys on the phone to dial a phone number. You could easily dial a number from the computer by generating the appropriate tones. Or if you are good at whistling, you can even whistle the phone number. Once upon a time "phone phreaks" used to make free calls by whistling the tones used by the phone company for internal routing. Needless to say, that is no longer possible.

L6.3 fdomain and tdomain

In the previous two exercises, you used the Fourier series to compute the time response $y[n]$ of an LTI system given the input $x[n]$ and frequency response function $H(j\omega)$. The frequency response function must be evaluated at frequencies from $-f_s/2$ to $f_s/2$. In Lab 5, you wrote code to generate a vector with these frequencies. For example, suppose you wish to evaluate the output of a first-order system with the frequency response function

$$H(j\omega) = \frac{A}{1 + j\omega\tau}.$$

Assuming x has an even number of elements N, the following code can be used:

```
ff = fs*[0: N/2-1 -N/2:-1]/N;
H = A./(1+j*2*pi*ff*tau);
X = fft(x);
Y = H.*X;
y = ifft(Y);
```

The messy part of the process is generating the vector **ff**. Indeed, our code does not work correctly for odd values of N. In Chapter 2 we explored a more elegant approach of using the `fdomain` function (`tdomain` is its inverse) that works for any value of N.

Look at `fdomain.m` and `tdomain.m` and see how they work. The M-file `fdomain` performs the FFT and also returns the vector **ff** of frequencies at which the frequency response must be evaluated. The M-file `tdomain` performs the IFFT and also returns the vector **tt** of times at which the output is sampled. Vector **ff** is handy for computing the frequency response, while vector **tt** is handy for plotting the result. These commands also shuffle the frequencies so that negative frequencies are at the beginning of the vector. Using these functions, the previous example code may be rewritten as

```
[X, ff] = fdomain(x, fs);
H = A./(1+j*2*pi*ff*tau);
Y = H.*X;
[y, tt] = tdomain(Y, fs);
```

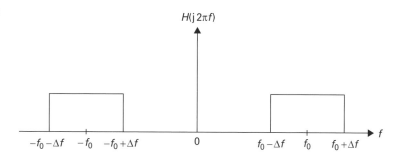

Fig. L6.1. Continuous-time frequency response of an ideal band-pass filter.

L6.4 Band-pass filters

An ideal continuous-time band-pass filter centered on frequency f_0 with width $2\Delta f$ has the continuous-time frequency response shown in Figure L6.1. All frequencies within the passband are passed unchanged and all other frequencies are completely attenuated.

Sketch the discrete-time frequency response of an ideal discrete-time band-pass filter with the same center frequency and width assuming the sampling rate f_s is greater than $2(f_0 + \Delta f)$.

Write a M AT L A B function H = bandpass(f0, deltaf, ff) that generates a vector containing the discrete-time frequency response function evaluated at frequencies given in the **ff** vector. **Hint**: you may find the commands <, >, and & useful.

Test your band-pass function using the sumcos function. Use a sampling rate of 1 kHz. Generate the response of a band-pass filter with a center frequency of 20 Hz and a width of 10 Hz (5 Hz on each side of center). Use the sumcos function to generate a vector

$$x(t) = 3\cos(2\pi 4t) + 7\cos(2\pi 18t)$$

with a duration of 0.5 seconds. If $x(t)$ is filtered through your band-pass filter, give an expression for the output $y(t)$. Make a five-panel plot with $x(t)$; the magnitudes of the input spectrum $X[k]$, the frequency response function $H[k]$, and the output spectrum $Y[k]$; and the output $y(t)$. Does the output match your expectation?

L6.5 DTMF decoding

To decode DTMF signals, you will need a filter bank with seven band-pass filters. By plotting the outputs and inspecting which filters respond to each part of the dialing, you will be able to determine which digits are being pressed.

Write another MATLAB file dtmfdecode(x, fs) that accepts a vector **x** representing the sound coming over the phone line. Your function should make an eight-panel plot. The first should be **x**. The next seven should be the outputs of band-pass filters centered around 697, 770, 852, 941, 1209, 1336, and 1477 Hz. You may choose an appropriate width of the passband. Use the fdomain and tdomain functions in your code.

Test your function on the (123)456-7890 phone number example you played earlier. Verify that you decode the proper tones.

L6.6 Forensic engineering

Inspector 59 is hot on the trail of a mutant terrorist. The Inspector learns that the suspect's accomplice will be calling the criminal from a pay phone. Our trusty Inspector cannot see the number being dialed but manages to tape the sound of the buttons being pressed from a distance. The recording is somewhat garbled by background noise. Now Inspector 59 needs your help to decipher the number that was called.

Copy mystery.mat file from the Lab 6 directory. This file is a digital recording of the number being called, sampled at 8 kHz. Read it into memory with the command load mystery.mat. This will create a vector called **mystery**. You can listen to it with the command soundsc(mystery, 8000). Use your dtmfdecode program to decode the mystery number. What is the number?

L6.7 Exercises

1. Provide a copy of your bandpass.m function.
2. Give an expression for $y(t)$.
3. Provide a five-panel plot testing the band-pass function.
4. Provide a copy of your dtmfdecode.m file.
5. Provide an eight-panel plot of the filter responses to the mystery number. What number was dialed?

Laboratory exercise 7: AM radio

L7.1 Objective

In this exercise, you will simulate an AM transmitter and a crystal receiver to demodulate an AM radio signal.

L7.2 Amplitude modulation

First you will simulate an AM transmitter. Ordinarily, AM radio operates in the 540–1600 kHz band. However, such a high sample rate would lead to vast amounts of data to process. For this exercise, we will investigate a system involving two stations transmitting on 30 kHz and 40 kHz carriers. Both stations are allocated a bandwidth of 5 kHz on either side of the carrier, just as in ordinary AM radio.

Suppose we have two stations. KAAA ("All A, all the time, never flat, never sharp") plays a continuous 440 Hz A note for its small but fiercely loyal band of listeners. KBRD plays a signal whose frequency varies from 300 Hz to 700 Hz in a sinusoidal manner.

Let the signal on KAAA be

$$x_1(t) = 0.8 \cos(2\pi(440)t).$$

Let the signal on KBRD be

$$x_2(t) = 0.9 \cos(2\pi(500t + 20\cos(10t))).$$

Generate these two signals for a duration of 2 seconds at a sampling rate of 100 kHz. Use soundsc to play x_1 and x_2. Do they sound like an A note and an oscillating chirp, respectively?

An AM station with an audio signal $x(t)$ operating at a carrier frequency f_c broadcasts a modulated signal

$$y_i(t) = (1 + x_i(t))\cos(2\pi f_c t).$$

Write a MATLAB program to generate $y_1(t)$ and $y_2(t)$, the modulated signals broadcast by the two stations. KAAA is assigned a 30 kHz carrier. KBRD is assigned a 40 kHz carrier. Use the same duration and sampling rate as you did for the $x(t)$ signals. Also create

$$y_3(t) = 0.08 y_1(t) + 0.3 y_2(t),$$

the sum of the two, representing the overall signal picked up at a receiver that is located different distances from the two transmitters and hence receives signals attenuated by different amounts. In this particular case, KBRD is coming in more strongly than KAAA.

Using the fdomain utility, make plots of the magnitude spectra of x_1, x_2, y_1, y_2, and y_3. Check that they match your expectations for an amplitude-modulated signal. For example, y_1 is shown in Figure L7.1. The plot is symmetric with three peaks in the positive frequencies and three peaks representing aliases of the negative frequencies. The center peak is the 30 kHz carrier. The two side lobes are at 29.56 kHz and 30.44 kHz caused by the 440 Hz tone.

L7.3 Demodulation

Now you will build a MATLAB function to model a crystal receiver. The crystal receiver has three parts. It first performs a band-pass operation to tune in one station while trying to exclude all others. It then rectifies the signal. Finally, it performs a low-pass filter to eliminate the carrier. Using your band-pass function from Lab 6, construct the frequency response H1 for a band-pass filter that passes frequencies between 25 kHz and 35 kHz (to tune in station 1). Next, the crystal receiver rectifies the result. Write a function y = rectify(x) that returns x if $x > 0$ and 0 otherwise. Finally write an H=lowpass(cutoff,ff) function similar to the bandpass function. It should generate the frequency response function for a low-pass filter passing frequencies below cutoff. Use it to generate the frequency response function H3 for a low-pass filter with a 5 kHz cut-off frequency.

Fig. L7.1. The magnitude spectrum of y_1 in the Lab 7 example.

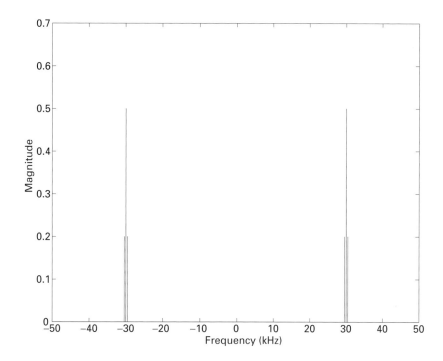

Put all of these functions together. To extract KAAA, band-pass filter y_3 with H1. Rectify the output. Finally, low-pass filter the result with H3 to obtain a signal z_1. Remember that the filtering takes place by multiplying in the frequency domain while rectification must be done in the time domain, so you will need to use the `fdomain` and `tdomain` commands several times to go back and forth between the domains. Extract KBRD with H2 (from 35 kHz to 45 kHz) and low-pass filter the result with H3 to obtain z_2, the signal for KBRD. Listen to the signals. Do they sound like x_1 and x_2?

L7.4 Pirate radio

After tracking the mutant terrorist back to the hideout at the American Pork Federation, Inspector 59 learns that the criminal is out driving a van loaded with gear to broadcast as a pirate radio station[1] and is busy sending subliminal messages to the public. The inspector digitizes the signal at a 100 kHz sampling rate. Unfortunately, there is another station also broadcasting below 50 kHz.

[1] That is, a station unlicensed by the FCC.

The Inspector's data is in the Lab 7 directory in a file called `pirate.mat`. Load it to obtain the vector called **pirate**. Invoke your MATLAB skills to determine what carrier frequency the mutant is broadcasting on and demodulate the subliminal message. Do not be confused by the other station that you may find.

If you encounter difficulties creating z_1 and z_2, you may want to look at the signals in both the time and frequency domains after the band-pass filtering, rectification, and low-pass filtering. Think through what each operation should do and check that your results match expectations.

L7.5 Exercises

1. Provide a five-panel plot with the spectra of the x and y signals.
2. Provide a copy of your `rectify.m` and `lowpass.m` files.
3. Provide a copy of your code to produce z_1 and z_2.
4. Do z_1 and z_2 sound like x_1 and x_2? If there are any differences, explain.
5. What frequency is the mutant broadcasting on? What is the subliminal message?

A Complex arithmetic

A1.1 Cartesian or rectangular form

A complex number z of real constants a and b can be expressed in *Cartesian* or *rectangular* form as

$$z = a + bj, \tag{A1.1}$$

where j is the complex unity ($j = \sqrt{-1}$), a is the real part of z, and b is the imaginary part of z. Thus,

$$a = \mathrm{Re}\,\{z\} \quad \text{and} \quad b = \mathrm{Im}\,\{z\}. \tag{A1.2}$$

Schematically, we can represent the complex number z as a point in a complex plane, where the horizontal axis corresponds to the real axis, and the vertical axis the imaginary axis (see Figure A1.1). The complex conjugate of z is denoted by $\bar{z}$, and is given by

$$\bar{z} = a - bj. \tag{A1.3}$$

Two complex numbers are said to be equal if and only if their real and imaginary parts are identical.

The addition and subtraction of complex numbers are defined by the following rule:

$$z_1 \pm z_2 = (a_1 + b_1 j) \pm (a_2 + b_2 j) = (a_1 \pm a_2) + (b_1 \pm b_2)j. \tag{A1.4}$$

The multiplication of complex numbers is defined by the rule

$$z_1 z_2 = (a_1 + b_1 j)(a_2 + b_2 j) = (a_1 a_2 - b_1 b_2) + (a_1 b_2 + b_1 a_2)j. \tag{A1.5}$$

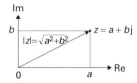

Fig. A1.1. Cartesian representation of a complex number.

The division of complex numbers is given by

$$\frac{z_1}{z_2} = \frac{a_1 + b_1 \mathrm{j}}{a_2 + b_2 \mathrm{j}} = \frac{(a_1 + b_1 \mathrm{j})(a_2 - b_2 \mathrm{j})}{(a_2 + b_2 \mathrm{j})(a_2 - b_2 \mathrm{j})} = \frac{a_1 a_2 + b_1 b_2}{a_2^2 + b_2^2} + \frac{a_2 b_1 - a_1 b_2}{a_2^2 + b_2^2} \mathrm{j},$$

$$(\text{A1.6})$$

where we have multiplied both the numerator and denominator of the quotient z_1/z_2 by the complex conjugate of z_2.

A1.2 Polar form

The same complex number z can also be represented schematically in *polar* form (see Figure A1.2), where r denotes the magnitude of z and θ the corresponding phase, in radians, measured counterclockwise from the horizontal or real axis. By inspection, we note that r and θ are simply given by

$$r = \sqrt{a^2 + b^2} \qquad \theta = \tan^{-1}\left(\frac{b}{a}\right). \tag{A1.7}$$

From calculus, recall that e^x can be expanded in a Maclaurin series as follows:

$$\mathrm{e}^x = 1 + x + \frac{x^2}{2!} + \frac{x^3}{3!} + \cdots. \tag{A1.8}$$

Letting $x = \pm \mathrm{j}\theta$, where θ is a real number, we can readily derive the Euler's identity

$$e^{\pm \mathrm{j}\theta} = \cos\theta \pm \mathrm{j}\sin\theta. \tag{A1.9}$$

By inspection, we can also write z as

$$z = r(\cos\theta + \mathrm{j}\sin\theta) = r\mathrm{e}^{\mathrm{j}\theta}. \tag{A1.10}$$

Equation (A1.10) is said to be the *polar* form of the complex number z. While either the Cartesian form (Eq. (A1.1)) or polar form (Eq. (A1.10)) can be used to represent a complex number, polar representation greatly

Fig. A1.2. Polar representation of a complex number.

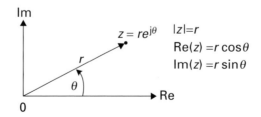

simplifies the multiplication and division of complex numbers. When two complex numbers are multiplied, the resultant is another complex number whose magnitude is simply the product of the magnitudes and whose phase is simply the sum of the phases; similarly, when two complex numbers are divided, the resultant complex number has a magnitude and phase which correspond to the quotient of the magnitudes and the difference between the phases:

$$z_1 z_2 = r_1 e^{j\theta_1} r_2 e^{j\theta_2} = r_1 r_2 e^{j(\theta_1 + \theta_2)}, \tag{A1.11}$$

$$\frac{z_1}{z_2} = \frac{r_1 e^{j\theta_1}}{r_2 e^{j\theta_2}} = \frac{r_1}{r_2} e^{j(\theta_1 - \theta_2)}. \tag{A1.12}$$

Finally, in expressing a complex number in polar form, it is important to keep track of the quadrant in which the complex number lies. For example, for positive, real constants a and b, while $\tan^{-1}(a/b)$ and $\tan^{-1}(-a/-b)$ are identical, $\tan^{-1}(a/b)$ lies in the first quadrant but $\tan^{-1}(-a/-b)$ lies in the third. To ensure the correct phase, it is often useful to plot the complex number in the Cartesian coordinates before calculating its corresponding phase.

B Constructing discrete-time signals from building blocks – least squares

In Chapter 2 we constructed continuous-time signals from building blocks or basis functions. An analogous approach can be used for constructing discrete-time signals. We wish to approximate the (possibly complex) discrete-time signal $x[n]$ by $\hat{x}[n]$ using a linear combination of the basis vectors $\phi_k[n]$,

$$x[n] \approx \hat{x}[n] = \sum_k a_k \phi_k[n], \tag{B1.1}$$

where the a_k are (possibly complex) constants. Instead of minimizing the integrated-squared-error (ISE) as we did for the continuous-time signals, for discrete-time signals, we minimize the summed-squared-error (SSE), defined as

$$\text{SSE} = \sum_n |x[n] - \hat{x}[n]|^2. \tag{B1.2}$$

The coefficients a_k that minimize the SSE satisfy the following orthogonality principle:

$$\sum_n (x[n] - \hat{x}[n]) \, \phi_k^*[n] = 0, \quad \text{for all } k, \tag{B1.3}$$

which can be proved using the same approach as that used for the continuous-time orthogonality principle and is left as an exercise for the reader. For K coefficients this yields K equations with K unknowns a_k, which can be readily solved using Gaussian elimination. The above formulation appears strikingly similar to a least-squares fit that is covered in a first course in linear algebra. In fact, as we will soon show, it is exactly the same concept.

Let us index the sequence $x[n]$ such that $0 \leq n \leq N - 1$ and the coefficients a_k such that $0 \leq k \leq K - 1$. Then Eqs. (B1.1) to (B1.3) become

$$x[n] \approx \hat{x}[n] = \sum_{k=0}^{N-1} a_k \phi_k[n], \tag{B1.4}$$

$$\text{SSE} = \sum_{n=0}^{N-1} |x[n] - \hat{x}[n]|^2, \tag{B1.5}$$

and

$$\sum_{n=0}^{N-1} (x[n] - \hat{x}[n]) \phi_k^*[n] = 0, \quad k = 0, 1, \ldots, K - 1. \tag{B1.6}$$

These expressions can be made more compact by using notations introduced in linear algebra. Let us first define the following vectors:

$$\mathbf{x} = [\, x[0] \quad x[1] \quad \ldots \quad x[N-1] \,]^T$$
$$\boldsymbol{\phi}_k = [\, \phi_k[0] \quad \phi_k[1] \quad \ldots \quad \phi_k[N-1] \,]^T, \quad k = 0, 1, \ldots, K - 1 \tag{B1.7}$$
$$\mathbf{a} = [\, a_0 \quad a_1 \quad \ldots \quad a_{K-1} \,]^T,$$

where $\mathbf{x}$ and $\boldsymbol{\phi}_k$ are of length $N - 1$, and $\mathbf{a}$ is of length K. Let us also define an $N \times K$ matrix, $[\Phi]$, whose columns are simply the basis vectors $\boldsymbol{\phi}_i$, as shown below:

$$[\Phi] = \begin{bmatrix} \phi_0[0] & \phi_1[0] & \cdots & \phi_{K-1}[0] \\ \phi_0[1] & \phi_1[1] & \cdots & \phi_{K-1}[1] \\ \vdots & \vdots & \cdots & \vdots \\ \phi_0[N-1] & \phi_1[N-1] & \cdots & \phi_{K-1}[N-1] \end{bmatrix}. \tag{B1.8}$$

Matrix $[\Phi]$ can also be expressed compactly as follows:

$$[\Phi] = \begin{bmatrix} \boldsymbol{\phi}_0 & \boldsymbol{\phi}_1 & \cdots & \boldsymbol{\phi}_{K-1} \end{bmatrix}. \tag{B1.9}$$

This compact notation allows the K equations given by the orthogonality principle to be written in matrix form as

$$[\Phi]^{*T} \mathbf{x} = [\Phi]^{*T} [\Phi] \mathbf{a}. \tag{B1.10}$$

Solving for the vector of constant coefficients, we obtain

$$\mathbf{a} = ([\Phi]^{*T}[\Phi])^{-1}[\Phi]^{*T}\mathbf{x}. \tag{B1.11}$$

Equation (B1.11) consists of a set of equations that are the standard normal equations for a least-squares fit. The reader is strongly encouraged to verify the results by expanding the product of matrices and comparing the resulting equations to those given by the orthogonality principle. Incidentally, these results are often presented without the complex conjugate of the transpose. That is because the Hermitian transpose always conjugates the elements of the matrix, and it is also the reason that M AT L A B does it, that is conjugates the elements, unless the user specifies M AT L A B not to by using the dot notation. Of course, if the matrix is real it makes no difference.

Lastly, just like the continuous-time case, if the basis vectors are orthogonal to one another, then the coefficients a_k can be determined individually, and additional orthogonal basis vectors can be added without recalculating the coefficients.

C Discrete-time upsampling, sampling and downsampling

We discussed the concept of upsampling in Section 1.9 as one of the discrete-time signal transformations. Clearly we can recover the original signal from the upsampled signal by removing the zeros added in the upsampling process. In this Appendix we will explore the effect of upsampling on the DFT of a sequence.

Consider a periodic sequence $x[n] = x[n + N]$. From our previous work, we know that expressions for IDFT and DFT are

$$x[n] = x[n + N] = \sum_{k=0}^{N-1} X[k]e^{jk(\frac{2\pi}{N})n} \tag{C1.1}$$

and

$$X[k] = X[k + N] = \frac{1}{N} \sum_{n=0}^{N-1} x[n]e^{-jk(\frac{2\pi}{N})n}. \tag{C1.2}$$

Let $y[n]$ be the sequence resulting from upsampling $x[n]$ by M, which can be described as

$$y[n] = x^{(M)}[n] = \begin{cases} x[n/M], & \text{for} \quad n = 0, \pm M, \pm 2M, \dots \\ 0 & \text{otherwise.} \end{cases} \tag{C1.3}$$

The periodic sequence $y[n]$ is also periodic, and can be expressed as

$$y[n] = y[n + MN] = \sum_{k=0}^{MN-1} Y[k]e^{jk(\frac{2\pi}{MN})n} \tag{C1.4}$$

and

$$Y[k] = Y[k + MN] = \frac{1}{MN} \sum_{n=0}^{MN-1} y[n]e^{-jk(\frac{2\pi}{MN})n}. \tag{C1.5}$$

Because $y[n]$ and $x[n]$ are related, our goal now is to determine the relationship between $Y[k]$ and $X[k]$. The upsampling process, also known as zero-stuffing, insures that

$$y[n] = 0, \quad \text{for } M(N-1) < n \leq MN - 1. \tag{C1.6}$$

Thus, Eq. (C1.5) becomes

$$Y[k] = \frac{1}{MN} \sum_{n=0}^{M(N-1)} y[n] e^{-jk(\frac{2\pi}{MN})n}. \tag{C1.7}$$

Letting $p = n/M$ or $n = Mp$, Eq. (C1.7) can be expressed as

$$Y[k] = \frac{1}{MN} \sum_{p=0}^{N-1} y[Mp] e^{-jk(\frac{2\pi}{MN})Mp}$$

$$= \frac{1}{M} \left\{ \frac{1}{N} \sum_{p=0}^{N-1} x[p] e^{-jk(\frac{2\pi}{N})p} \right\} \tag{C1.8}$$

or

$$Y[k] = \frac{1}{M} X[k] = Y[k+N]. \tag{C1.9}$$

This means that $Y[k]$ is periodic with period N and consists of the sequence $X[k]$ repeated M times and scaled by $1/M$.

As an example, consider a signal $x[n]$ with period $N = 6$ and DFT $X[k] = [3, 6, 9, 0, 9, 6]$. Figure C1.1 shows $X[k]$, $x[n]$, $y[n]$, and $Y[k]$ for $M = 4$. Note that $Y[k]$ consists of $X[k]$ repeated four times, and thus $Y[k] = Y[k+4]$.

Now suppose instead of upsampling we sample a discrete-time signal. We start with a periodic sequence $x[n] = x[n+N]$ and define the sampled sequence:

$$x_s[n] = \begin{cases} x[n], & \text{for} \quad n = 0, \pm M, \pm 2M, \dots \\ 0 & \text{otherwise.} \end{cases} \tag{C1.10}$$

To insure that $x_s[n] = x_s[n+N]$, we require N/M be a positive integer. In practice, we can pad the discrete signal to satisfy this requirement. Note that $x_s[n]$ can be written as

$$x_s[n] = x[n]s[n], \quad \text{where } s[n] = s[n+M] = \sum_{p=-\infty}^{\infty} \delta[n - pM]. \tag{C1.11}$$

Expressing $s[n]$ in terms of its Fourier coefficients allows us to write

Fig. C1.1. Discrete-time upsampling with $M = 4$.

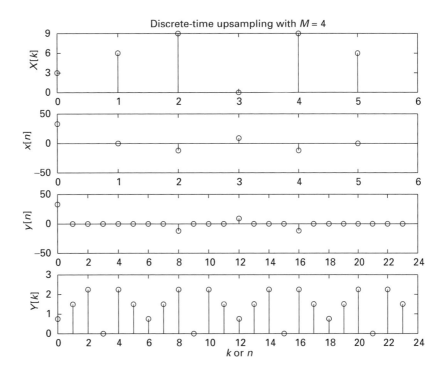

$$s[n] = s[n + M] = \sum_{m=0}^{M-1} a_m e^{jm\left(\frac{2\pi}{M}\right)n}, \tag{C1.12}$$

where

$$a_m = \frac{1}{M}, \quad \text{for } m = 0, 1, \ldots, M. \tag{C1.13}$$

Finally, expressing $x[n]$ in terms of its Fourier coefficients yields

$$
\begin{aligned}
x_s[n] &= \left(\sum_{k=0}^{N-1} X[k] e^{jk\left(\frac{2\pi}{N}\right)n} \right) \left(\frac{1}{M} \sum_{m=0}^{M-1} e^{jk\left(\frac{2\pi}{M}\right)n} \right) \\
&= \frac{1}{M} \sum_{m=0}^{M-1} \left(\sum_{k=0}^{N-1} X[k] e^{j2\pi\left\{ \left(k + \frac{N}{M}m\right)\right\}n} \right).
\end{aligned}
\tag{C1.14}
$$

While the last expression looks a bit intimidating, it simply means that $X_s[k]$, the DFT of $x_s[n]$, consists of $X_s[k]$, the DFT of $x[n]$, repeated m times at intervals of N/M and scaled by $1/M$. The following example will help to clarify this and, more importantly, illustrate the criteria that must be satisfied to allow $x[n]$ to be reconstructed from $x_s[n]$.

Fig. C1.2. Discrete-time
sampling with $N = 24$ and
$M = 2$.

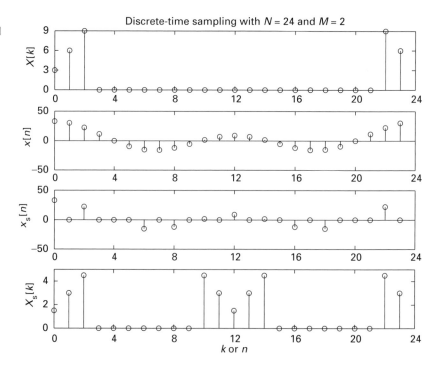

Consider a periodic discrete-time signal with period $N = 24$ and DFT
as follows:

$$X[k] = [\,3, 6, 9, 0, 0, 0, 0, 0, 0, 0, 0, 0, 0, 0, 0, 0, 0, 0, 0, 0, 0, 0, 9, 6\,].$$

$$(C1.15)$$

Figures C1.2 through C1.5 show the results of sampling the sequence for
$M = 2, 3, 4, 6$. For $M = 2, 3, 4$ we can recover the original DFT from the
DFT of the sampled signal by using a digital filter to remove the extra
terms (there is no overlap). For $M = 6$, however, we do have overlap (it
is our old friend aliasing!), and we cannot recover the DFT of the original
signal. Furthermore, note that for $M = 2, 3, 4$ we need keep only every
second, third, and fourth term, respectively, since if we know M we can
simply upsample to replace the zeros and then filter the result to recover
the original sequence, that is we can keep the signal $y[n] = x[Mn]$. This
process is called *downsampling*. In practice, there will usually be some
aliasing, and engineering judgment must be used to determine how much
is acceptable.

Fig. C1.3. Discrete-time sampling with $N = 24$ and $M = 3$.

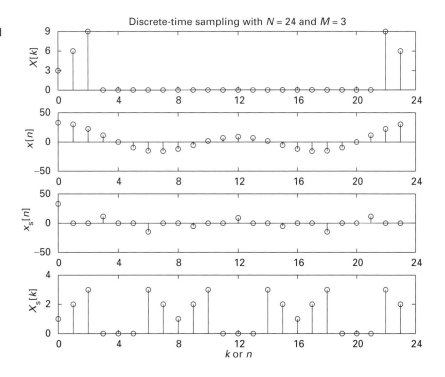

Fig. C1.4. Discrete-time sampling with $N = 24$ and $M = 4$.

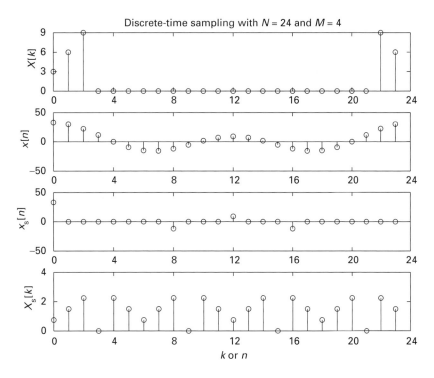

Fig. C1.5. Discrete-time sampling with $N = 24$ and $M = 6$.

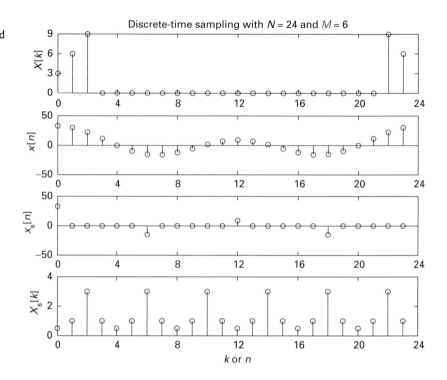

Now that you understand the fundamentals you can refer to more advanced texts[1] to see how the processes of upsampling, sampling, and downsampling can be used to implement so-called multi-rate processing. The idea is to process and keep the minimum amount of data required.

[1] For example, Oppenheim, Willsky, and Young, *Signals and Systems, Second Edition*, Prentice Hall, 1997.

Index

dual tone multifrequency (DTMF) touchtones,
406
decoding exercise, 407–408
duty cycle, 46, 51

ears *see* Gibbs phenomenon, 48
`eig` command, 209
eigensolutions, 202
eigenvalue problem, 203, 204
eigenvalues, 202–203, 204, 209–210
eigenvectors, 202, 204–205, 209, 210
electrical systems, 158, 165–166, 341–342
analogy with mechanical systems, 158,
165–166, 170, 172, 342
combination of elements, 172–178
elements and variables, 158–165
governing equations, 167–172
inputs, 166–167
energy loss, 129, 192
see also dissipation of energy
envelopes, 52, 53, 313–314, 386–387
equations of motion, 120, 132
equilibrium, 140
see also stability
equilibrium value *see* steady state value
Euler's formula, 7, 414
even functions, 285
excitation frequency, 217, 245, 320, 322
`exp` command, 367–368
expansion, of signals, 16

Faraday's Law, 161
farads, 164
fast Fourier transform (FFT), 59, 341, 342
automobile suspension problem, 402–403
block diagram of use, 270
data acquisition system, 110–112
examples, 62–69
laboratory exercise, 392–393
see also `fft` command
FBD *see* free body diagram
`fdomain` function, 67–69, 100, 271–272,
397–398, 406, 410, 411
feedback, 299, 359
FFT *see* fast Fourier transform
`fft` command, 59, 60, 62–69
`fftshift` command, 100, 272
filter order, 283
`find` command, 374
finite impulse response filters, 299–305, 342,
355

first order differential equations,
184–185
first-order systems, 342, 349
flux, magnetic, 161
folding, 92, 94–95, 99, 346–347
forced response, 183, 204–211
`format` command, 370–371
Fourier series, 5–6, 42–43, 344
and inputs, 221–222, 270
complex form, 43, 44–45
finite approximations, 47–49
other forms, 45
real form, 44–45
free body diagram, 120
free response, 183, 202–204
frequency, of signal, 4, 9
frequency response function, 188–189, 219,
351–353
and discrete vs. continuous-time systems,
291–299
and inputs, 221–222
and transfer functions, 359
automobile suspension problem,
400–401
block diagram for use, 270
gain, 222
impedance method, 242–249
in discrete-time systems, 281–291
in second-order systems, 232–238
laboratory exercise, 394–399
linear time-invariant systems, 219–221
magnitude, 222
numerical methods, 269–270
phase, 222
polar notation for, 222
FRF *see* frequency response function
fundamental frequency, 47
Für Elise project, 385–386

gain factor, 224
Gauss' Law, 164
Gibbs phenomenon, 48, 49, 302
governing equations, 120, 342
electrical systems, 167–168, 342
mechanical systems, 132–140
GPS receivers, 355
Gram–Schmidt process, 78
gramophones, 3
gravity, 135, 139
ground nodes, 160
ground voltage, 160